MICROALGAL PRODUCTION
for Biomass and High-Value Products

MICROALGAL PRODUCTION

for Biomass and High-Value Products

Edited by

Stephen P. Slocombe

Scottish Association for Marine Science (SAMS)
Oban, UK

John R. Benemann

CEO, MicroBio Engineering, Inc.
San Luis Obispo, California, USA

CRC Press
Taylor & Francis Group
Boca Raton London New York

CRC Press is an imprint of the
Taylor & Francis Group, an **informa** business

Front cover (top to bottom): Tubular photobioreactor for the production of *Haematococcus pluvialis*, a source

of astaxanthin at BGG, Yunnan Province, China; The 20 hectare raceway pond system for Spirulina production

in California, USA (Earthrise Nutritionals, LLC); A paddle wheel mixed raceway pond in a greenhouse

cover; Column reactors used for microalgae production for aquaculture.

Photo credits: Jianguo Liu, Institute of Oceanology, Qingdao, China; and Amha Belay, Earthrise Nutritionals,

LLC.

CRC Press
Taylor & Francis Group
6000 Broken Sound Parkway NW, Suite 300
Boca Raton, FL 33487-2742

First issued in paperback 2021

Version Date: 20160405

ISBN 13: 978-1-03-209792-3 (pbk)
ISBN 13: 978-1-4822-1970-8 (hbk)

Library of Congress Cataloging-in-Publication Data

Names: Slocombe, Stephen P., editor. | Benemann, John R., editor.
Title: Microalgal production for biomass and high-value products / [edited by] Stephen P. Slocombe and John R. Benemann.
Description: Boca Raton, FL : CRC Press, 2016. | Includes bibliographical references and index.
Identifiers: LCCN 2015040117 | ISBN 9781482219708 (alk. paper)
Subjects: LCSH: Microalgae. | Microalgae--Biotechnology. | Biomass energy.
Classification: LCC QK568.M52 M537 2016 | DDC 579.8--dc23
LC record available at http://lccn.loc.gov/2015040117

Visit the Taylor & Francis Web site at
http://www.taylorandfrancis.com

and the CRC Press Web site at
http://www.crcpress.com

Contents

Preface

In 2000, the Saudi Oil Minister Sheikh Yamani predicted that renewable and other technologies will usurp the fossil fuels that currently drive the world economy: "Thirty years from now there will be a huge amount of oil—and no buyers" (Fagan 2000). The implication was that oil will be left in the ground. Now at the midpoint (2015), this appears prescient in the prediction of excess oil production capacity, but not, thus far at least, concerning the growth of renewable energy sources generally and transportation biofuels in particular. Corn, sugarcane, and soybean oil are still the main sources for liquid biofuels, comprising mainly ethanol and biodiesel, and amount to less than 2% of global transportation fuels (EIA 2012). Furthermore, many independent experts consider these to be in competition with food and feed production, and generally view them as not truly renewable. More sustainable biofuel sources that do not compete with food production are being developed, including microalgal biofuels.

During the last decade, there has been a great and increasing interest in microalgae production, propelled by the dual drivers of liquid biofuels and CO_2 greenhouse gas abatement. Liquid biofuels are the main goal of ongoing research and development (R&D) in this field, as they are preferable to gaseous or solid fuels on several accounts, with a particular interest by the aviation industry. However, in the past few years, the focus of microalgae R&D has expanded to feed, food, fertilizers, chemicals, and high-value specialty products, as well as wastewater treatment. To be considered renewable, microalgae processes must demonstrate a substantial net benefit in CO_2 emission reduction, based on a life cycle assessment (LCA), and not deplete other nonrenewable resources, such as phosphorous.

The vision for microalgae is for high productivity cultivation processes that convert solar energy into large amounts of biomass, which would yield tens of thousands of liters of ethanol, biodiesel, green jet fuel, and other advanced biofuels, per hectare per year. This is many times the volume of biodiesel or ethanol produced by conventional crops. In addition, biofuels from microalgae could potentially be low in cost and, perhaps most importantly, would use wastewater, seawater, or brackish water, thus avoiding competition with food production.

Microalgae are also attractive because their production requires a concentrated source of CO_2, which in most projections is provided from fossil power plants. Although any source of CO_2 will do, including those of a biogenic origin (from waste decomposition, food and beverage industries, etc.), small to medium-scale sources (<100 MWe) would be favored over larger ones. Other fuels derived from microalgal biomasses, such as biogas or ethanol, can also be considered, including animal feed and chemicals, if the products result in significant CO_2 emission abatement based on an LCA of the process. Of course, economics will be the main driver in the future development of microalgae processes and products. And a major factor in production costs is productivity.

Presently, there are insufficient data on the actual productivity of microalgae cultures either in terms of tons of biomass or liters of biofuel produced per hectare per year. Most data are obtained on smaller scales or for shorter time periods and then extrapolated to large scales and annual production. Much more research is required to realize the promise, even to fully understand the challenges, of microalgae biofuel production. An immense interest during the past few years in microalgal production for biofuels and greenhouse gas abatement, and the substantial R&D support from governments, corporations, venture funds, individual investors, and entrepreneurs, is now yielding a bonanza of basic and applied research advances. These advances promise to make the visions of large-scale microalgae production a reality, even before 2030.

This book provides an overview of this exciting field from a variety of perspectives by experts and practitioners in this field. No single volume can encompass all the aspects and advances that are being made, but the text does provide a guide on how laboratory and applied research are leading to advances in commercial production. It brings together the latest advances of interest to those already working in this field, while also providing a gateway to this exciting new technology for those who are just beginning to learn about the promise of microalgae for a sustainable future.

Stephen P. Slocombe
John R. Benemann
Editors

REFERENCES

EIA. 2012. International energy statistics. U.S. Energy Information Administration, Washington, DC. http://www.eia.gov/cfapps/ipdbproject/IEDIndex3.cfm (accessed December 19, 2015).
Fagan, M. 2000. Sheikh Yamani predicts price crash as age of oil ends. http://www.telegraph.co.uk/news/uknews/1344832/Sheikh-Yamani-predicts-price-crash-as-age-of-oil-ends.html (accessed June 25, 2015).

MATLAB® is a registered trademark of The MathWorks, Inc. For product information, please contact:

The MathWorks, Inc.
3 Apple Hill Drive
Natick, MA 01760-2098 USA
Tel: 508-647-7000
Fax: 508-647-7001
E-mail: info@mathworks.com
Web: www.mathworks.com

Acknowledgments

It was a pleasure to work with the authors involved on this project, and we thank them wholeheartedly for their excellent work. We also thank those who kindly provided an internal review for one or more chapters. We are grateful for the helpful advice shown by the staff at CRC Press/Taylor & Francis, particularly to Ashley Weinstein; Linda Leggio; Arun Aranganathan; Randy Brehm, who mentored the process from the beginning; and Bob Trigiano for the original invitation and encouragement to write this book. For providing comments for one or more chapters, we are indebted to Martina Strittmatter, Gary Caldwell, Alexei Solovchenko, Maria Heute-Ortega, Linda O'Higgins, Malcolm Slocombe, John Lefley, and Thomas Butler for their time and expertise. SPS thanks his wife, QianYi, his sons Barnaby and Felix, and his parents for their support and inspiration. JRB thanks his partner, Kate Ang, for keeping him focused.

Editors

Stephen P. Slocombe, PhD, is an experienced plant molecular biologist who has been working on microalgal biotechnology at the Scottish Association for Marine Sciences (SAMS) in Oban, United Kingdom, since 2009. He holds a BSc (Hons) (1986) in combined biochemistry and genetics from the University of Leeds and a PhD in botany from the University of Leicester (1991) on the subject of Crassulacean acid metabolism. He has published extensively on biotechnology-related topics such as the regulation of lipid metabolism, particularly through the investigation of mutants in fatty acid breakdown and fatty acid transport, which ultimately led to the production of triacylglycerols in leaves. He has also studied novel fatty acid synthesis pathways in Solanaceae for their potential in improved physical properties of biofuels. At SAMS, he oversaw the screening of the microalgal CCAP (Culture Collection of Algae and Protozoa) collection and is currently working on improving oil production and other biotechnological applications in marine microalgae, using molecular and physiological approaches.

John R. Benemann, PhD, earned a BS in chemistry and a PhD in biochemistry, both from the University of California (UC) Berkeley, followed by postdoctoral work at UC, San Diego, and then in an independent research position in the Department of Civil Engineering, UC, Berkeley, until 1980. He founded EnBio Inc, and SeaAg, Inc.; two small companies in the business of higher-value microalgae products and aquaculture, respectively. During the 1980s, he was an associate professor in the Department of Applied Biology, Georgia Institute of Technology (Atlanta). Dr. Benemann has worked as an independent consultant and for the past 5 years as CEO of MicroBio Engineering Inc., a research and consulting engineering company focused on wastewater treatment and algal technologies, which he founded with Tryg Lundquist, California Polytechnic State University, San Luis Obispo. He is also cofounder and director of the not-for-profit Institute for Environmental Management Inc., which for more than 20 years has developed the "Controlled Landfill Bioreactor," and other technologies that deal with solid waste. In addition, Dr. Benemann is the cofounder and director of the Algae Biomass Organization, a trade organization, and member of the Executive Committees of the International Society for Applied Phycology and the European Algae Biomass Association. He has consulted for and advised several U.S. government agencies and international organizations, and is a frequent conference speaker.

Contributors

F. Gabriel Acién
Department of Chemical Engineering
University of Almería
Almería, Spain

Roberto E. Armenta
Mara Renewables Corporation
Dartmouth, Nova Scotia, Canada

John Beardall
School of Biological Sciences
Monash University
Clayton, Victoria, Australia

Patrick V. Brady
Geoscience Research and Applications
    Group
Sandia National Laboratories
Albuquerque, New Mexico

David E. Brune
Department of Food Systems and
    Bioengineering
University of Missouri
Columbia, Missouri

Laura T. Carney
Heliae Development, LLC
Gilbert, Arizona

William Chang
Department of Biology
Brooklyn College
Brooklyn, New York

Aaron Chavis
Marine Sciences Laboratory
Pacific Northwest National Laboratory
Sequim, Washington

Boris Chubukov
Marine Sciences Laboratory
Pacific Northwest National Laboratory
Sequim, Washington

André Coleman
Hydrology Group
Pacific Northwest National Laboratory
Richland, Washington

Luís A. Costa
A4F, Algae for Future
Lisboa, Portugal

Braden Crowe
Marine Sciences Laboratory
Pacific Northwest National Laboratory
Sequim, Washington

John G. Day
Culture Collection of Algae and Protozoa
Scottish Association for Marine Science
Oban, United Kingdom

Scott Edmundson
Marine Sciences Laboratory
Pacific Northwest National Laboratory
Sequim, Washington

Jose Maria Fernández-Sevilla
Department of Chemical Engineering
University of Almería
Almería, Spain

Diana B. da Fonseca
A4F, Algae for Future
Lisboa, Portugal

Luís T. Guerra
A4F, Algae for Future
Lisboa, Portugal

Samuel Hobbs
Marine Sciences Laboratory
Pacific Northwest National Laboratory
Sequim, Washington

Andy Huang
Department of Biology
Brooklyn College
Brooklyn, New York

Michael Huesemann
Marine Sciences Laboratory
Pacific Northwest National Laboratory
Sequim, Washington

Floral Joseph
Department of Cell Biology
State University of New York
    Downtown Medical Center
Brooklyn, New York

Todd W. Lane
Department of Systems Biology
Sandia National Laboratories
Livermore, California

Jianguo Liu
Institute of Oceanology
Chinese Academy of Sciences
Qingdao, Shandong Province, People's
    Republic of China

and

Yunnan Alphy Biotechnology Co., Ltd.
Chuxiong, Yunnan Province, People's
    Republic of China

Robert C. McBride
Triton Algae Innovations
San Diego, California

Mark P. McHenry
School of Engineering and Information
    Technology
Murdoch University
Perth, Western Australia, Australia

Sofia H. Mendonça
ALLMA
Lisboa, Portugal

Navid R. Moheimani
School of Veterinary and Life Sciences
Murdoch University
Perth, Western Australia, Australia

Emilio Molina Grima
Department of Chemical Engineering
University of Almería
Almería, Spain

Koenraad Muylaert
Laboratory of Aquatic Biology
Catholic University of Leuven Kulak
Kortrijk, Belgium

João C. Navalho
A4F, Algae for Future
Lisboa, Portugal

Peter Neofotis
Department of Biology
Brooklyn College
Brooklyn, New York

and

The Graduate Center
City University of New York
New York, New York

Robert Nielsen
ExxonMobil Corporate Strategic
    Research
Annandale, New Jersey

Juergen E.W. Polle
Department of Biology
Brooklyn College
Brooklyn, New York

and

The Graduate Center
City University of New York
New York, New York

Saul Purton
Institute of Structural and Molecular
    Biology
University College London
London, United Kingdom

John A. Raven
The James Hutton Institute
University of Dundee
Dundee, United Kingdom

and

Plant Functional Biology and Climate
  Change Cluster
University of Technology
Sydney, New South Wales, Australia

Michael Ross
Department of Microbial and Molecular
  Biology
Scottish Association for Marine Science
Oban, United Kingdom

Edgar T. Santos
A4F, Algae for Future
Lisboa, Portugal

Gregory Schwartz
BioResource and Agricultural Engineering
  Department
Cal Poly State University
San Luis Obispo, California

Joana G. Silva
A4F, Algae for Future
Lisboa, Portugal

Val H. Smith
Department of Ecology and
  Evolutionary Biology
University of Kansas
Lawrence, Kansas

Andrew Spicer
Algenuity
Eden Laboratory
Bedfordshire, United Kingdom

Michele S. Stanley
Department of Microbial and Molecular
  Biology
Scottish Association for Marine Science
Oban, United Kingdom

Zhiyong Sun
Mara Renewables Corporation
Dartmouth, Nova Scotia, Canada

Vincent J. Tocco
Marine Sciences Laboratory
Pacific Northwest National Laboratory
Sequim, Washington

John P. van der Meer
Pan-American Marine Biotechnology
  Association
Halifax, Nova Scotia, Canada

Peter Waller
Department of Agricultural and
  Biosystems Engineering
The University of Arizona
Tucson, Arizona

Joseph Weissman
ExxonMobil Corporate Strategic
  Research
Annandale, New Jersey

Mark Wigmosta
Hydrology Group
Pacific Northwest National
  Laboratory
Richland, Washington

Litao Zhang
Institute of Oceanology
Chinese Academy of Sciences
Qingdao, Shandong Province, People's
  Republic of China

QianYi Zhang
Department of Microbial and Molecular
  Biology
Scottish Association for Marine
  Science
Oban, United Kingdom

Yong Zhang
Yunnan Alphy Biotechnology Co., Ltd.
Chuxiong, Yunnan Province, People's
  Republic of China

Introduction

Stephen P. Slocombe and John R. Benemann

THE MICROALGAE INDUSTRY

Well before the recent surge of interest in microalgae for biofuels, in the 1950s, the green alga *Chlorella* was studied with a view to producing human food, the focus of the first pilot plant for microalgae biomass production (Burlew 1953). By the early 1960s, a small commercial microalgal industry was developed in Japan and later in Taiwan for the production of *Chlorella* as a human food "supplement." The biomass was produced in circular ponds of up to about 0.1 ha, and was sold as tablets or green powder, first in Japan and now worldwide. Today, close to 10,000 tons of *Chlorella* biomass are produced annually, roughly equally divided between autotrophic production in open ponds using sunlight and CO_2 (often supplemented with acetate); and heterotrophic processes in which the algae are grown in the dark in fermenters using sugar as both a carbon and energy source (Chapters 13 and 14).

Bulk (i.e., ton quantity) selling prices for *Chlorella* biomass for food supplements range between \$20 and \$30/kg but vary greatly depending on the producer, production process, volume, market conditions, customers, and other factors. For example, products derived from fermenters and photobioreactors (PBRs) are cleaner and fetch premium prices compared to open pond cultures, mainly produced in China (Chapter 13). It must be noted that reliable information on production volumes and selling prices is not available, and thus, estimates of commercial *Chlorella* production are based on personal knowledge of the industry.

This also applies to the other three microalgae species, which are currently produced commercially in relatively large volume using sunlight and CO_2 (autotrophic growth): *Spirulina* (*Arthrospira platensis*), again mainly used for food supplements, though it is also a source of a food colorant, phycocyanin; *Dunaliella salina*, extracted for beta-carotene; and *Haematococcus pluvialis*, a source of astaxanthin, another carotenoid (Chapter 12).

Since the early 1970s, *Spirulina* has been produced commercially and is now the market leader, with about twice the production of *Chlorella* but selling at a somewhat lower bulk price. *Spirulina* is produced only autotrophically (i.e., using only sunlight and CO_2 as energy and carbon sources) and almost exclusively in large paddle wheel mixed raceway-type ponds. There is a growing, though in aggregate still small, cottage industry producing *Spirulina* for local markets and even for personal use. A major recent development was the approval by the European Union (EU) and the United States for the food colorant phycocyanin, a blue protein, which is present in over 10% of *Spirulina*, that is extracted from the biomass. A major *Spirulina* producer, Earthrise Nutritionals, LLC, in Irvine, California, recently commissioned a \$10 million plant to produce phycocyanin.

Dunaliella and *Haematococcus* are also cultivated autotrophically and extracted for their natural beta-carotene and astaxanthin, respectively; these carotenoids are being found in relatively large amounts in these microalgae and, again, sold mainly as human nutritional products (nutraceuticals), for their antioxidant and other beneficial properties. Their total production is small in terms of tonnage, but these products sell (in bulk) on the order of $1,000–$10,000/kg for beta-carotene and astaxanthin, respectively, on a pure basis. There is only one major producer for beta-carotene, BASF (Ludwigshafen, Germany), is a German chemical company that only a few years ago acquired approximately 1000 ha of the hypersaline *Dunaliella* production ponds in Australia. BASF is currently expanding to an even larger scale of production in Saudi Arabia in a joint venture with the National Aquaculture Group (NAQUA).

There are now more than a dozen *Haematococcus* astaxanthin producers, with many more start-ups, resulting in a transition from a seller to a buyer's market, with a resulting drop in price. The ultimate goal for microalgae carotenoid production is to replace the lower-priced synthetic carotenoids and expand the palette of products to include lutein, zeaxanthin, fucoxanthin, and other carotenoids, with existing or developing markets.

Another significant commercial application for microalgae is in aquaculture, where small amounts—on the order of a few kilograms—are produced at shrimp, bivalve, and fish hatcheries and nurseries. These supply the live feed required at the larval and juvenile stages or are fed to zooplankton, which is then fed to the animals. Several microalgae species, in particular, diatoms and flagellates, are produced for these purposes, and there are a number of small-scale producers supplying the aquaculture industry with live algae concentrates or freeze-dried powder. Some additional products from microalgae have found niche markets in cosmetics, food, and other specialty markets, but production of these is still on a relatively small scale.

A major advance in commercial algae production has been in heterotrophic production of sugars in the dark (Chapter 14). Martek (Columbia, Maryland), the market leader (acquired by DSM in Heerlen, the Netherlands, in 2011), pioneered the fermentative production of the dinoflagellate *Crypthecodinium cohnii* to produce DHA as an ingredient in infant formulas, which is now a several hundred million dollar per year market. Recently, Solazyme (San Francisco, California), a new start-up venture in the United States, expanded the applications of *Chlorella* fermentation into specialty oil, cosmetics, food, and even fuels.

Commercial microalgae production has, over the last five decades, steadily expanded, though thus far only into niche, low-volume, high-value products. As the sector grows and research advances, lower-cost production processes will allow markets to grow, resulting in larger production plants, achieving economies of scale and further market expansion. To draw an analogy from the cultivation of microalgae: commercial production at this point has only advanced from the lag phase into the early logarithmic phase. Continuing exponential growth in algal products and production will result in a major future industry—algaeculture—producing millions of tons of algal biomass for food, feed, fuels, chemicals, and much more.

This growth trend is accelerated due to the major advances in microalgae biotechnology. In particular, the many "omics"—genomics, proteomics, lipidomics, and so on—are now leading to a more holistic understanding of both the complexities and power of biological processes generally and microalgae specifically (Chapter 6). With about 10,000 researchers and engineers currently engaged in the research and development of this technology and industry, in all aspects, from biology to production, rapid advances can be anticipated. The greater than 10-fold increase in R&D in this field during the past decade has mostly been driven by private investors, companies, and governments interested in microalgae biofuels. This interest is now expanding into new areas, from commodity animal feeds to pharmaceuticals, from fertilizers to wastewater treatment, as well as many other applications. Next, a review of the historical development of microalgae production and its potential is discussed, providing a background and perspective to the chapters that follow.

MICROALGAE R&D: A HISTORY

Microalgae have been consumed as a traditional food source in many countries, most notably around the highly alkaline Lake Chad in Africa, where natural blooms of *Spirulina* were collected, dried, and sold in local markets, leading to the "discovery" of this novel food source only 50 years ago (Ciferri 1983). Another cyanobacterium, *Nostoc* sp., which grows as small grape-like balls in freshwater, has been consumed as traditional human food in many countries (Guo et al. 2015).

Scientific research into microalgae began in the late nineteenth century with its initial cultivation in the laboratory, in particular, from the fast-growing *Chlorella* species. This became the subject of the first studies on photosynthetic efficiencies during the early twentieth century (Warburg 1919). These early studies suggested a very high photosynthetic efficiency, with only 4 photons required to reduce one CO_2 molecule. Later research concluded that the correct figure was at least twice that and closer to 10 photons per CO_2 fixed into carbohydrates (see Chapters 1 and 2). The idea that microalgae have very high productivities, much higher than conventional crop plants, became one of the allures of microalgae and led to a proposal to use microalgae as a source of vegetable oil, fuel, and food, dating back to the 1940s when it was found that some microalgae accumulate high levels of such lipids (Harder and von Witsch 1942).

In the 1940s, after World War II, global food shortages were predicted and it was toward this goal that the first efforts in microalgal mass culture research were initiated, culminating in the first pilot plant—two plastic bags approximately 100 m² in a raceway configuration—to cultivate *Chlorella* (Burlew 1953). Many of the problems that are currently the focus of R&D, and reported in the chapters in this book, were already recognized at that time: biomass productivity, algal grazers and pathogens, strain selection and maintenance, harvesting and processing of the algal biomass, and the engineering design of low-cost cultivation systems. Open raceway ponds were also developed at about the same time for growing microalgae for wastewater treatment (Oswald and Gotaas 1957).

The use of CO_2 from the waste gases of power plants to grow microalgae was proposed and subjected to an initial techno-economic analysis in 1960 for large-scale production of biogas, and then electricity production from waste-grown algae (Oswald and Golueke 1960). During the 1970s, there was a revival in the concept of algal biofuel in response to the oil crisis, and research was initiated on this wastewater–biogas concept (Benemann et al. 1978, 1980). In 1980, the U.S. Department of Energy initiated the "Aquatic Species Program" (ASP), shifting the focus from wastewater and biogas to brackish water sources in the Southwest United States, and the production of algal oils (lipids) for transportation fuels. The ASP carried out extensive screening for new strains, some early work on algal genetics, and considerable outdoor work with open ponds, as reviewed in the ASP close-out report (Sheehan et al. 1998). A major Japanese government program, costing several hundred million dollars, more than 10-fold of that spent on the ASP, was carried out during the 1990s. It focused mainly on microalgal CO_2 fixation to reduce greenhouse gases, but no report on this research was published.

DEVELOPMENT OF COMMERCIAL PRODUCTION

As noted, in about 1960, a small commercial industry for *Chlorella* production started in Japan, soon expanding to Taiwan, using open circular ponds (about 0.1 ha in size) and batch cultivation, often combined with a short stage during which acetate was added (i.e., "mixotrophic" growth) prior to harvest. *Chlorella* is also produced by heterotrophic growth in fermenters, mainly in Korea. Since about 2000, it has been produced in a large array of tubular photobioreactors (about 1 ha) in Germany (now owned by Roquette, in Lestrem, France). Recently, another such system started in Portugal (ALGAFARM/ A4F in Lisboa, Portugal, see Chapter 13). *Chlorella* was initially sold mainly as a food "supplement" in Japan, but by the 1980s, markets for *Chlorella* and *Spirulina* "nutraceuticals" had also developed in the United States, Europe, and elsewhere.

The first *Spirulina* plant started operating in the early 1970s, near Mexico City, by Sosa Texcoco, S.A., utilizing about a 10 ha section of a large alkaline evaporation pond that was used to manufacture sodium carbonate and salt. This project was inspired by the discovery of French researchers for the use of *Spirulina* as a human food source in Lake Chad, Africa. The *Spirulina* from this plant was initially sold mainly to Japan to be mixed with *Chlorella*, but soon it developed its own markets and today, *Spirulina* outsells *Chlorella* by over twofold. The success of the Mexican plant encouraged the cultivation of *Spirulina* in purpose built paddle wheel mixed raceway ponds, first in the late 1970s in Thailand by a Japanese company (Dainippon Ink and Chemicals, or DIC, Tokyo); and by the early 1980s, in the United States by Earthrise Nutritionals, LLC (Irvine, California, later acquired by DIC) and Cyanotech Corporation (Kailua-Kona, Hawaii). Other countries also initiated *Spirulina* production in the 1990s, in particular, India. By the mid-2000s, China became the major global microalgae producer, now with about two-thirds to three-quarters of the total output, for both *Chlorella* and *Spirulina*. Almost all of these are produced using paddle wheel mixed ponds, typically 0.2–0.8 ha in size.

Commercial production of *Dunaliella salina* began in Australia in the 1980s as a source of natural beta-carotene (pro-vitamin A), using about a thousand hectares of very large (up to 100 ha each) unmixed ponds, with very low productivity but even lower operating costs (Borowitzka and Moheimani 2013). Smaller plants, using a similar process and also raceway ponds, are also operating in several countries but are not significant producers of natural beta-carotene. This provitamin/antioxidant algal product is superior to synthetic beta-carotene, the major commercial product.

Similar to other vegetable beta-carotene sources, it contains both the *cis-* and *trans-* isomers, while the synthetic product only contains the *trans* isomer (which has lower antioxidant activity) (Ben-Amotz and Levy 1996).

This is also the case for astaxanthin, another carotenoid produced commercially since the early 2000s by the autotrophic culture of *Haematococcus pluvialis* (see Chapter 12). The first commercial plants were operated in Ketura, Israel by Algatechnologies, Ltd., using tubular reactors, and in Hawaii by Cyanotech Corporation using open raceway ponds and Fuji Chemicals (Kamiichi-machi, Japan), using over a thousand dome-type PBRs. A plant with completely enclosed tanks using artificial illumination has operated for over a decade in Sweden, bought some years ago by Fuji Chemicals, which recently built such a plant in Moses Lake in Washington State. Now, more than a dozen producers have started to produce algal astaxanthin using these technologies, with many more at the start-up stage. The largest plant in the world now operates in China, with over 1000 km of tubular PBRs (BGG, Kunming, Yunnan Province).

During the 1990s, production of the long-chain omega-3 fatty acid docosahexenoic acid (DHA) by the heterotrophic dinoflagellate *Crypthecodinium cohnii* grown in fermenters was developed in the United States by Martek Corporation (acquired by DSM). An essential fatty acid for brain development, this is sold as an ingredient in human infant formula (Horrocks and Yeo 1999). Production of DHA and EPA (eicosapentaenoic acid) by microalgae (and algae-like thraustrochytrids) in fermenters is now a booming business with many new entrants into this market, aiming to replace fish oil-derived EPA and DHA in both the human nutritional and aquaculture and animal feed markets (Chapters 4 and 14). Algal DHA is now the largest microalgae product in sales, at several hundred million dollars, although *Spirulina* and *Chlorella* are nearing this mark and algal carotenoids are catching up.

In brief, the markets for human nutritional products from microalgae continue to grow worldwide, and are approaching a billion dollars in combined bulk sales; multiplied over 10-fold by the time the products reach the individual consumers. However, a lack of reliable production and consumption figures, let alone sales prices, does not allow accurate estimates of production costs or markets sizes. Also, markets often fluctuate rapidly. For example, in the mid-1990s, when the Sosa Texcoco operation in Mexico closed, the sudden drop in supply, with only Earthrise and Cyanotech remaining as producers, resulted in *Spirulina* prices doubling, from the low teens to over $25/kg. This encouraged an expansion in production by these U.S. companies and the emergence of several new producers. By the early 2000s, total production had not only recovered but was actually well above earlier volumes, and prices dropped again. Over the past 10 years, *Spirulina* production has increased from under about 5,000 tons to about 15,000 tons today, suggesting a 15% annual growth rate.

The case of *Haematococcus pluvialis* astaxanthin provides a more recent example. With only a handful of producers and limited supplies, but with strong demand, this product was selling for over $10 million/ton (on a pure basis)— about 10-fold the price of natural beta-carotene from *Dunaliella salina*. However, increasing supplies of astaxanthin are coming online, and prices are already experiencing a downward trend.

Another market that is attracting increased attention is the aquaculture of mollusks, fish, and crustaceans, where the larval stages often require live microalgae.

This industry and its integration of algal growth with production and recycling are discussed in Chapter 9.

RECENT DEVELOPMENTS: MICROALGAE BIOFUELS, CO$_2$ ABATEMENT, AND HIGHER-VALUE PRODUCTS

The greater interest in microalgae during the past decade has been its promise to both reduce CO$_2$ emissions by power plants while at the same time producing liquid biofuels to replace oil, which until recently, was thought by many to already be approaching a peak of production. A signal event was the announcement in July 2009 of a joint venture project between ExxonMobil (Irving, Texas), and Synthetic Genomics, Inc. (San Diego, California), the company founded by Craig Venter, of the human genome fame. The initial 5-year project was to spend a reported $300 million, and it was projected that a commercial process could be developed within a decade. By the time of the ExxonMobil announcement, several dozen algae biofuel start-ups were already underway, some well capitalized by large venture funds. For example, the venture fund of Bill Gates, along with others, invested over $200 million in Sapphire Energy (San Diego, California). Along with an additional grant of $50 million from the U.S. Department of Energy, Sapphire built the largest algae biofuel plant in the world in New Mexico. With 40 ha of open paddle wheel raceway ponds, this was modeled on Earthrise Nutritional's *Spirulina* farm in California.

Many other large private investments, amounting from tens to even hundreds of millions of dollars, have gone into more than a dozen start-up microalgae biofuel companies in the United States. This, along with hundreds of millions of dollars in U.S. government funding to universities, government laboratories, and small companies, often as part of research consortia. The European Union, Japan, Korea, and many other countries have also supported large numbers of research consortia, including private companies to develop microalgae for biofuels and CO$_2$ abatement. It can be roughly estimated that about a billion dollars a year has been invested in such projects globally over the past 7 or 8 years, with 10,000 researchers and engineers now working on such R&D projects.

However, the very low value of fossil fuels made the near-term economics of microalgae biofuels problematic, even before the precipitous drop in oil prices in mid-2014. Indeed, even before the ExxonMobil/Synthetic Genomics collaboration was announced in July 2009, there were some notable failures. In March 2009, GreenFuel Technologies, Inc. (Phoenix, Arizona), the earliest of the algal biofuels start-ups, went out of business after 5 years of operation with a total loss of $70 million. In March 2013, the CEO of ExxonMobil pronounced that microalgae biofuels were still "25 years" in the future, and dialed down the algal biofuels project—though it continues. By that time, the fact that algal biofuels would not be ready for commercial production for some time had become too obvious to ignore, and most, though not all, biofuel start-ups, had shifted their focus from biofuels to higher-value products, mostly human nutritionals.

For example, several of the early biofuel start-up ventures now focus on astaxanthin, the highest value microalgal product, until recently selling for $10 million per ton, though, as noted earlier, prices are now falling as many new producers are entering the market. Others now aim to replace fish meal and fish oil, which have more than doubled in price over the past decade to more than $2000/ton. Some marine microalgal oils closely resemble fish oil composition (see Chapter 4), making this an attractive market. However, the algal biomass would have a lower value than fish meal, because of its lower protein content and lower-quality amino acid composition, though it would still command considerably higher prices than soybeans, selling for under $500/ton. Presently, the focus is mainly on replacing purified fish oil sold as nutritional supplements, the prices of which are manifold higher than fish oil sold for animal feed. This is mostly due to the high cost of purifying crude fish oil to an acceptable quality for human consumption. The greatest current interest is in *Nannochloropsis*, a microalga that contains about 4%–5% dry weight EPA and is a subject of intensive research for large-scale cultivation for biofuels.

Another approach which is being pursued is to fractionate algal biomass into various components, from high-value EPA and carotenoids for human consumption to specialty animal feed, with the residuals used for bulk animal feeds and biofuels— the "biorefinery" concept. However, for animal feed, the cost of the fractionation would likely make selling the entire algal biomass more profitable. For example, lutein, a carotenoid, which is currently extracted from marigold petals and sold as an ingredient for chicken feed (to color eggs and skins), could be produced with microalgae, but it would be more profitable for the whole algal biomass to be sold without extracting the lutein. For the human nutraceutical markets, only small amounts of biofuels—such as biogas—would be derived from such biorefineries, helping to reduce the carbon footprint (greenhouse gas emissions) of the process. Although not a significant economic driver, this is of increasing importance for all enterprises.

THE NEXT STEP: FROM HIGH-VALUE PRODUCTS TO COMMODITIES

Microalgae production of specialty animal and aquaculture feed, which are intermediate in value and market size between human nutritional products and commodity feed and biofuels, would provide the next step in the development of this technology. By requiring larger-scale production plants, economies of scale would significantly decrease production costs. However, this alone will not be sufficient—even greater cost reductions must be achieved through technological advancements in the entire production process, starting with the algal strains being cultivated to the processing of the biomass. In particular, improved algal strains will be required that exhibit higher productivities (see Chapters 1 through 6), and also can be protected against invasions and infections by other algae, grazers, fungi, and so on (Chapters 7 and 8). The use of CO_2 from power plants can also be viewed as a benefit, but implementation will be a significant cost (see Chapter 10). Advances in harvesting and dewatering must also be implemented to reduce overall production costs (see Chapter 11).

Similar technology advances will be required in microalgal culture in fermenters, in particular the use of lower-cost reactors. These will be more susceptible to contamination, thus requiring more robust strains (see Chapter 14). Such advances could provide the "stepping-stone" to move from the current very high-value niche nutraceuticals markets to the production of lower-cost specialty feed, with the ultimate goal of commodity feeds and fuels.

Microalgae have several major advantages over terrestrial crops, in particular, their ability to use seawater and brackish waters is of great importance where fresh-water resources are limiting. They can also remediate wastewater (see Chapter 9). Perhaps more importantly, they have a potential for much higher productivities than crop plants. Of course, there are also drawbacks that must be overcome, beginning with the undeveloped nature of this technology, which remains to be proven beyond the current handful of algal species that are being produced commercially on a small scale at a high cost. A great deal more research is required.

The algal biomass industry is still at a relatively early stage of development. With an estimated annual 30,000 tons of dry weight biomass produced globally, roughly equally divided between autotrophic and heterotrophic production, this is still a very small industry, compared with the agriculture, fermentation, or even seaweed (macroalgae) industries. For a comparison, global production of palm oil, for instance, amounts to over 50 million tons annually, and even linseed oil, a minor product, stands at half-a-million tons (FAOSTAT 2013). Fish oil is nearly a one million ton per year product (Shepherd and Jackson 2012) and animal feed is approaching a billion tons (Alltech 2014). About 25 million tons (wet weight) of commercial seaweed are produced annually (FAO 2014). Global biodiesel production only amounts to about 0.5% of the world total crude oil production (EIA 2012).

However, as noted earlier, the growth of the microalgae industry is accelerating: *Spirulina* biomass production—the largest component in the industry—has been expanding for over a decade at an estimated annual growth rate of 15% (i.e., doubling every 5 years). If such a market-driven growth were applied to all commercial microalgae production, autotrophic and heterotrophic, by 2030 the total output of algal biomass would reach a quarter million tons, and by 2050 the industry would be making a major impact in commodities markets, both in animal feed and fuel.

The question is how can the microalgae industry grow and reach the 100-fold higher production levels of output required to compete in the same league as other commodity crops? This will require a more than 10-fold reduction in overall costs, necessitating major improvements in algal biomass productivity and major cost reductions at all stages of the production process, along with diversification into new markets in the value product value-product sector of commodity markets. These topics, which are dealt with in detail in the chapters that follow, are discussed briefly next.

BIOMASS PRODUCTIVITY AND PHOTOSYNTHETIC EFFICIENCY

Biomass productivity, reflects the overall photosynthetic efficiency achieved by the algal strains cultivated in the production system, whether a closed or open system (Chapter 1). It depends on the genetic, thus physiological, characteristics of the algal cells and how these respond to the environment encountered during cultivation. Ambient temperature, and its diurnal fluctuations, is a major factor that affects over-all photosynthetic efficiency (Chapter 2). Strains must be selected for growth under prevailing temperature conditions. Another factor is O_2, algal cultivation systems accumulate high concentrations (several-fold air saturation levels) and algal strains must be selected to be resistant to O_2 inhibition. The fluctuations in light received by the algal cells, due to mixing of the ponds or PBRs and mutual shading, is another, though poorly understood, parameter affecting overall photosynthetic efficiency. Again, strains must be selected that have adapted to the light regime in the production systems. Selection protocols would need to utilize the algal cultivation system itself, in which the most productive strains would outcompete and ultimately displace the less efficient ones (Chapter 1).

Some factors are, however, not dependent on the selection of algal strains. For example, CO_2 supply increases productivity over 10-fold, as the transfer of CO_2 from the atmosphere into ponds is very limited (Chapter 2). Another example: nitrogen should be supplied as ammonia or urea; nitrate, apart from being more expensive, would reduce productivity by using up about 10% of the photosynthetic reductant during assimilation (Chapter 1).

Further increases in photosynthetic efficiency, beyond those achievable by the selection of strains that have adapted to the environment of the cultivation system, will require more advanced techniques, including genetic modifications. Microalgae have evolved to maximize productivity of the individual cell, rather than the whole culture. They evolved to use light at low levels, by assembling large light-capture pigment arrays ("antenna"), but are not efficient at high light levels, resulting in most of the light actually absorbed being wasted (Chapter 2). This can be overcome by reducing antenna size, which would minimize dissipation (i.e., wastage) of light (photons) at midday, leading to more efficient energy conversion by the culture as a whole (Chapter 1). This approach could lead to a doubling of photosynthetic efficiency (Chapter 2).

Gains in productivity are not as likely achievable by increasing the rate of photosynthesis itself, as this would entail elevating both electron transport in the photosynthetic apparatus and the levels of enzymes required for carbon fixation (Chapter 2). Instead, alterations to the Calvin cycle components and respiratory pathways, to minimize energy wastage, are suggested as a more viable option (Chapter 1). In the long term, this might be accomplished by introducing completely artificial pathways that are more efficient. Another approach that has been suggested is to boost the CO_2 supply to the algal cells, through more active carbon concentrating mechanisms (CCMs), though it is not clear if this is actually a limiting factor during algal cultivation.

SCREENING AND SELECTION OF SUPERIOR STRAINS

There are still a limited number of algal species and strains that are currently being used industrially. As suggested earlier, screening and selection for new strains, better adapted to the cultivation environment, would lead to improved biomass productivity and increased outputs of the products of interest, an even more important goal than increased productivity alone. There is also a strong case for continued investigation of marine and extremophile environments to obtain both more robust strains and to avoid competition with agriculture. The very wide phylogenetic diversity of the hundreds of thousands of algal species and essentially infinite strains found in the biosphere argues for the potential of this approach. Chapters 3 and 4 provide examples of such screens for superior algal strains.

A protocol for isolating strains from environmental samples and the selection of locations for such collections are presented in Chapter 3. The samples are filtered to remove grazers, followed by immediate plating on agar in the field, followed by growth in liquid media and flow cytometry (FACS cell sorting) for isolation of clonal colonies. These then underwent several further rounds of screening to identify high lipid producers. This approach yielded a subset of promising strains for large-scale testing from an input of 5000 environmental isolates (Chapter 3). The testing of one of these strains is described in Chapter 5.

A slightly different approach is described for the screening of marine strains from the CCAP collection at SAMS (Chapter 4). This screen entailed a primary screen for growth in saline media followed by a secondary screen with replicate cultures to assess biomass content and yields. The focus was on compositional analysis including for fatty acids, elemental C and N, proteins, and carbohydrates. This generated a subset of top strains for producing biofuels (oil or carbohydrate), feeds, aquaculture, and high-value fatty acids. The outcome from these and other screens was that the highest biomass and oil productivities were from strains of green algae (*Scenedesmus*, *Desmodesmus*, *Chlorella*) and Eustigmatophyceae (marine *Nannochloropsis*).

A major bottleneck is testing the candidate strains at their actual scales in specific locations—a costly and time-consuming exercise. Chapter 5 shows that the outdoor cultivation at scale could be effectively modeled from lab-scale measurements of growth parameters and climate data. This could allow for better laboratory screening of strains suited for large-scale cultivation. This chapter presents a model for the batch growth for a *Chlorella sorokiniana* strain (isolated during the screening work described in Chapter 3), which was generated from data obtained from laboratory growth experiments and can accurately predict batch cultivation growth in outdoor ponds at a site in Arizona.

GENETIC IMPROVEMENTS

Model strains that emerge from screening efforts could then be used to generate metabolomic, genomic, and related "big data" sets, which would inform the genetic modifications needed for increasing biomass productivity, a requirement for reducing costs. Genetic modification of microalgae is reviewed in Chapter 6, and aspects relevant to higher oil production by marine strains are discussed in Chapter 4. In Chapter 6, it is argued that the limited options and knowledge for sexual breeding in most algal species, combined with the shortcomings of mutagenesis, favor using genetic engineering technologies in strain improvement R&D. Regulatory approval for GM algae, particularly in the EU, is a major uncertainty however, given that containment is uncertain. Advances in molecular biology that allow genetic modifications without the introduction of foreign genes suggest an avenue to overcome objections to GM strains (see Chapter 6). Genetic technologies to produce high-value and novel products were demonstrated using model algae such as *Chlamydomonas* and *Phaeodactylum*, which are not production strains, except for high-value pharmaceuticals grown at a small scale, under highly controlled conditions. For lower-value products, this work needs to be transferred into existing or novel top production strains.

Chapter 4 describes some successful inroads already being made into increasing oil content in microalgae by reducing partitioning of photosynthesis products into starch and by suppressing factors regulating oil mobilization. Regulatory factors that control oil production directly, must still be identified.

CROP PROTECTION: AVOIDING LOSSES TO GRAZERS, PATHOGENS, AND OTHERS

A major impediment to the advancement of the algal industry, in particular for open ponds, are weeds, pests, and pathogens that all too often decimate algal cultures. There are myriads of these: invading wild algae, grazing zooplankton and protozoa, such as amoebae and ciliates, fungi—in particular, the chytrids, lytic bacteria, and viruses all of which can decimate algal cultures (Chapters 7 and 8). Crop protection methods for dealing with such pests and pathogens are discussed in Chapter 7. Even relatively simple chemical treatments, such as high ammonia or low concentrations of hypochlorite can be effective. Other crop protection options include the development of pesticides and fungicides, biological control agents, the cultivation of extremophiles—which have few competitors and pests—or the development of consortia of algal strains that could be more stable and resistant than single strains.

The cornerstone of any crop protection strategy is early detection of the pathogenic agent, to increase the chances of an intervention being successful, see Chapter 8. Molecular methods offer the possibility of rapid diagnosis of pathogens combined with high-throughput analyses. Ultimately, this would mean routine and highly automated sampling and extraction of nucleic acid for PCR-based methods (or microarray techniques), which essentially provide bar-coded information on pathogens, or next-generation (next-Gen) sequencing, which provides a more complete analysis of the entire microbiome. Both approaches can also yield quantitative data, assuming an even extraction of nucleic acid. Hence, transient increases in pathogen nucleic acid can be correlated with events such as culture crash, improving the identification of the causative agent, with several examples of this given from different phyla in Chapter 8. Future developments could include using similar methods to monitor distress signals in the target strain. Complementary approaches to molecular approaches could include remote sensing of changes in spectral properties of ponds in response to pathogens or detecting engineered early-warning signals from modified strains (Chapter 7).

NUTRIENT AND CO$_2$ SUPPLY

In intensive microalgal cultivation, the only factors that should limit productivity should be those that are not controllable, that is, light and temperature. A major requirement is to avoid nutrient deficiencies, with nitrogen and phosphorous generally the major, but far from the only, nutrients required (see Chapter 1). The bulk of nitrogen fertilizer supplied in agriculture is produced from fossil fuel by the Haber–Bosch process, yet excess nutrients (e.g., fertilizer runoff) are an increasing source of pollution from agriculture and livestock husbandry, and remediation measures are required. Processes using microalgal biomass production to both recover fertilizer values and produce valuable co-products are being developed (Chapter 9). Seasonal variation in output must be taken into account, however, and complete water remediation may require a two-stage method to take advantage of the high nutrient-removal capacity at low concentrations in microalgae. In aquaculture, partitioned pond systems have been shown to improve fish productivity and also provide significant algal biomass output at the same time through sedimentation or collection of fish excreta that can be used for biofuels/biogas or as fertilizer.

The single largest nutrient required for algal biomass production is CO_2. As noted earlier, unlike the situation for higher plants, with their large leaf surface (leaf area index) and leaf architecture, CO_2 from air is insufficient to support more than about a tenth of the algal growth achieved with CO_2 fertilization. CO_2 supply is both a resource issue—a suitable source is required, preferably at 10% or higher concentration (e.g., the flue gas from a coal-fired power plant), and an economic issue—as the distribution and transfer of such CO_2 is a major cost. Although algae cultures allow for flue gas, that is, fossil CO_2, utilization, in principle this is no different from pulling CO_2 from the air. CO_2 supply to algal cultures is somewhat complex, requiring a balancing of the requirements of pH control and loss of CO_2 during and after transfer into the cultures, but is also relatively well understood (Chapter 10).

HARVESTING

Next to the growth ponds themselves, the harvesting/dewatering process is one of the largest costs in microalgae production. Settling is a relatively low-cost process, but few of the microalgae species being produced commercially or of interest in biofuels production will settle, at least not fast enough, for a practical process. These, however, require some chemical or physical process, such as chemical flocculation or centrifugation, costly both in capital and operating costs, and energy inputs (Chapter 11). Photobioreactors and shallow cascade ponds, have an advantage here over raceway ponds—with typically 10-fold, or higher, cell density and concomitant decreases in harvesting cost. Autoflocculation induced by pH or magnesium/calcium coprecipitation is a potentially promising process, but only where water chemistry favors this. Bioflocculation, in which the algal cells spontaneously flocculate and settle when removed from the growth system, has the greatest potential, but must be developed for specific cases and algal strains (Chapter 11).

THE FUTURE OF THE MICROALGAE INDUSTRY

The algal biomass industry, for applications from high-value human foods to replacements for fossil fuels, is only in its initial, formative stages. It has the potential to diversify into many existing and, most likely, new markets. This is assured by the diversity of microalgae, with their large variety of products of interest and metabolic capabilities.

For example, the green alga *Haematococcus pluvialis* currently produces the most valuable algal product, natural astaxanthin, rendering production highly profitable. One advantage is that large nonmotile cells contain the product astaxanthin, and this allows for easy harvesting by settling of cells. But it is also very challenging due to the relatively complex life cycle of this alga and the susceptibility of the cultures to infections (Chapter 12). It is likely that in a few years, production costs will decrease by an order of magnitude, allowing competition with synthetic astaxanthin used to color farm-raised salmon—a $200 million per-year industry.

Another example is *Chlorella*, which has been grown commercially since the 1960s in open ponds. The use of closed systems, that is, PBRs, offers a production process system that is more controlled and allows for a cleaner and, thus, higher-quality product (Chapter 13). *Chlorella* is also currently produced heterotrophically in fermenters, and these now allow application of GMO techniques to produce novel oils and products (Chapter 14). Indeed, growth in fermenters is the other major approach to algal production, though in this they would need to compete with other such processes using yeast or bacteria. One area of research is to reduce costs by avoiding reactor sterilization (Chapter 14).

The three basic production processes discussed herein—open ponds, closed photo-bioreactors, and fermenters—are all being developed in many different designs and configurations, some are already being used at the industrial scale. Balancing the costs of production with what the markets will bear, becomes the major issue. Each approach has its strengths and limitations: open systems are lower in cost but more vulnerable to pathogens; closed systems are more expensive but more controlled. Fermentation requires inexpensive cheap sugars or other substrates and expensive fermentation systems. Each technology will likely gravitate toward specific products and markets. The long-term goal of mass production of lower-value products such as biofuels and chemical feedstocks will probably be reached via intermediate-value products, such as specialty animal feeds.

Taken together, research and development must continue and the industry must take advantage of the great advances being made in all areas by basic and applied research at academic and government laboratories. There are still many challenges to overcome before microalgae production can become a large bioindustry, with millions of tons of biomass and products being produced, generating tens of billions of dollars in revenue. The chapters that follow provide a signpost to the various routes for achieving this objective.

REFERENCES

Alltech. 2014. Global feeds survey summary. http://www.alltech.com/sites/default/files/alltechglobalfeedsummary2014.pdf (accessed December 20, 2015).

Ben-Amotz, A. and Y. Levy. 1996. Bioavailability of a natural isomer mixture compared with synthetic all-*trans* beta-carotene in human serum. *Am. J. Clin. Nutr.* 63:729–734.

Benemann, J. R., B. L. Koopman, J. C. Weissman, D. M. Eisenberg, and P. Goebel. 1980. Development of microalgae harvesting and high rate pond technologies in California. In *Algae Biomass: Production and Use*, eds. G. Shelef and C. J. Soeder, pp. 457–496. Amsterdam, the Netherlands: Elsevier.

Benemann, J. R., B. L. Koopman, J. C. Weissman, D. M. Eisenberg, and W. J. Oswald. June 1978. An integrated system for the conversion of solar energy with sewage-grown microalgae. Report, Contract D (0-3)-34, U.S. Department of Energy, Washington, DC, SAN-003-4-2.

Borowitzka, M. A. and N. R. Moheimani. 2013. Open pond culture systems. In *Algae for Biofuels and Energy*, eds. M. A. Borowitzka and N. R. Moheimani, Vol. 5, pp. 133–152. Developments in Applied Phycology. Dordrecht, the Netherlands: Springer.

Burlew, J. S. 1953. *Algae Culture: From Laboratory to Pilot Plant.* Washington, DC: Carnegie Institution of Washington.

Ciferri, O. 1983. *Spirulina*, the edible microorganism. *Microbiol. Rev.* 47:551–578.

EIA. 2012. International energy statistics. U.S. Energy Information Administration, Washington, DC. http://www.eia.gov/cfapps/ipdbproject/IEDIndex3.cfm (accessed December 20, 2015).

FAO. 2014. The state of world fisheries and aquaculture: Opportunities and challenges. Rome, Italy. http://www.fao.org/3/a-i3720e.pdf (accessed December 20, 2015).

FAOSTAT. 2013. Crops processed. Food and Agriculture Organization of the United Nations Statistics Division. Rome, Italy. http://faostat3.fao.org/browse/Q/QD/E (accessed December 20, 2015).

Guo, M., G.B. Ding, S. Guo et al. 2015. Isolation and antitumor efficacy evaluation of a polysaccharide from *Nostoc commune* Vauch. *Food Funct.* 6:3035–3044.

Harder, R. and H. V. von Witsch. 1942. Über Massenkultur von Diatomeen. *Ber. Deut. Bot. Ges.* 60:146–152.

Horrocks, L. A. and Y. K. Yeo. 1999. Health benefits of docosahexaenoic acid (DHA). *Pharmacol. Res.* 40:211–225.

Oswald, W. J. and C. G. Golueke. 1960. Biological transformation of solar energy. *Adv. Appl. Microbiol.* 2:223–262.

Oswald, W. J. and H. B. Gotaas. 1957. Photosynthesis in sewage treatment. *Trans. Am. Soc. Civil. Eng.* 122:73–105.

Sheehan, J., T. Dunahay, J. Benemann, and P. Roessler. 1998. A look back at the U.S. Department of Energy's Aquatic Species Program: Biodiesel from algae. Close-out Report. Golden, CO. http://www.osti.gov/servlets/purl/15003040-tW7nZs/native/ (accessed December 20, 2015).

Shepherd, C. J. and A. J. Jackson. 2012. Global fishmeal and fish oil supply—Inputs, outputs, and markets. http://www.seafish.org/media/594329/wfc_shepherd_fishmealtrends.pdf (accessed December 20, 2015).

Warburg, O. 1919. Über Die Geschwindigkeit Der Photochemischen Kohlensäurezersetzung in Lebenden Zellen. *Biochem. Zeitschr.* 100:230–270.

1 Algal Photosynthesis and Physiology

John A. Raven and John Beardall

CONTENTS

1.1 INTRODUCTION

Algae, herein referring to microalgae including cyanobacteria, occur in almost all environments, exhibiting a wide range of specific growth rates and productivities. The diversity of algae includes genotypes that can, through evolution, grow over large ranges of mean and extreme values of flux in photosynthetically active radiation (PAR) of both low and high concentration of dissolved nutrients and of wide ranges in temperature (Raven and Geider 2003). They can also deal with the influences of both abiotic factors, such as salinity and mixing in the water column, and biotic factors, such as grazers and parasites/parasitoids, including viruses (Falkowski and Raven 2007). Biotechnological applications require algae adapt to the light, nutrient, and temperature conditions in the cultivation system, allowing relatively high growth rates of individual cells and overall high culture productivity. In addition, such biotechnological processes must yield the desired outputs, from oils for use as fuels to nutritional supplements, such as carotenoids, polysaccharides, and other cell wall components and even the whole organisms as biomass for use in feed and food. The need for high culture productivities as well as the desired product composition and quality imposes significant limitations on the algal species and strains to be cultivated (Borowitzka 2013; Raven and Ralph 2015).

Within the range of genetically determined attributes of microalgae, there is considerable phylogenetic variation, for example, in light-harvesting and photoprotective pigments and in biomass composition. The genetic potential extends also to determining the extent of phenotypic acclimation to variations in environmental conditions on timescales similar to, or less than, the relatively short generation time of these organisms, from a few hours to days. Organisms isolated from a natural environment and placed in a production system will respond both genotypically, through selection of new more competitive variant strains (adaptation), and phenotypically, by physiological acclimation to such novel environments for optimal utilization of limiting resources (Rosen 1967). However, such selection and acclimation might not be appropriate for biotechnological applications, where the goal is to maximize production either of biomass or of selected cell constituents, such as oils, carotenoids, and polysaccharides. Isolation and screening for particular strains from nature, followed by genetic selection under the cultivation process and by genetic modification, could help to achieve these biotechnological goals to produce the desired products in large amounts and at high productivity. In addition to the cultivation process, the algal biomass must also be harvested, dried, or processed to extract the products of interest and the culture media and unused nutrients recycled, for such a process to be as self-contained as possible (Borowitzka 2013; Raven and Ralph 2015).

To produce specific products, such as oils from microalgae, maximizing the production rate of oil involves minimizing the biomass allocated to all other functions while maintaining or increasing carbon assimilation rates for biomass increase, especially oil production. This requirement underpins the following discussion of algal physiology and photosynthesis, bearing in mind that pathways that achieve a high overall conversion of substrate to product frequently also exhibit lower energetic efficiencies (Odum and Pinkerton 1955; Raven 1984a,b; Bar-Even et al. 2010, 2011, 2012; Fabris et al. 2012; Beardall and Raven 2013; Flamholz et al. 2013; Raven and Ralph 2015). To achieve a "minimum allocation" of energy and other resources to the production of biomass other than the desired product involves a greater energy input in operating the pathways that produce the desired product (e.g., lipids, etc.). Balancing the requirements for catalytic biomass and product formation is a key issue addressed herein, and photon energy harvesting and transformation is the starting point in such a discussion.

1.2 PHOTON HARVESTING AND TRANSFER TO REACTION CENTERS

The absorption of photons in photosynthesis involves pigments (chromophores) bound to specific apoproteins in light-harvesting pigment complexes. These apoproteins alter the absorption properties of the chromophores compared to their free state and facilitate the very high efficiency of absorbed photon excitation energy (exciton) transfer to, ultimately, a reaction center pigment complex, where the transformation of exciton energy into chemical energy takes place. Note that the reaction centers absorb photons at somewhat longer wavelengths than the light-harvesting pigment complexes; this is necessary for the efficient transfer of exciton energy from light harvesting to reaction center complexes (Falkowski and Raven 2007; Raven et al. 2013, 2014; Raven and Ralph 2015). In all oxygenic photosynthetic organisms (except the cyanobacterium *Acaryochloris* and its relatives, which use chlorophyll *d*), reaction centers use the red absorption peak of chlorophyll *a* at 680 nm in photosystem II (PSII) and at 700 nm in photosystem I (PSI).

The range of light-harvesting pigments (Table 1.1) covers the entire visible spectrum. Table 1.1 also shows that there is variation in the specific absorption coefficients of the pigments, which indicates the rate of absorption of photons at a given incident photon flux density at the relevant absorption maximum. Integrating the specific absorption coefficient over the entire absorption spectrum of the pigment allows calculation of its effectiveness in absorbing photons and the photons required for synthesizing the pigment–apoprotein complex. Such calculations show that a dominant factor in variations in resource costs (i.e., energy, nitrogen in protein and, through the need for RNA in protein synthesis, phosphorus) of photon absorption is the quantity of apoprotein associated with each molecule of the pigment (Table 1.1) (Raven 1984a,b, 2013a,b; Raven et al. 2013). The apoproteins of light-harvesting pigment–protein complexes can be the most abundant proteins in photosynthetic cells acclimated and/or adapted to low PAR levels (Table 1.1) (Raven 1984a,b; Raven et al. 2013; Raven and Ralph 2015).

A diversity of light-harvesting pigments allows complete coverage of the 400–700 nm wavelength range, and altering their relative quantities would adjust the absorption spectrum of the organism to that of the incident light quality. The argument as to the utility of a great diversity of pigments applies most clearly to very small cells where there is a minimal self-shading in individual algal cells (the package effect) and the absorption properties of pigment–apoproteins are distinctly shown (Raven 1984a,b, 1994, 1996). Larger cells have more pigments in the

TABLE 1.1

Range of Specific Absorption Coefficients of the Chromophores and the Mass of Apoproteins per mol Photosynthetic Chromophore, of Oxygenic Photosynthetic Organisms and the Range of Photosynthetic Chromophore Apoproteins as a Fraction of Total Cell Protein of Microalgae

Metric	Value and Units	References
Mean (400–700 nm) specific absorption coefficient of photosynthetic chromophores	1212–3367 $m^2 mol^{-1}$ chromophore	Raven (1984a)
Mass of apoprotein per mol photosynthetic chromophore	2.0–15.7 kg protein per mol chromophore	Raven (1984a)
Mass of apoprotein of photosynthetic chromophore per mass of total cell protein	0.04–0.4 kg apoprotein per kg total cell protein	Raven et al. (2013)
Mass of RuBisCO protein per mass of total cell protein	0.02–0.16 kg RuBisCO per kg total cell protein	Raven et al. (2013)
Mass of ribosomal protein per mass of total cell protein	0.09–0.21 kg ribosomal protein per kg total cell protein	Raven et al. (2013)

Notes: For comparison, the fraction of cell protein contributed by ribulose bisphosphate carboxylase/oxygenase (RuBisCO) protein and by ribosomal protein is also shown.

optical path through the cell, so there is a larger package effect, resulting in reduced impact of individual pigment species on the overall absorption spectrum than in smaller cells. This argument applies to individual cells and thus is amenable to natural selection for best adaptation to different light quality.

The overall process of oxygen production from water by photosystem II (PSII) and reduction of carbon dioxide to the carbohydrate level by NADPH from photosystem I (PSI) (and ATP) has deep evolutionary roots, based in the photosynthetic proteobacteria and Chloroflexaceae with a PSII-like reaction center and the Chlorobiaceae and Heliobacteriaceae with a PSI-like reaction center (Falkowski and Raven 2007). However, their PSII-like photosystems have very long wavelength absorption by reaction centers, between 870 and 890 nm, whose energy does not generate a sufficiently high redox potential to oxidize water and produce oxygen (Falkowski and Raven 2007). It has been suggested that the wavelength range used by oxygenic organisms could be increased by replacing the existing PSI by the PSI-like photosystem from a photosynthetic green sulfur bacterium absorbing at longer wavelengths than PSI or, indeed, the PSI-like *Chlorobium* photosystem (Blankenship et al. 2011). Assuming that the *Chlorobium* photosystem, but with a reaction center using photons of a much longer wavelength (lower energy content) than the original, can be engineered into oxygenic photosynthetic organisms and that $NADP^+$ can be reduced by this introduced PSI-like photosystem, there would be a smaller redox span between the two photoreactions (Figure 3B of Blankenship et al. 2011). This suggests that fewer protons would be pumped, and hence ADP phosphorylated, per electron transferred from water to $NADP^+$. Such extensions of the wavelengths used for oxygenic photosynthesis would be of more use in terrestrial plants than in aquatic plants since even a shallow water body absorbs more infrared solar radiation than does the atmosphere. Provided that the algal culture is not deep, the absorption of infrared photons by water would still leave a significant fraction for absorption by bacteriochlorophylls. In either terrestrial or aquatic habitats, a significant fraction of the photons absorbed by the introduced photosystem would be in the blue region of bacteriochlorophyll absorption, in competition with the blue absorption peak of the chlorophyll of the original photosystem II, similar to the case with the chlorophylls of the original photosystem I. Further investigation of this interesting possibility is needed.

An additional point about the use of 680 and 700 nm for the respective photochemical reactions of the two photosystems, PSII and PSI, of oxygenic photosynthesis is that the wavelengths used are just on the long wavelength side of the wavelength of maximum photon emissions from the sun, as moderated by passage through the atmosphere. This means, for an organism on land or at the ocean surface, that over half of the photons emitted from the sun that penetrate the atmosphere can be used in photosynthesis, including some of the UVA. However, full energy per photon content of shorter wavelengths is not used in photosynthesis: all photon energy is collected by the reaction centers at 680 and 700 nm, amounting to a loss of about 20%–21% of the actual photon energy in the shorter wavelengths reaching the earth (Williams and Laurens 2010; Raven and Donnelly 2013). The use of the longest wavelengths compatible with the photochemistry required for the overall photosynthetic reaction might only be coincidentally related to the potential use of over half of solar PAR wavelengths if oxygenic photosynthesis evolved at depth in a water body, where most red (and almost all infrared) photons have been absorbed by the overlying water. The reason for this assumption of a deep origin of oxygenic photosynthesis is that, before the global oxygenation event and the origin of a stratospheric ozone shield absorbing UVB, photosynthetic organisms would have occurred at depths in the water where much of the damaging UVB is absorbed. A counterargument comes from the recent work on the environmental correlates of the evolution of cyanobacteria, the clade of photolithotrophic bacteria in which oxygenic photosynthesis evolved (Blank and Sanchez-Baracaldo 2010; Sanchez-Baracaldo et al. 2014). This work showed that the extant relatives of basal cyanobacteria are all from low-salinity habitats and that the most basal extant cyanobacterium is *Gloeobacter* (Blank and Sanchez-Baracaldo 2010; Sanchez-Baracaldo et al. 2014), which occurs on aerially exposed rock surfaces (Mareš et al. 2013; Saw et al. 2013). If this were the habitat of the earliest-evolving cyanobacteria, then they would have been exposed to red as well as shorter wavelengths of PAR, which can be used via the phycobilins as well as by chlorophyll *a* and carotenoids. This accords with the hypothesis that photochemistry takes place at the longest wavelengths compatible with overall energetics, allowing the use of as many of the incident solar photons as possible, including those in the red region that is not available deep in water bodies. However, this begs the question of how UVB was dealt with by these organisms; an extracellular UVB screening compound such as scytonemin (found in many extant cyanobacteria) could have been involved.

A final point about the light-harvesting machinery is the number of light-harvesting pigment molecules per PSI or PSII reaction center (one of the uses of the term photosynthetic unit and the meaning used here). Natural selection can overendow the organism with light-harvesting machinery relative to the mean PAR in the natural environment, for example, the diel change in PAR as a function of solar altitude and variations due to entrainment in vertical water movements in the upper mixed layer. Decreasing the photosynthetic unit size would, unless there is a compensating increase in the number of reaction centers per cell, mean less pigment per cell and hence a lower photosynthetic rate at low irradiances because less PAR is absorbed. At high irradiances, a smaller photosynthetic unit size (such as occurs for photosystem II, but not photosystem I, in mutant strains of algae) decreases the potential for photodamage from excess PAR plus UVB, relative to what can be used in autotrophic carbon dioxide assimilation. This, in turn, means that there is the potential for downregulating photoprotective machinery such as photochemical quenching or the variants, using both photosystems or only photosystem II, of the water–water cycle or non-photochemical quenching, including the xanthophyll cycles (Raven 2011; Raven and Ralph 2015). This sparing of the need for photoprotection, arising from fewer light-harvesting pigments per photosynthetic unit (defined earlier), requires that there also be no decrease in the number of such photosynthetic units per cell; otherwise, the capacity for carbon dioxide assimilation would be limited on a cell basis by restrictions on interphotosystem redox reactions and proton pumping and hence on regeneration of ribulose-1,5 bisphosphate and on ribulose-1,5-bisphosphate carboxylase/oxygenase (RuBisCO) activity. As we have noted, reduced pigment content will restrict photosynthetic rate at low irradiances. Thus, in assessing such

modifications to the photosynthetic unit, it is necessary to take into account the sinu-
soidal variation in solar radiation through the daily photoperiod. While decreasing
the photosynthetic unit size can limit photoinhibition around midday for cells near
the surface of the pond or reactor, it would also decrease the capacity for photosynthesis at the lower irradiances near dawn and dusk.

1.3 PHOTOCHEMISTRY AND DOWNSTREAM REACTIONS IN THE THYLAKOID MEMBRANE LEADING TO THE PRODUCTION OF NADPH, ATP, AND OXYGEN

The two photosystems (Figure 1.1) have different maximum photon yields. PSI has
a maximum photon yield approaching 1, that is, for each photon whose energy is
transferred to the photosystem I reaction center, 1 electron is transported from P_{700},
the reaction center chlorophyll a, to the primary acceptor, a special chlorophyll a, to
produce P_{700}^+ and chl a^- (Falkowski and Raven 2007; Raven et al. 2014; Raven and
Ralph 2015). PSII has a maximum photon yield of not more than 0.8, that is, for each
photon whose energy is transferred to the photosystem II reaction center, a maximum of 0.8 electrons from P_{680}, the reaction center chlorophyll a, is passed on to the
primary acceptor pheophytin a to produce P_{680}^+ and pheophytin a^- (Falkowski and
Raven 2007; Raven et al. 2014; Raven and Ralph 2015) (Table 1.2). The 4-electron
reaction of oxidizing two water molecules to produce one oxygen at the oxidizing
end of PSII and reducing two NADP$^+$ at the reducing end of photosystem I requires
4 photochemical reactions of PSII, with a photon requirement of 4/0.8 (=5), and 4
photochemical reactions of PSI, with a photon requirement of 4/1 (=4), giving a total
of 9 photons absorbed by the light-harvesting pigments (Raven et al. 2014; Raven and

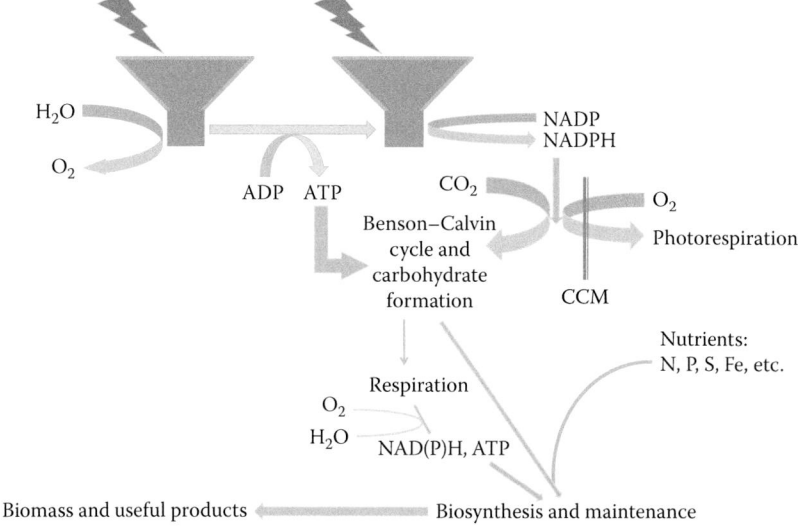

FIGURE 1.1 Conceptual diagram of the flow of energy and elements into biomass during
microalgal growth.

TABLE 1.2

Computed Minimum Photon Costs (as Absorbed Photons) of Assimilation of Carbon Dioxide into Carbohydrate by Oxygenic Photolithotrophic Microorganisms with Diffusive Entry of Carbon Dioxide to the RuBisCO–Benson–Calvin Cycle from the Present Atmosphere, or from the Present Atmosphere Using Carbon-Concentrating Mechanisms, or Using Diffusive Entry to the 3-Hydroxypropionate Bi-Cycle from the Present Atmosphere

Mechanism of Conversion of External Carbon Dioxide to Intracellular Carbohydrate in Microalgae	Minimum Cost of Absorbed Photons per mol Carbon Dioxide Converted to Carbohydrate	References
Diffusive entry of carbon dioxide to RuBisCO from the present atmospheric concentration of carbon dioxide	9.92–9.96 mol photons per mol carbon dioxide	Raven et al. (2014)
Entry of inorganic carbon using one of a range of carbon dioxide concentrating mechanisms, assuming a leakage of carbon dioxide equal to the net inorganic carbon assimilation rate	9.5–11.0 mol photons per mol carbon dioxide	Raven et al. (2014)
Diffusive entry of carbon dioxide to RuBisCO from a saturating concentration of carbon dioxide	9.0 mol photon per mol carbon dioxide	Raven et al. (2014)
Diffusive entry of carbon dioxide to the 3-hydroxypropionate bi-cycle from a saturating concentration (slightly higher than the present atmospheric concentration) of carbon dioxide	9.0 mol photon per mol carbon dioxide, plus one surplus mol ATP	Raven et al. (2012, 2014)

Notes: For comparison, values are given for diffusive entry to the RuBisCO–Benson–Calvin cycle from saturating carbon dioxide and for a (hypothetical) organism transformed to replace the RuBisCO–Benson–Calvin cycle with the 3-hydroxypropionate bi-cycle (the only known high carbon dioxide affinity, oxygen insensitive pathway of autotrophic carbon dioxide assimilation [Bar-Even et al. 2010, 2011, 2012; Raven et al. 2011, 2012, 2014]).

Ralph 2015) (Table 1.2). Each electron moving from water to NADP+ is associated with the energetically uphill movement of 3 protons from the plastid stroma (cytosol in cyanobacteria) to the thylakoid lumen, so that 12 protons are removed from the stroma per 4 electrons moved to NADP+ and per 9 photons absorbed (Falkowski and Raven 2007; Raven et al. 2014; Raven and Ralph 2015).

Oxidation reduction and associated proton pumping in the thylakoid reactions of photosynthesis are temporally and energetically downstream of the light reactions. As far as present knowledge allows such a generalization, these downstream reactions are conserved among oxygenic photosynthetic organisms in terms of stoichiometry and energetic efficiency. This similarity occurs despite some substitutions of redox catalysts, that is, plastocyanin or cytochrome c_6 as electron donors to the oxidizing end of photosystem I (P_{700}^+) and ferredoxin or flavodoxin as electron donors to the ferredoxin/flavodoxin/NADP+ oxidoreductase.

These substitutions may be completely (phylo)genetically determined (e.g., some algae only have plastocyanin, while some others have only cytochrome c_6). Alternatively, they can relate to genome–environment interactions, so that algae that can produce both plastocyanin (containing copper) and the iron-containing cytochrome c_c (=cytochrome cc) can have ratios of expression of the two proteins depending on the relative availability of the two metals. Furthermore, algae that can produce both ferredoxin (containing iron) and flavodoxin (contains no metals), express flavodoxin under iron-limiting conditions but otherwise express only ferredoxin (Raven et al. 1999; Blaby-Haas and Merchant 2012).

Returning to the stoichiometry and energetics of the proton pumping per electron transferred from water to NADP+, there does not seem to be any mechanistic or energetic possibility of increasing the proton/electron ratio in this noncyclic electron transport. The energetic argument relies in part on the proton/ATP stoichiometry of the ATP synthase and the *in vivo* free energy of hydrolysis of the resulting ATP.

Before considering the possibilities of modifying the stoichiometry of the ATP synthase, we examine the proton/electron ratio of cyclic electron transport driven by photoreaction I. The maximum possible proton/electron ratio of this cyclic electron flow is 4, with a ratio of 2 for the ferredoxin/NADP+ to PQ segment by analogy with the (mitochondrial matrix located) NADH/UQ segment of the mitochondrial electron transport chain (Falkowski and Raven 2007; Raven and Ralph 2015) and also 2 for the PQH_2 to cytochrome *f* segment by analogy with this segment of noncyclic electron flow and the UQH_2 to cytochrome *c* segment of the mitochondrial redox chain. It must be emphasized that this proton/electron ratio of 4 for cyclic electron flow round photoreaction I has not been demonstrated, but it is a potential aim for improving the energetic efficiency of this reaction in cases where photo-produced ATP is needed in excess of that produced by noncyclic electron transport and proton pumping, such as in powering carbon dioxide concentrating mechanisms (CCMs). In addition, there are also more direct uses of the proton gradient, for example, in bicarbonate to carbon dioxide conversion, catalyzed by the carbonic anhydrase Cah3 in the equilibrium dictated by the low pH maintained by the proton pump. The low pH is maintained despite removal of protons by transfer back to the stroma with the ATP synthase and, in *Chlamydomonas reinhardtii*, in conversion of bicarbonate to carbon dioxide in competition with ATP synthesis (Raven 1997; Raven et al. 2014). The reduced ferredoxin/NADPH to PQ segment of cyclic electron flow is also used in catalyzing the energized removal of hydroxyl ions generated in the conversion of carbon dioxide, which has entered the cytosol of cyanobacteria by diffusion through proteinaceous channels in the plasma membrane, into bicarbonate that is then consumed by carboxysomes. Such efforts directed toward achieving the theoretical maximum might address any shortfall in the proton/electron ratio of the reduced ferredoxin/NADPH(NADH) to PQ(UQ) segment with the measured proton/electron ratio being sometimes less than is predicted from the structural biology (Raven et al. 2014; Raven and Ralph 2015).

Turning to the CF_0/CF_1 ATP synthase in the thylakoid membrane, the theoretical maximum proton/ATP ratio is set by the ratio of proton-conducting *c* channels in the membrane-located portion to the alpha–beta adenylate-binding component in the stroma-exposed (cytosol exposed in cyanobacteria) portion, with 3 ADP bound and 3 ATP released per rotation of the rotor component. The plastid ATP synthases examined have 14 c subunits, and cyanobacteria have 13–15 c subunits, so the proton/ATP synthesis ratio predicted from the structural biology is 14:3 (=4.67) for plastids and 13:3 (=4.33), 14:3 (=4.67), or 15:3 (=5.0) for the cyanobacteria (Raven et al. 2014; Raven and Ralph 2015). However, the observed proton/ATP ratio for the plastid ATP synthase is only 3.9 $\pm$ 0.3, that is, significantly less than 4.67. This suggests that only 12 of the 14 proton channels are used in the synthesis from each rotation of the rotor and the phosphorylation of 3 ADP. This has energetic implications: for a constant value of free energy of hydrolysis of ATP of 55 kJ mol^{-1}, the free energy difference of protons across the thylakoid membrane must be more than 13.75 kJ mol^{-1} protons for a 12:3 stoichiometry, rather more than the expected 11.79 kJ mol^{-1} protons (Raven et al. 2014; Raven and Ralph 2015). This has implications for the energetics of the redox reactions, which drive the proton pumps that generate the proton gradient that powers the ATP synthetase. It is particularly relevant to look at the analogous redox reactions in the inner mitochondrial membrane, in which the ATP synthetase has 10 c subunits per complex and a predicted proton/ATP ratio of 3.33 but a measured proton/ATP ratio of 2.9 $\pm$ 0.2. The proton gradient across the inner mitochondrial membrane must then be in excess of 18.33 kJ mol^{-1} protons rather than the 16.5 kJ mol^{-1} proton based on the structural biological c subunit count per rotor, that is, 10. It seems that the free energy of hydrolysis of ATP in the mitochondrial matrix is less than in the cytosol, with an energy "top-up" from the energized ATP–ADP exchange and phosphate import at the inner mitochondrial membrane (Raven 1984b; Falkowski and Raven 2007; Nichols and Ferguson 2013). Accordingly, the required proton free energy difference across the inner mitochondrial value may be only 3/4 of the value quoted earlier for the proton/ATP ratio of 2.9, that is, 3/4 of 18.3 kJ mol^{-1} or 13.75 kJ mol^{-1}. This is the same as the required value for the proton gradient at the thylakoid membrane of plastids. Accordingly, the energetics of the plastidial and the mitochondrial redox-driven proton pump are identical. For noncyclic electron transport and proton pumping, the relevant comparison is between the PQH_2 to cytochrome f segment in plastid thylakoids and UQH_2 to cytochrome c_1 segment of the inner mitochondrial membrane. For the highest possible proton/electron ratio of cyclic electron flow in plastids, the comparison is between the ferredoxin/NADP$^+$ to PQ segment and the NADH (in the matrix) to UQ segment of the mitochondrial electron transport chain (Raven and Ralph 2015).

A final point that must be considered in any suggestion of increasing the energetic efficiency of ATP synthesis is an additional role for the pH difference (but not the electrical potential difference) component (Falkowski and Raven 2007) of the proton free energy difference across the thylakoid membrane. Excessive excitation of PSII can produce a very high pH difference because there is a larger redox supply of energy driving the proton pumps with no increased possibility of the use of the proton gradient in additional ADP phosphorylation. The stromal pH is kept essentially constant, and the lumenal pH decrease under very high irradiances initiates non-photochemical quenching of excitation (including the xanthophyll cycles) that restricts photodamage to PSII (Raven 2011).

To conclude the consideration of the thylakoid components of photosynthesis, the present minimum photon cost of noncyclic photophosphorylation is 9 photons to oxidize two water molecules, reduction of 2 $NADP^+$, and the phosphorylation of three ADP (Table 1.2). The partial reactions are shown in the following equations:

$$2\ H_2O_{(lumen)} + 2\ NADP^+_{(stroma)} + 12\ H^+_{(stroma)} + 9\ photons\ (400–700\ nm)$$
$$\rightarrow O_{2(lumen)} + 2\ NADPH_{(stroma)} + 2\ H^+_{(stroma)} + 12\ H^+_{(lumen)} \tag{1.1}$$

$$3\ ADP_{(stroma)} + 3\ phosphate_{(stroma)} + 12\ H^+_{(lumen)}$$
$$\rightarrow 3\ ATP_{(stroma)} + 3\ H_2O_{(stroma)} + 12\ H^+_{(stroma)} \tag{1.2}$$

Combining Equations 1.1 and 1.2:

$$2\ H_2O_{(lumen)} + 2\ NADP^+_{(stroma)} + 3\ ADP_{(stroma)} + 3\ phosphate_{(stroma)} + 9\ photons$$
$$(400–700\ nm) \rightarrow O_{2(lumen)} + 2\ NADPH_{(stroma)} + 2\ H^+_{(stroma)} + 3\ ATP_{(stroma)} + 3\ H_2O_{(stroma)} \tag{1.3}$$

Most of the NADPH and ATP generated in noncyclic photophosphorylation are used in CO_2 assimilation via the Benson–Calvin cycle, which is considered in the next section. At CO_2 saturation, the cofactor stoichiometry of CO_2 assimilation into carbohydrate, represented as (CH_2O), is shown in the following equation:

$$2\ NADPH_{(stroma)} + 2\ H^+_{(stroma)} + 3\ ATP_{(stroma)} + 3\ H_2O_{(stroma)} + 1\ CO_{2(stroma)}$$
$$\rightarrow 1\ (CH_2O)_{(stroma)} + 1\ H_2O_{(stroma)} + 2\ NADP^+ + 3\ ADP_{(stroma)} + 3\ phosphate_{(stroma)} \tag{1.4}$$

The sum of these reactions is shown in the following equation, where the asterisk indicates oxygen originating in water:

$$2\ H_2O^*_{(lumen)} + 1\ CO_{2(stroma)} + 9\ photons\ (400–700\ nm)$$
$$\rightarrow 1\ O_2^*_{(lumen)} + 1\ (CH_2O)_{(stroma)} + 1\ H_2O_{(stroma)} \tag{1.5}$$

1.4 CO₂ ASSIMILATION USING THE BENSON–CALVIN CYCLE: APPARENT INEFFICIENCIES AND THE MEANS OF COUNTERING THEM

The Benson–Calvin cycle uses ribulose-1,5-bisphosphate carboxylase/oxygenase (RuBisCO) as the carboxylase, carrying out the reaction shown in the following equation:

$$1\ CO_2 + 1\ H_2O + 1\ ribulose\text{-}1,5\text{-}bisphosphate^{2-} \rightarrow 2\ 3\text{-}phosphoglycerate^{2-} + 2\ H^+ \tag{1.6}$$

The regeneration of one ribulose-1,5-bisphosphate^{2-} from the two 3-phosphoglycerate^{2-} with the production of 1 (CH_2O) uses 2 NADPH + 2 H$^+$ plus 3 ATP.

Equation 1.6 represents the reaction found in the absence of O_2 or at very high CO_2/O_2 ratios. RuBisCO also has an oxygenase activity, expressed in the presence of oxygen, when there is insufficient carbon dioxide to outcompete the oxygen and thus allow expression of the oxygenase activity, as shown in the following equation:

$$1\ O_2 + 1\ \text{ribulose-1,5-bisphosphate}^{2-}$$
$$\rightarrow \text{3-phosphoglycerate}^{2-} + \text{2-phosphoglycolate}^{2-} + 2\ H^+ \quad (1.7)$$

RuBisCO evolved well before the global oxygenation event about 2.3 billion years ago (Falkowski and Raven 2007; Raven et al. 2012), with RuBisCO involved in autotrophy in some chemolithotrophs and also in some anoxygenic photolithotrophs. Under these conditions, the oxygenase activity was not expressed, just as in anoxic habitats today. The origin of photolithotrophy in cyanobacteria (all of which use the Benson–Calvin cycle of autotrophic carbon dioxide fixation) in freshwater habitats (Blank and Sanchez-Baracaldo 2010) led to first local buildup of free oxygen and hence local occurrence of the oxygenase activity. The start of the global oxygenation event occurred when cyanobacteria started proliferating into the ocean. Coupled with burial of some of the organic carbon produced, the equivalent oxygen was available to start oxygenation at the surface ocean and the atmosphere, and ultimately, more than a billion and a half years later, the whole ocean. The relatively high carbon dioxide at the time of the global oxygenation event would have limited the oxygenase activity, but there was (geologically) soon a low carbon dioxide episode indicated by an ice age, although the carbon dioxide would not have to be as low as that occurring in more recent ice ages, because the young sun was fainter than is the case today (Sanchez-Baracaldo et al. 2014).

Responses to oxygenase activity include conversion of 2-phosphoglycolate to glycolate by 2-phosphoglycolate phosphatase, followed by glycolate excretion. This would lead to a negative carbon balance of photosynthesis in cyanobacteria in today's air-equilibrated water bodies, as a result of the kinetics of the oxygenase and carboxylase activity of the Form IA and Form IB RuBisCOs found in cyanobacteria (Raven 1984b; Raven et al. 2012). All cyanobacteria, and other oxygenic photosynthetic organisms, express photorespiratory carbon oxidation cycles that convert two 2-phosphoglycolate into one 3-phosphoglycerate (and hence triose phosphates) with the loss of one carbon dioxide and an input of ATP and NADPH for the metabolism of glycolate, as well as the ATP and NADPH used by the oxygenase rather than the carboxylase activity (Raven et al. 2012). Raven et al. (2014) showed that with diffusive entry of carbon dioxide from the medium to a RuBisCO, with the kinetics giving the smallest known ratio of oxygenase to carboxylase activities in an air-equilibrium solution (Tcherkez et al. 2006), the minimum photon cost of photosynthesis is 9.92–9.96, the range depending on the photorespiratory carbon oxidation cycle used (Raven et al. 2014). However, most RuBisCOs involve higher photon costs (Table 1.2).

All extant cyanobacteria, most algae, and a minority of embryophytes ("higher plants") have CCMs that increase the carbon dioxide concentration at the active site of RuBisCO. This saturates (or almost saturates) the carboxylase activity and greatly decreases the oxygenase activity of RuBisCO. There is a great range of mechanisms for these CCMs. The minimum photon cost in a number of well-characterized cases is discussed by Raven et al. (2014), who include the costs of leakage of carbon dioxide from the intracellular pool at a rate equal to the rate of photosynthesis and give a range of photon costs of photosynthesis of 9.5–11.0 (Table 1.2), that is, spanning the minimum costs of the most energetically economical photorespiratory carbon oxidation cycles following diffusive carbon dioxide flux from the medium to RuBisCO.

These additional energy costs for using the RuBisCO–Benson–Calvin cycle in air-equilibrated growth media (Table 1.2) could be avoided by the use of an alternative autotrophic pathway for carbon dioxide fixation that uses carboxylases without an oxygenase function and which have ATP costs equal to, or less than, that of the carbon dioxide–saturated RuBisCO–Benson–Calvin cycle. The reductant costs of converting carbon dioxide to carbohydrate are, of course, independent of the pathway used. Alternative carboxylases often have relatively low affinities for carbon dioxide (as substrate or, for those using bicarbonate, after conversion to equilibrium carbon dioxide at the appropriate intracellular pH value), but some have affinities equal to, or rather higher than, those of the highest-affinity RuBisCOs (Raven et al. 2012). Finally, the specific reaction rate on a mass of enzyme–protein basis of even the fastest-working RuBisCOs is less than that of some of the alternative autotrophic carboxylases (Raven et al. 2012). This means that the fraction of total cell protein taken up by RuBisCO is high, although this fraction is significantly decreased by the presence of CCMs (Losh et al. 2013; Raven 2013a,b) (Table 1.1).

Replacing the RuBisCO–Benson–Calvin cycle with an alternative autotrophic carbon dioxide assimilation pathway, with no oxygenase activity and a high affinity for carbon dioxide, would mean that no photorespiratory carbon oxidation cycle or CCM would be needed. This would decrease the ATP requirement for operating the pathway and also decrease the energy and nutrients needed to synthesize the enzymes in the pathway. Of the alternative carbon dioxide assimilation pathways found in nature, the only example that fulfills these requirements and, crucially, is not inhibited by oxygen and has a lower energy (absorbed photon) cost than the RuBisCO–Benson–Calvin cycle is the 3-hydroxypropionate bi-cycle (Bar-Even et al. 2010, 2011, 2012; Raven et al. 2012) (Table 1.2). There is also the possibility of synthetic cycles produced using enzymes from a number of other pathways (Bar-Even et al. 2010, 2011, 2012). Any such replacement would need retention of those enzymes of the Benson–Calvin cycle (reductive pentose phosphate cycle) required for the production of the erythrose-4-P or pentose phosphates used in biosynthesis. These enzymes are also found in the oxidative pentose phosphate pathway. Such replacement of the Benson–Calvin cycle has not yet been achieved.

1.5   ASSIMILATION OF OTHER RESOURCES NEEDED FOR PHOTOLITHOTROPHIC GROWTH

The abiotic factors most frequently limiting the growth of algae in nature are frequently limited by deficiency or excess of PAR (considered in Sections 1.2 and 1.3), too high or too low temperature, and the elemental resources such as nitrogen, phosphorus, and iron (Raven et al. 1999; Raven and Geider 2003; Falkowski and

Raven 2007). Microalgae have a range of transporters with a range of affinities for inorganic combined nitrogen (i.e., non-N_2) and organic nitrogen species (Falkowski and Raven 2007), as well as for inorganic phosphate (Raven 2012, 2013a). Organic phosphates are made available by extracellular phosphatases that access phosphate esters and (in some cyanobacteria) uptake of organic phosphonates with intracellular production of inorganic phosphate (Raven 2012, 2013a). Iron uptake from oxic environments involves extracellular ferric iron, using siderophores (iron-binding peptides produced by cyanobacteria and other bacteria) that bind the iron, and the ferric iron–siderophore complex is taken up by the cells. Eukaryotic microalgae generally have extracellular ferric iron reductases, though the transporter that moves iron into the cell uses ferric iron generated by a copper-containing ferrous iron oxidase, thereby increasing the specificity of the transporter for iron (Blaby-Haas and Merchant 2012; Marchetti et al. 2012; Raven 2013c). Combined nitrogen levels limit the productivity of nondiazotrophic algae over much of the world's ocean as a result of the failure of diazotrophy by (mainly) cyanobacteria to fix dinitrogen at a rate adequate to overcome the losses of combined nitrogen by sedimentation and, especially, denitrification (Falkowski and Raven 2007). Restrictions on diazotrophy by marine cyanobacteria include the availability, in various parts of the ocean, of the large iron and/or phosphorus requirements of diazotrophs (Sohm et al. 2011; Raven 2012, 2013a,b; Sanchez-Baracaldo et al. 2014; Weber and Deutsch 2014).

In the biotechnological use of algae in cultures, the supply of nitrogen, phosphorus, and iron can generally be arranged so as not to be growth limiting, and with saturating concentrations of the relevant chemical species of these elements, the energy costs of uptake should be minimal. This contrasts with most natural environments in which microalgae evolved where nutrients are typically present at very low concentrations, and there can be a significant energetic cost in active nutrient uptake.

An additional energy cost is incurred by a change in redox state of the element between the form supplied and a more reduced form that is assimilated during growth. No redox change is involved in the assimilation of phosphate (inorganic or as esters), and the ability of some cyanobacteria to use exogenous organic phosphonates involves their uptake followed by cleavage and oxidation in producing phosphate. The small quantity of iron required means that any required reduction of ferric to ferrous iron requires very little energy. By contrast, the commonly used nitrogen source, nitrate, requires significant energy inputs. We assume a carbon/nitrogen ratio is equal to the Redfield ratio of 106:16 by atoms, or 6.625. For nitrate reduction to ammonium entirely in the photophase, the reduction of each nitrate requires four NAD(P)H that, with 4.5 photons needed per $NADP^+$ reduced, means 18 absorbed photons. Per carbon dioxide assimilated means 2.72 more absorbed photons than the values listed earlier for photons per carbon assimilated into carbohydrate, that is, an increased photon cost of up to 30%. Further photon costs are incurred if nitrate is reduced in the dark (e.g., Needoba and Harrison 2004) since carbon dioxide has to be assimilated into carbohydrate with further energy input to produce a stored polysaccharide or lipid that can then be respired to regenerate NAD(P)H used in nitrate reduction.

1.6   LOSS OF REDUCED CARBON RELATED TO GROWTH AND MAINTENANCE

Conversion of organic carbon to carbon dioxide in respiratory reactions is inevitable in algal growth and maintenance. Even in the light, where the production of reductant and ATP used in biosynthesis can, in principle, be produced by the thylakoid reactions of photosynthesis rather than in the respiration of stored carbon (Raven 1976a–c), respiratory processes are needed to produce carbon skeletons essential for biosynthesis. Examples are production of the C4-dicarboxylic acids needed to produce the aspartate family of amino acids and pyrimidines and the C5 2-oxoglutarate needed to produce the glutamate family of amino acids and (with the exception of mitochondria of *Euglena*) tetrapyrroles such as chlorins of chlorophylls and hemes of cytochromes, algal hemoglobin, and catalase. This biosynthetic use of tricarboxylic acid cycle intermediates requires anaplerotic C3 + C1 carboxylations to maintain the concentration of the tricarboxylic acid cycle intermediates, partly offsetting respiratory carbon dioxide production. For biosynthesis in the dark, the tricarboxylic acid cycle clearly also involves carbon skeleton production, but also generation of ATP, while the oxidative pentose phosphate pathway generates NADPH. Maintenance processes include the synthesis of replacements for damaged proteins (mainly recycling undamaged amino acids from the damaged proteins) and recouping leaked solutes. The main respiratory product used in maintenance is ATP that, as stated earlier, can, in principle, be produced by the thylakoid reactions of photosynthesis in the photoperiod. In the dark, respiratory processes are needed for ATP generation for maintenance. Minimizing carbon loss in respiration by restricting biosyntheses and other growth-related processes to the photophase could minimize carbon loss in respiration and thus decrease the photon requirement, an argument similar to that used earlier for photoreduction of nitrate. However, limiting biosynthesis to the photophase means that the enzymes and other catalysts used in biosyntheses are only used for about half of the diel cycle, thus increasing resource (photons; elements such as carbon, nitrogen, phosphorus; and iron) costs of the catalysts relative to a more even use over the full light-dark cycle (Raven 1976a–c, 2013a,b).

Maximizing the use of respiratory energy in the growth and maintenance can be achieved by avoiding inefficiencies, such as the absence of 2-oxoglutarate dehydrogenase from the tricarboxylic acid cycle of *Euglena* and many cyanobacteria, and its replacement by a less energy-efficient bypass (Raven and Ralph 2015). Another inefficiency that could be avoided is the absence, in dinoflagellates, of the NADH (in the mitochondrial matrix) to UQ proton-pumping redox reaction, which eliminates 2 (possibly 1.5) of the usual 5 (possibly 4.5) protons pumped per electron and transferred from NADH (in the matrix) to oxygen. A third inefficiency, potentially avoided by avoiding or limiting the use of alternative oxidase activity, is the loss of the 3 protons pumped per electron moving from UQH_2 to oxygen via cytochrome *c* and cytochrome oxidase. However, balancing NAD(P)H production, ATP production and carbon skeleton production for reductive biosynthesis may need some alternative oxidase activity. There seems to be no possibility of increasing the protons per electron of electron transfer from matrix-located NADH to oxygen using cytochrome oxidase or decreasing the protons per ATP of the mitochondrial ATP synthase (see the discussion under thylakoid reactions of photosynthesis; Section 1.3). As to replacing parts of the photosynthetic pathway by a more "efficient" pathway (as with replacement of the RuBisCO–Benson–Calvin cycle), there is the suggestion of a "synthetic" version of glycolysis to replace the original Emden–Meyerhoff–Parnas pathway (Flamholz et al. 2013). It must be remembered that if the energetically efficient pathways operate closer to thermodynamic equilibrium, it could incur higher resource costs of enzyme synthesis that must be considered (see Section 1.1).

1.7 CONCLUSIONS

1. *For autotrophic cultures of algae and cyanobacteria, a major resource cost of photon absorption is the apoprotein associated with each molecule of pigment.* The apoproteins of light-harvesting pigment–protein complexes can be the most highly expressed proteins in photosynthetic cells acclimated and/or adapted to low PAR levels, thereby presenting cells with high costs in the energy, nitrogen, and phosphorus (via RNA) necessary for their biosynthesis.

2. *Acclimation/adaptation to low PAR levels can lead to an excess of light-harvesting machinery relative to the mean PAR in the natural environment, leading to the potential for photodamage.* At high irradiances, a smaller photosynthetic unit size decreases the potential for photodamage from excess PAR plus UVB, and there is the potential for downregulating photo-protective machinery such as photochemical quenching. On the other hand, this restricts the photosynthetic rate at low irradiances; the variation in solar radiation through the photoperiod must be taken into account in determining the effects of such changes in the context of algal cultures using natural light. *While decreasing photosynthetic unit size can limit photoinhibition around midday, such a strategy would also decrease photosynthetic capacity near dawn and dusk when photon flux is lower.*

3. *The two photosystems have different photon yields.* Photosystem I has a maximum yield close to 1, whereas Photosystem II has a maximum yield of 0.8. Thus, the 4-electron reaction oxidizing two water molecules to produce one oxygen at Photosystem II and reducing two $NADP^+$ at Photosystem I requires 4 photochemical reactions of Photosystem II, with a photon requirement of 4/0.8 (=5), and 4 photochemical reactions of photosystem I, with a photon requirement of 4/1 (=4), for a total of 9 photons absorbed by the light-harvesting pigments, also resulting in the phosphorylation of 3 ADP. Most of the NADPH and ATP generated in noncyclic photophosphorylation are used in CO_2 assimilation via the Benson–Calvin cycle using ribulose-1,5-bisphosphate carboxylase/oxygenase (RuBisCO) as the carboxylase.

4. *The oxygenase activity of RuBisCO leads to additional energy costs.* All oxygenic photosynthetic organisms express photorespiratory carbon oxidation cycles, which convert two 2-phosphoglycolate into one 3-phosphoglycerate (and hence triose phosphates) with the loss of one carbon dioxide and an input of ATP and NADPH for the metabolism of glycolate, as well as the ATP and NADPH used in the Benson–Calvin cycle. Diffusive entry of carbon dioxide

from the medium to a RuBisCO, with the kinetics giving the smallest known ratio of oxygenase to carboxylase activities in an air-equilibrated solution, leads to a *minimum* photon cost of photosynthesis of 9.92–9.96, depending on the photorespiratory pathway used, and most RuBisCOs have kinetics that incur higher photon costs.

5. *CCMs suppress photorespiration—at an energy cost.* All present-day cyano-bacteria and most algae have CO_2 concentrating mechanisms (CCMs), which increase the carbon dioxide concentration at the active site of RuBisCO and greatly improve carboxylase activity and decreases the oxygenase activity of RuBisCO. There is a great range of mechanisms for these CCMs, which all come with an additional energy cost: the minimum photon cost of photosyn-thesis in organisms expressing CCMs ranges from 9.5 to 11.0.

6. *Nitrogen assimilation can be energetically costly.* Additional energy cost is incurred if the redox state of the nutrient element differs between the form supplied and the form that is assimilated during growth. Thus, for nitrate reduction to ammonium in the light, the reduction of each nitrate requires four NAD(P)H that, with 4.5 photons needed per $NADP^+$ reduced, requires 18 photons. This means 2.72 additional photons per carbon dioxide in addi-tion to those needed for carbon assimilation into carbohydrate (an increase of 30%). Further photon costs are incurred if nitrate is reduced in the dark as NAD(P)H is then derived from respiration of organic C.

7. *Respiration leads to CO_2 release.* Respiratory conversion of organic carbon to CO_2 is inevitable in algal growth and maintenance. Even in the light, respiratory processes are needed to produce carbon skeletons essential for biosynthesis. The biosynthetic use of tricarboxylic acid cycle intermediates requires anaplerotic C3 + C1 carboxylations to maintain the concentration of the tricarboxylic acid cycle intermediates, partly offsetting respiratory carbon dioxide production.

8. *Possibilities for modification of respiratory losses.* Maximizing the use of respiratory energy in the growth and maintenance may be achieved by avoiding the inefficiencies found in some species or possibly by introducing "new" pathways through metabolic engineering. There seems however to be no possibility of increasing the protons per electron of electron transfer from matrix-located NADH to oxygen using cytochrome oxidase or decreasing the protons per ATP of the mitochondrial ATP synthase.

9. *Options for improvement of CO_2 assimilation pathways.* Replacing the Benson–Calvin cycle with another autotrophic carbon dioxide assimilation pathway could decrease both the ATP requirement for operating the pathway and also the energy and nutrients needed to synthesize the enzymes required. Introduction of carboxylases with a lack of oxygenase activity and a high affinity for carbon dioxide would result in no photorespiratory carbon oxida-tion cycle, and hence, no CCM would be needed. The only known example that fulfills these requirements is the 3-hydroxypropionate bi-cycle, though there is also the possibility of creating synthetic cycles using enzymes from a number of other pathways. Such replacement of the Benson–Calvin cycle has not yet been achieved.

10. *In conclusion, in algal biotechnology, the supply of nitrogen, phosphorus, and iron can generally be arranged so as not to be growth limiting, making light the limiting factor in productivity.* The earlier reviewed energetics of photosynthesis establishes the minimum photon requirements for production of algal biomass, which depends on the organisms, light environment, nutrients supply, and other environmental conditions. This provides a clear way forward for efforts to engineer improvements in microalgal biomass production.

ACKNOWLEDGMENTS

Discussions with Professor Peter J. Ralph have been very helpful. The University of Dundee is a registered Scottish charity, No. SC 015096.

REFERENCES

Bar-Even, A., E. Noor, N. E. Lewis, and R. Milo. 2010. Design and analysis of synthetic carbon fixation pathways. *Proc. Natl. Acad. Sci. USA* 107:8888–8894.

Bar-Even, A., E. Noor, and R. Milo, 2012. A survey of carbon fixation pathways through a quantitative lens. *J. Exp. Bot.* 63:2325–2342.

Bar-Even, A., E. Noor, Y. Savir et al. 2011. The moderately efficient enzyme: Evolutionary and physicochemical trends shaping enzyme parameters. *Biochemistry* 50:4402–4404.

Beardall, J. and J. A. Raven. 2013. Limits to phototrophic growth in dense cultures: CO_2 supply and light. In *Algae for Biofuels and Energy*, eds. M. A. Borowitzka and N. R. Moheimani, pp. 91–97. Berlin, Germany: Springer.

Blaby-Haas, C. E. and S. S. Merchant. 2012. The ins and outs of algal metal transport. *BBA Bioenergetics* 1825:1531–1552.

Blank, C. E. and P. Sanchez-Baracaldo. 2010. Timing of morphological and ecological innovations in the cyanobacteria—A key to understanding the rise of atmospheric oxygen. *Geobiology* 8:1–23.

Blankenship, R. E., D. M. Tiede, J. Barber et al. 2011. Comparing photosynthetic and photovoltaic efficiencies and recognising the potential for improvement. *Science* 332:805–809.

Borowitzka, M. A. 2013. Energy from microalgae: A short history. In *Algae for Biofuels and Energy*, eds. M. A. Borowitzka and N. R. Moheimani, pp. 1–15. Berlin, Germany: Springer.

Fabris, M., M. Matthijs, S. Rombauts, W. Vyverman, A. Goosens, and G. J. E. Baarts. 2012. The metabolic blueprint of *Phaeodactylum triconutum* reveals a eukaryotic Entner-Doudoroff glycolytic pathway. *Plant J.* 20:1004–1014.

Falkowski, P. G. and J. A. Raven. 2007. *Aquatic Photosynthesis*, 2nd edn. Princeton, NJ: Princeton University Press.

Flamholz, A. M., E. Noor, A. Bar-Even, W. Liebermeister, and R. Milo. 2013. Glycolytic strategy as a tradeoff between energy yield and protein cost. *Proc. Natl. Acad. Sci. USA* 110:10039–10044.

Losh, J. L., J. I. Young, and F. M. M. Morel. 2013. Rubisco is a small fraction of total protein in marine phytoplankton. *New Phytol.* 198:52–58.

Marchetti, A., D. M. Schruth, C. A. Durkin et al. 2012. Comparative metatranscriptomics identifies molecular bases for the physiological responses of phytoplankton to varying iron availability. *Proc. Natl. Acad. Sci. USA* 109:E317–E325.

Mareš, J., P. Hrouzek, R. Kaňa, S. Ventura, O. Strunecký, and J. Komárek. 2013. The primitive thylakoid-less cyanobacterium *Gloeobacter* is a common rock-dwelling organism. *PLoS One* 8:e66323.

Needoba, J. A. and P. J. Harrison. 2004. Influence of low light and light: Dark cycle on NO_3^- uptake, intracellular NO_3^-, and nitrogen isotope discrimination by marine phytoplankton. *J. Phycol.* 40:505–516.

Nichols, D. G. and S. T. Ferguson. 2013. *Bioenergetics.* Amsterdam, the Netherlands: Academic Press.

Odum, H. T. and R. C. Pinkerton. 1955. Time's speed regulator: The optimum efficiency for maximum output in physical and biological systems. *Am. Sci.* 43:52–58.

Raven, J. A. 1976a. Division of labour between chloroplasts and cytoplasm. In *The Intact Chloroplast*, ed. J. Barber, pp. 403–443. Amsterdam, the Netherlands: Elsevier.

Raven, J. A. 1976b. The quantitative role of "dark" respiratory processes in heterotrophic and photolithotrophic plant growth. *Ann. Bot.* 40:587–602.

Raven, J. A. 1976c. Transport in algal cells. In *Transport in Cells and Tissues Encyclopedia of Plant Physiology*, New Series, eds. U. Lüttge and M. G. Pitman, pp. 129–188. Berlin, Germany: Springer.

Raven, J. A. 1984a. A cost-benefit analysis of photon absorption by photosynthetic unicells. *New Phytol.* 98:593–625.

Raven, J. A. 1984b. *Energetics and Transport in Aquatic Plants.* New York: AR Liss.

Raven, J. A. 1994. Photosynthesis in aquatic plants. In *Ecological Studies*, Vol. 100: *Ecophysiology of Photosynthesis*, eds. E. D. Schultze and M. M. Caldwell, pp. 299–318. Berlin, Germany: Springer.

Raven, J. A. 1996. The bigger the fewer: Size, taxonomic diversity and the range of pigments in marine phototrophs. *J. Mar. Biol. Assoc. UK* 76:211–217.

Raven, J. A. 1997. CO_2 concentrating mechanisms: A direct role for thylakoid lumen acidification? *Plant Cell Environ.* 76:147–154.

Raven, J. A. 2011. The cost of photoinhibition. *Physiol. Plant.* 142:87–104.

Raven, J. A. 2012. Protein turnover and plant RNA and phosphorus requirements in relation to nitrogen fixation. *Plant Sci.* 188–189:25–35.

Raven, J. A. 2013a. The evolution of autotrophy in relation to phosphorus requirements. *J. Exp. Bot.* 64:4023–4046.

Raven, J. A. 2013b. RNA function and phosphorus use by photosynthetic organisms. *Front. Plant Sci.* 4:1–13.

Raven, J. A. 2013c. Iron acquisition and allocation in stramenopile algae. *J. Exp. Bot.* 64:2119–2127.

Raven, J. A., J. Beardall, and M. Giordano. 2014. Energy costs of carbon dioxide concentrating mechanisms in aquatic organisms. *Photosynth. Res.* 121:111–124.

Raven, J. A., J. Beardall, M. Giordano, and S. C. Maberly. 2012. Algal evolution in relation to atmospheric CO_2: Carboxylases, carbon concentrating mechanisms and carbon oxidation cycles. *Philos. Trans. R. Soc. Lond. B* 367:493–507.

Raven, J. A., J. Beardall., A. W. D. Larkum, and P. Sanchez-Baracaldo. 2013. The influence of photosynthesis on genome function. *Philos. Trans. R. Soc. Lond. B* 368:20120264.

Raven, J. A. and S. Donnelly. 2013. Brown Dwarfs and Black Smokers. The potential for photosynthesis using the radiation from low-temperature black bodies. In *Habitability of Other Planets and Satellites*, ed. J.-P. Devera, pp. 269–284. Berlin, Germany: Springer.

Raven, J. A., M. C. W. Evans, and R. E. Korb. 1999. The role of trace elements in photosynthetic electron transport in O_2-evolving organisms. *Photosynth. Res.* 60:111–149.

Raven, J. A. and R. D. Geider. 2003. Adaptation, acclimation and regulation of photosynthesis in algae. In *Photosynthesis in Algae*, eds. A. W. D. Larkum, S. E. Douglas, and J. A. Raven, pp. 385–412. Dordrecht, the Netherlands: Kluwer.

Raven, J. A., M, Giordano., J, Beardall., and S.C. Maberly. 2011. Algae and aquatic plant carbon concentrating mechanisms in relation to environmental change. *Photosynth. Res.* 109:281–296.

Raven, J. A. and P. J. Ralph. 2015. Enhanced biofuel production using optimality, pathway modification and waste minimization. *J. Appl. Phycol.* 27:1–31.

Rosen, R. 1967. *Optimality Principles in Biology.* London, UK: Butterworths.

Sanchez-Baracaldo, P., A. Ridgwell, and J. A. Raven 2014. A Neoproterozoic revolution in the marine nitrogen cycle. *Curr. Biol.* 24:652–657.

Saw, J. H. W., M. Schatz, M. V. Brown et al. 2013. Cultivation and complete genome sequencing of *Gloeobacter kilaueenis* sp. nov., from a lava cave in Kilauea Caldeira, Hawai'i. *PLoS One* 8:76376.

Sohm, J. A., E. A. Webb, and D. G. Capone. 2011. Emerging patterns of marine nitrogen fixation. *Nat. Rev. Microbiol.* 9:499–508.

Tcherkez, G. G., G. D. Farquhar, and T. J. Andrews. 2006. Despite slow catalysis and confused specificity, all ribulose bisphosphate carboxylases may be nearly perfectly optimized. *Proc. Natl. Acad. Sci. USA* 103:7246–7251.

Weber, T. and C. Deutsch. 2014. Local versus basin-sale limitation of marine nitrogen fixation. *Proc. Natl. Acad. Sci. USA* 111:8741–8746.

Williams, P. J. le B. and L. M. L. Laurens. 2010. Microalgae as biodiesel and biomass feedstocks: Review and analysis of the biochemistry, energetics and economics. *Energ. Environ. Sci.* 3:553–590.

2 Algal Growth Kinetics and Productivity

Joseph Weissman and Robert Nielsen

CONTENTS

2.1 INTRODUCTION: WHAT IS GROWTH KINETICS?

Growth kinetics, a set of mathematical models and analyses, is used to design microbial growth and productivity experiments, predict their outcomes, and interpret the results. The central concept in growth kinetics is the "specific growth rate," defined as the rate of increase of cell mass per unit cell mass. One can make an analogy to economics, in which specific growth rate is interest rate and cell mass is capital. The maximum specific growth rate is attained when all of the nutrients required for growth are available in excess thus allowing the biomass of the population to grow exponentially as the algal cells replicate at the fastest rate of replication that the organism can achieve, within the physical and chemical environment in which it is cultivated (e.g., pH, temperature, salts).

A major objective of growth kinetic studies is to predict, for any set of conditions, the biomass production rate, that is, productivity. Here, the focus is specifically on microalgae grown photoautotrophically (e.g., on light energy and CO_2). Productivity can be expressed in various units, such as grams of biomass produced per liter of a culture medium per hour, or, for microalgae grown photoautotrophically (on light and CO_2), per square meter (of ground area) per day, or even as tons per hectare per year. Biomass in this analysis is the ash-free dry weight (AFDW) of the organic matter produced by the algal cells. To calculate the efficiency of producing biomass, the heat of combustion of the organic matter (which in some cases includes cellular excretion products) is divided by the energy content of the absorbed photons. This accounts for variations in cellular composition observed under different growth regimes. Note that metrics based on cell numbers, often used in algal studies, may not translate into such productivity units because any conversion factor of cell size to biomass will depend on growth rates and environmental conditions.

Maximum specific growth rates are often measured during batch growth. After inoculation at low biomass density, a batch growth curve starts with an initial lag, during which the cells adjust to the growth environment. Once this adjustment is accomplished, the cells reproduce for some time at their maximum specific growth rate under the specific environmental conditions at hand. Eventually, after the cellular biomass increases to a given concentration, one nutrient starts to limit growth and the specific growth rate declines, with the rate of decline depending on the limiting nutrient. Most growth kinetic models attempt to describe the specific growth rate in terms of the environment and the limiting nutrient. This turns out to be a formidable task as it requires a detailed knowledge of cellular metabolic pathways and their regulation, the transport and utilization of nutrients, and the influence of the physicochemical composition of the growth medium on these processes. Consequently, many simplifying assumptions must be made when modeling growth kinetics.

Generally, growth kinetic models are simplified to the determination of the instantaneous specific growth rate as a function of the availability of the growth-limiting nutrient. In order to complete the analysis, the yield of biomass per unit of the growth-limiting nutrient is required. For autotrophic microalgae being grown for biofuels, the key substrate will be light energy, typically solar photons, and the focus in microalgal growth kinetics is on light as the growth-limiting nutrient. This turns out to be a more complex problem than for the dissolved nutrients as the availability of light is not uniform throughout the culture but depends on the position of each individual algal cell within it. It usually also depends on the light history of the cell, that is, its movement within the light field.

The results of growth experiments are compared to the mathematical growth models in order to analyze the data and, if required, to modify the model. Good experimental design, data analysis, and modeling are all critical to understanding algal growth, but even so, only relatively simple growth models can be experimentally tested at present. This chapter focuses on light as a limiting nutrient to growth and photosynthesis.

A very large literature on studies and models of microalgal growth kinetics has accumulated over the past several decades (see Béchet et al. 2013). These are of varying degrees of complexity, but highlight the great importance of microalgal growth and photosynthesis in ecosystems, from oceans to freshwater environments to the impacts of pollution including global warming. The intent here is not to thoroughly review this vast field, but rather to extract from prior work those advances that the authors believe are most relevant to understanding how algal growth kinetics and photophysiology determine photosynthetic productivity. First, a specific example of experimental growth kinetic data and analysis is presented.

2.2 GROWTH AND PRODUCTIVITY OF *Chaetoceros muelleri*

A growth kinetic analysis is exemplified using the set of experimental results shown in Figure 2.1 through Figure 2.4 for the diatom *Chaetoceros muelleri*. In Figure 2.1, the daily biomass productivity (gram ash-free dry weight [AFDW]$\cdot$m$^{-2}\cdot$day^{-1}) is shown as a function of average specific growth rate for small outdoor pond cultures (15 cm deep, 1.4 m^2, paddle wheel mixed) diluted continuously during the daylight hours or semicontinuously once each morning. Here, the average specific growth rate is imposed by the rate of dilution of the cultures. The control was a pond diluted semicontinuously at 40% per day, throughout the approximate 3-month duration of the experiment. There was not one daily biomass density, or specific growth rate, in each of these cultures as these changed over the course of the day. The precision of productivity measurements was ±5%.

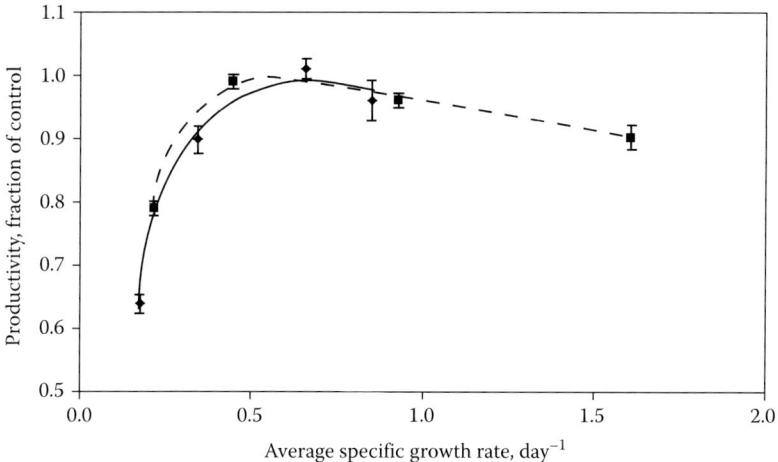

FIGURE 2.1 Productivity as a function of average specific growth rate. Outdoor semicontinuous and continuous cultures of *Chaetoceros muelleri* grown in 1.4 m^2 ponds. Dashed line, semicontinuous; solid line, continuous. Control pond productivity varied day to day from about 30 to 38 g$\cdot$m$^{-2}\cdot$day^{-1}. (Weissman, J. C. and Goebel, R. P., U.S. DOE Aquatic Species Program, 1986, unpublished.)

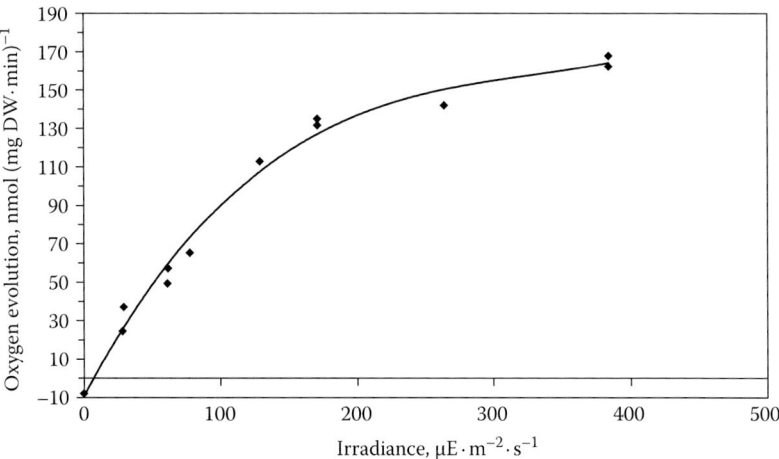

FIGURE 2.2 Photosynthesis–irradiance curve for *Chaetoceros muelleri* grown in the laboratory at 200 $\mu E \cdot m^{-2} \cdot s^{-1}$ diluted semicontinuously at 50% per day. (Weissman, J. C. and Goebel, R. P., U.S. DOE Aquatic Species Program, 1986, unpublished.)

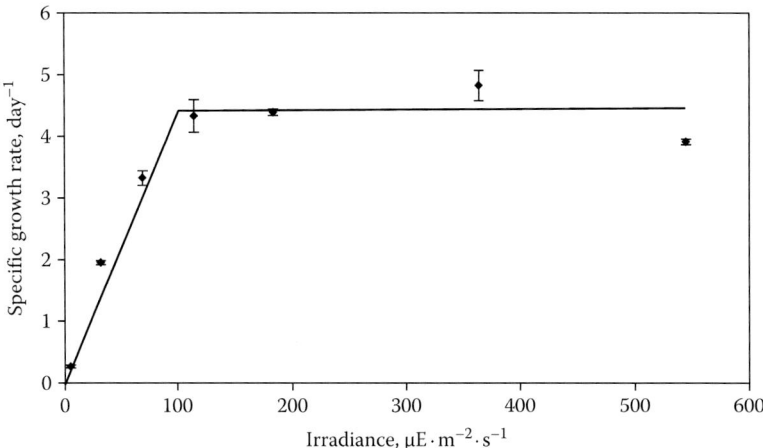

FIGURE 2.3 Specific growth rate of optically thin cultures of *Chaetoceros muelleri* as a function of applied irradiance. (Weissman, J. C. and Goebel, R. P., U.S. DOE Aquatic Species Program, 1986, unpublished.)

The continuous cultures were sampled numerous times during the period of dilution, and the samples composited to give an average biomass. This was multiplied by the dilution rate (equal to the average specific growth rate), which is the total volume that was continuously removed (and replenished) from the pond each day, divided by the volume of the pond to give the volumetric productivity (g AFDW $\cdot$ L^{-1} $\cdot$ day^{-1}). This was translated into areal biomass productivity by multiplying by the volume to area ratio of the culture (150 L $\cdot$ m^{-2} in this case). For each point on the graph, the average pond

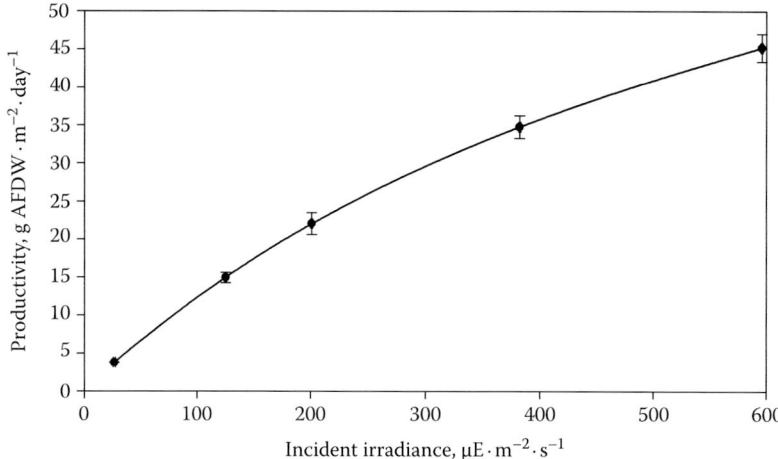

FIGURE 2.4   Maximum productivity of dense cultures of *Chaetoceros muelleri* as a function of incident irradiance. (Weissman, J. C. and Goebel, R. P., U.S. DOE Aquatic Species Program, 1986, unpublished.)

density may be calculated by taking the areal productivity and dividing by the number of liters removed per day (the corresponding dilution rate on the graph times $150 \, L \cdot m^{-2}$). The highest density was approximately $700 \, mg \cdot L^{-1}$, and the lowest was approximately $220 \, mg \cdot L^{-1}$. To obtain specific growth rate in $hour^{-1}$ from the units of day^{-1}, as shown on the graph, requires the illuminated hours per day, which was about 13 hours in midsummer at the location of this experiment. Overnight losses are accounted for in the biomass measurements (see Section 2.6.1 for a discussion of overnight losses). The continuous cultures in these experiments were sufficiently dense to absorb virtually all sunlight that was incident on the ponds, even at the highest rate of dilution imposed.

The semicontinuously diluted cultures had volume removed each morning. For these cultures, the average specific growth rate is not equal to the dilution rate. It is estimated by the natural log of the ratio of the biomass immediately before dilution to that right after dilution on the previous day. At the lowest volume removal per day (about 20%), the specific growth rates for semicontinuously diluted ponds and continuously diluted ponds were similar at about $0.22 \, day^{-1}$ for the former versus $0.18 \, day^{-1}$ for the latter. At large dilutions per day, the difference in specific growth rates were large: at 80% per day, the average specific growth rate of semicontinuously diluted cultures was twice that of the continuously diluted cultures (1.61 versus $0.8 \, day^{-1}$). In the semicontinuous case, the algal biomass needed to increase fivefold each day to maintain a steady density, from approximately 45 to $225 \, mg \cdot L^{-1}$. At this low initial density, about 50% of the incident light was transmitted through the pond culture without being absorbed before reaching the bottom of the pond. However, since the bottom was white, additional light was absorbed upon reflection.

A central objective of a growth kinetic analysis is to explain and predict the gross features of these curves using conceptual and/or mathematical models. Such models should, for example, explain why there is a maximum in the relationship between productivity and the imposed specific growth rate using a relationship between photo-synthetic rate (described next) and irradiance. Only the simplest model will be discussed in this chapter, but the challenge of developing models based on greater physiological detail will be addressed.

Often, the photosynthesis–irradiance curve (P–I curve) is used to describe specific growth rate in terms of irradiance. As shown in Figure 2.2, it is a saturation curve, typical of biological growth rate measurements in relation to a nutrient level, except there is often a decline in photosynthetic rate at irradiances higher than those tested in the experiment shown. Photosynthetic rate is measured by taking a sample from a (usually dense) algal culture, either the one being modeled or the one in a presumably similar physiological state, diluting it until it is optically thin (less than 10% light attenuation), and then placing it in a chamber equipped with an O_2 electrode (typically 2 mg chlorophyll a per liter is a suitable concentration for a 1 cm light path chamber). The sample is exposed to a constant irradiance (light intensity) and, after allowing some time for stabilization, the rate of oxygen evolved is measured. The experiment is repeated as a function of a series of irradiances, as well as in the dark to obtain a dark respiration rate. Alternatively, radioactive carbon dioxide uptake can be used to measure photosynthetic rate. The photosynthetic rate may be expressed in terms of the chlorophyll content of the cuvette or the biomass content. Thus, the units are oxygen (carbon dioxide) either per unit chlorophyll per unit time or per unit cell mass per unit time. The former case is often labeled P, the latter Q.

The mathematical form of the P–I curve may be determined by best fit to the data of a P–I experiment performed as described. Alternatively, it can be derived from a microscopic model of photosynthesis (at the level of the photosynthetic unit [PSU]) for a given physiological state (absorption cross section, maximum electron transport rate). To obtain the P–I curve, this model of light utilization by a PSU must be integrated over the cell containing many PSUs absorbing at different irradiances, as light is attenuated across the cell. Therefore, even a P–I curve determined from an optically dilute suspension of cells represents an integration of the photosynthesis by many PSUs absorbing at different irradiances. Attenuation across a cell may be great. For instance, a 30 cm deep, 100 mg $AFDW \cdot L^{-1}$ suspension of low-light adapted cells (highly pigmented: chlorophyll at 2% of AFDW, optical cross section of 5 $m^2 \cdot (g\ chl\ a)^{-1}$) will absorb most of the light incident on the surface of the suspension, for example, >90%. If the cells are 10 μm in diameter, it is easily calculated that this suspension contains only about three cells on average in a volume with a 30 cm depth and cross-sectional area equal to that of the cell. Thus, each cell must absorb about 50% of the light. The P–I curve determined from an empirical or mechanistic model is integrated over the depth (volume) of the culture, using a model for the attenuation of the incident light within the culture (or with depth) to derive the overall culture productivity for a given specific growth rate or biomass density (which, as per Figure 2.1, should have a one-to-one correspondence under similar conditions). The objective is to derive the results in Figure 2.1 from data such as in Figure 2.2.

Several important parameters are defined by this simple P–I curve:

1. The initial slope is proportional to the maximum efficiency at which the algal cells convert photon energy into assimilated carbon dioxide (albeit measured as evolved oxygen). It can be represented mathematically

by an absorption cross section ($m^2 \cdot g^{-1}$ biomass or chlorophyll, see discussion in the following text) times the actual quantum efficiency O_2 evolved, or CO_2 fixed, per photon absorbed.

2. The maximum photosynthetic rate (P_{max}), where oxygen evolution becomes independent of irradiance at least until photoinhibition sets in at irradiances higher than in this experiment, causing the rate to decline.
3. The saturating irradiance, I_s, the intersection of the initial slope, and P_{max} (about $150 \, \mu E \cdot m^{-2} \cdot s^{-1}$ in Figure 2.2).
4. The compensation point, the light intensity at which the rate of photosynthesis (or growth) is zero, that is, where photosynthesis balances with respiration. This is often quite low and difficult to determine. In Figure 2.2, the intercept of the P–I curve is at about $10 \, \mu E \cdot m^{-2} \cdot s^{-1}$.
5. The dark respiration rate, that is, the rate of O_2 uptake in the dark. This is one measure of maintenance metabolism. Another, which includes maintenance metabolism in the light as well, is discussed in Section 2.6.1.
6. The light intensity at which photosynthesis starts to be inhibited—photoinhibition. For this strain, and diatoms in general, photoinhibition is observed only at relatively high-light intensities, typically well above the maximum light intensity used in these experiments.

Myers and Graham (1971) already questioned whether the photosynthetic rates measured in such a laboratory experiment can actually represent what occurs in a dense culture including in actual outdoor biomass production, as in Figure 2.1. In the O_2 electrode measurements, the rates of O_2 evolution depend greatly on the exact experimental conditions. For example, one common experimental method is to increase light intensity stepwise, starting with zero light (dark respiration measurement). This, however, typically gives different results than if the experimental protocol is reversed, which is going stepwise from high to low light, or if each point in such a P–I curve is measured with a fresh sample from the culture as recommended in Myers and Graham (1971). In brief, the steady-state rates of photosynthesis are affected when cultures are perturbed in being taken from one environment and placed in a new one, even if the environmental conditions are similar.

Photosynthetic rate is a type of specific growth rate measurement and can be converted into units of reciprocal time (specific growth rate is biomass/biomass/unit time), with some important assumptions. If CO_2 assimilation is used to measure the photosynthetic rate, then carbon concentration units are converted to carbon mass units, which are then converted to the same mass units that were used to normalize the photosynthetic rate, using known or assumed values. If the photosynthetic rates were originally measured as oxygen evolution, concentrations of oxygen are converted to carbon using some assumption for the ratio of CO_2 taken up to oxygen evolved (the inverse of what is called the photosynthetic quotient). One of the main assumptions in using photosynthetic rates to calculate specific growth rates is that cells are photosynthesizing at constant rate throughout their cell cycle.

One common form used to model the P–I curve is the rectangular hyperbola. In this case, $I_s = I_k$, where I_k is defined as the irradiance at which the rate is half maximal. Since the rectangular hyperbola has its greatest curvature at the origin (lowest light), it is not consistent with most P–I curves, which have a prolonged region of linear increase of rate at low irradiance.

Figure 2.3 shows another form of growth/light intensity relationship, the specific growth rate versus the irradiance relationship, or µ–I curve. Here, the specific growth rates of dilute algal cultures (to avoid self-shading) are plotted against the irradiance to which they are acclimated. This is accomplished most easily by diluting, turbidostatically or semicontinuously, optically thin cultures exposed to a steady light level (irradiance). For turbidostatic dilution, the specific growth rate is equal to the dilution rate. For semicontinuous dilution, the specific growth rate is the natural logarithm of the ratio of the biomass density before dilution to that just after the previous dilution, divided by the time interval between dilutions. The biomass density may be measured by one or several methods (organic carbon, chlorophyll, cell volume or number) as long as the method is sensitive enough to measure low biomass density (AFDW is generally not sensitive enough for this application). Oceanographers and limnologists measure µ–I curves in stratified water columns by incubating bottles to measure gross photosynthetic rate (and respiration in dark controls) as a function of depth (and thus, light intensity). As may be expected, these alternative techniques give rather different results.

The data in Figure 2.3 were all obtained at a temperature of 28°C–30°C and at pH 7.5 ± 0.1 with all nutrients saturating. Each point represents a different physiological state. The maximum specific growth rate may be different from the maximum photosynthetic rate (again, after conversion to a specific growth rate) measured in short-term experiments. Maximum specific growth rate is not only determined by the rate of photosynthesis, but also by the cell cycle. It is possible for a cell cycle to constrain growth below the maximum photosynthetic rate. On the other hand, a strain is always able to photosynthesize at a rate at least equal to its specific growth rate. Note that here there is no clear compensation point; acclimated cultures show positive growth even at very low irradiances. The lowest rate 0.3 day^{-1} was measured at 7 $\mu E \cdot m^{-2} \cdot s^{-1}$ and is 7% of the maximum rate. There may be several reasons why the compensation point on a P–I curve is different from the one on a µ–I curve, including the use of different illumination for each type of experiment, different methods of measuring light, different physiological state, and they may not be measuring the same physiological processes.

A major difference between the P–I curve and the µ–I curve is that the former represents the response of a single physiological state to irradiance, while the latter is a set of physiological states each determined by irradiance. It is beyond any current techniques to model how state variables like chlorophyll content per unit carbon (and other physiological parameters like cell size, proximate composition, per unit carbon levels of the photosynthetic units and metabolic enzymes for carbon fixation and reduction) change with irradiance. Changes in such variables must be measured for each strain and environmental variable, in addition to light.

If the light absorbed by the dilute suspension representing a single (µ, I) point is measured (based on the emission spectrum of the light and the absorption spectrum of the suspension), then the quantum requirement (or inversely the quantum efficiency) for incorporating carbon into the biomass can be calculated and compared to the limit set by our understanding of the microscopic mechanism of photosynthesis (e.g., a quantum requirement of approximately 10–12 quanta per CO_2 fixed, depending on the biochemical composition of the biomass, or 9–11 quanta per O_2 evolved).

For a stratified water column, the μ–I curve may be used along with the attenuation of irradiance with depth to obtain the column integrated productivity. However, it is not generally correct to use the μ–I curve as the basis to estimate column productivity in a mixed water column unless intermixing of layers is slow. Each point represents a distinct physiological state with different optical properties. In dense mechanically mixed mass cultures, the cells are low light acclimated.

There is another type of curve—that of maximum biomass productivity versus incident light intensity (Figure 2.4). Here, the maximum biomass productivity, measured at each incident growth irradiance, is plotted against the irradiance. These experiments were conducted under daylight fluorescent lamps, in 1 L Roux flasks, diluted semicontinuously at the optimal rate (previously determined to be 40%–60% per day) at pH 7.5 and at 28°C–30°C. A major test of the power of algal growth kinetics modeling is to derive this curve. Note that unlike the earlier cases, this is not a saturation curve but, rather, a curve of diminishing return: after a constant initial slope in subsaturating light, the productivity increases less and less for a given increase in incident irradiance, as light saturation sets in. It is not possible, until we develop much better models, to extract the saturating irradiance from this curve. Note also, that if this experiment had been carried out at a higher light intensity, typical of full sunlight, a maximum productivity of about 60 $g \cdot m^{-2} \cdot day^{-1}$ would have been achieved, much lower than the extrapolation of productivity measured at low light to such a high light. This illustrates a major constraint on the productivity of plant and algal photosynthesizes: light saturation.

2.3   SIMPLE MODEL OF PHOTOSYNTHESIS

2.3.1   PHOTOSYNTHETIC MECHANISM

With this experimental introduction to the relationship between light and photosynthesis in algal cultures, a review of the basics of photosynthesis is useful, in order to clarify the nature of two key parameters in light limited growth kinetics: the saturating irradiance (I_s) and the maximum rate of photosynthesis (P_{max}). We will demonstrate how experimental measurements of μ_{max} set an upper limit on the turnover time of the photosynthetic apparatus.

The experiments of Emerson and Arnold (1932) elucidated many of the fundamental facts of photosynthesis by applying short flashes of high-intensity light to algal cultures and measuring the resulting O_2 evolution per flash. Thousands of molecules of chlorophyll are involved in the evolution of 1 molecule of oxygen, about 2000–3000 chlorophyll molecules in their experiments. In order to maintain the maximal amount of oxygen per flash, a temperature-dependent period of time in the dark was required between successive flashes of high-intensity light (dark times were limited by the apparatus to no less than 30 ms). These facts imply that chlorophyll is not involved in a normal enzymatic reaction with oxygen, and that light absorption and the regeneration of the state allowing maximum light utilization (maximum oxygen per flash) are different events with the former being a photochemical process (fast, not dependent on temperature) and the latter a more conventional biochemical reaction (as seen from the temperature dependence). Although it took many years to formulate a mechanistic model of the photosynthetic apparatus, the fundamental basis did not change: light absorption and stable biochemical storage of light energy are two distinct processes, and the ability to utilize the absorbed light rapidly saturates with increasing amounts of applied light.

A complication is that two different but connected photochemical processes, with somewhat different action spectra in red light, are involved in O_2 production and CO_2 fixation. These insights eventually led to the now-familiar Z-scheme model (Hill and Bendall 1960), which consists of photosynthetic units (PSUs) with two types of photosystems, called the lower redox potential photosystem (PSII) and the positive redox photosystem (PSI), operating in series. Each contains a light-harvesting apparatus for photon capture, called antennae, composed of hundreds of pigment molecules (in the case of the green algae, or chlorophytes, mostly chlorophyll *a* and *b*), and a central reaction center. The latter contains a specialized chlorophyll *a* dimer, which mediates the initial charge separation. The photons captured by the light-harvesting pigments (excitons) are transformed into chemical energy at the reaction centers (traps). The two distinct photosystems are separated by electron transport proteins that bridge between the reducing end of the lower redox potential photosystem (PSII) and the oxidizing end of the more positive redox photosystem (PSI). Together, these PSUs oxidize (at PSII) water and "lift" the electron to the level of the powerful biological reducing agent NADPH (at PSI), which is used in CO_2 fixation. As electrons are removed from the ultimate electron donor, water, at the oxidizing side of PSII, the protons from the water molecules are released in the cavity (the lumen) formed by the thylakoid membrane, the structure in which the PSUs and ancillary proteins and pigments are embedded. Some of the electron transporters (the lipid soluble quinones) also transport protons, from outside to inside, thus further conserving some of the light energy in an electrochemical gradient. Thus, one proton released into the lumen from water oxidation and two translocated across the thylakoid membrane into the lumen yields three protons for each electron transported from oxygen to the reducing side of PSI. A membrane spanning ATPase uses this gradient to form ATP by channeling the protons back across the membrane. Thus, the whole process is integrated via the cavity enclosing membrane on a larger scale than any one PSU. The quinones carry electrons from PSII to another macromolecular membrane-bound protein complex called cytochrome b_6f, via a complicated cycle, the Q cycle (Mitchell 1975). From cytochrome b_6f, a water-soluble electron carrier, plastocyanin, bridges the gap to PSI through diffusion in the lumen. Although the photosystems act primarily in series, there is some cyclic flow of electrons initiated solely due to photon absorption by PSI. This cyclic flow translocates hydrogen ions and generates some extra ATP, when needed, but at the cost of additional photons. The turnover time of a PSU, that is, the time it takes to transport an electron from PSII to PSI, is key in determining the rate-limiting step in photosynthesis, to be discussed in the following text.

PSUs in many microalgae (and plants) can share the light-harvesting antenna such that if, for example, one PSII reaction center is closed, the exciton may find its way to the reaction center of another PSII and become trapped (participate in water oxidation). Part of the light-harvesting antenna can even be shared between PSII and PSI in state transitions, presumably to balance the absorption of light as dynamically required by downstream metabolic needs for NADPH and ATP.

A central problem of photosynthesis by microalgae, and also higher plants, is that at high-light intensities the rate of photons arriving at the cells near the surface of a culture, or leaf, and absorbed by the light-harvesting pigments is much faster than the rate at which metabolic pathways can use the reductant and energy. This is reflected by the saturating light intensity (e.g., Figure 2.2), which is the parameter that characterizes the balance between the rate of photon absorption and utilization by the photosynthetic apparatus (e.g., production and subsequent utilization of NADPH and ATP).

2.3.2 LIGHT ABSORPTION AND LIGHT SATURATION

The light absorbed by a cell (or pigment, or antenna) at a particular wavelength is determined by its optical or absorption cross section at that wavelength and the irradiance it experiences. The term also applies to the average absorption from a multi-wavelength light source weighted by the number of photons from the light source absorbed within each small range of wavelengths. The absorption cross section is typically expressed as an area per unit pigment or biomass or per cell. Generally, it is the absorption of light by a suspension of cells that is measured, but in simple spectrophotometers, the decrease in light reaching the detector is the sum of both absorption and scatter. The latter can be obviated by the use of an integrating sphere spectrophotometer. The optical cross section per unit of biomass multiplied by the amount of biomass in the suspension determines the rate of attenuation of light through the suspension, for example, the extinction coefficient in the Beer–Lambert's law.

Photosynthesis in a cell can be viewed, simplistically, as a large number of photosynthetic units operating in parallel to provide electrons (carried by NADPH) and energy (ATP) to reductant- and energy-requiring metabolic pathways, principally CO_2 reduction. The rate of photon absorption by a photosystem or PSU can be estimated from the irradiance incident on it, the absorption cross section per unit chlorophyll mass (or number), and the mass (or number) of chlorophylls in the unit. For example, with some simplifying assumptions (including that the number of chlorophylls and the absorption cross section for PSI and PSII are the same) at steady-state, full sunlight (about 2000 $\mu E \cdot m^{-2} \cdot s^{-1}$ incident irradiance), at an optical cross section of 0.005 $m^2 \cdot mg^{-1}$ chlorophyll, at about 900 mg chlorophyll/mmole chlorophyll, and at an antenna size for each of the photosystems of about 300 chlorophyll molecules (all typical for chlorophytes acclimated to subsaturating light), by converting to μmoles, one can calculate that each PSU absorbs $2000 \cdot 0.005 \cdot 900 \cdot 2 \cdot 300$, or about 5400 photons per second (2700 photons per second per photosystem). This amounts to a potential for a steady-state average rate of about 1 electron every 0.37 ms, transported through the intersystem electron transport system.

The actual rate of electron transport sets a limit on what fraction of the absorbed photons could be used subsequently for carbon fixation. A more stringent limit may be set by the rate at which carbon fixation is capable of using four-electron equivalents. Since this is a steady-state analysis, it does not matter which particular step is limiting the rate most; all steps must run at that pace. If photons are absorbed at a rate faster than electrons can be transported, the reaction centers are "closed" and the photonic excitation is dissipated as heat and fluorescence. If 5 ms is assumed for electron transport turnover time, the rate of photon absorption at 2000 $\mu E \cdot m^{-2} \cdot s^{-1}$ may exceed the maximum rate of electron transport by a factor of about 13.5 (5 ms/0.37 ms). Stated another way, the incident irradiance would need to be reduced by a factor of 13.5, that is, to about 150 $\mu E \cdot m^{-2} \cdot s^{-1}$ to bring the rate of light absorption into balance with this rate of electron transport. This is called the saturating irradiance, and for this example, it is evident in Figure 2.2.

A significant approximation is inherent in using the simple model of the photosynthesis described earlier to estimate the electron transport rate and then the saturating irradiance. Due to the attenuation of light in a cell, not all PSUs are exposed to the same irradiance. This complicates the analysis by lowering the average absorption per photosystem, but does not change it conceptually. The error is substantially less for small cells with less attenuation of light across a single cell.

Continuing with this example, a corresponding μ_{max} can be calculated as the rate of carbon fixation (per sec) per unit carbon in the biomass and assuming again that the turnover time of electron transport is 5 ms. Using the assumptions earlier for the size of the PSU, 1 mol O_2 is evolved per 600 mol chlorophyll per 20 ms (5 ms × 4 electrons transferred). Also using a photosynthetic quotient of 1.15 O_2 per CO_2 (a realistic number for algal cells growing on ammonia), changing 1 mol C to 12 g C and 1 mol chlorophyll to 900 g chlorophyll and using a chlorophyll content of 6% of biomass carbon gives a maximum growth rate of

$$\mu_{max} = (1 \cdot (1/1.15) \cdot 12 \cdot 1000 \cdot 3600)/(20 \cdot 600 \cdot 900 \cdot (100/6))$$
$$= 0.21 \text{ hour}^{-1} \text{ (3.3 hours doubling time) or } 5.0 \text{ day}^{-1}$$

This is about 17% higher than the measured value of μ_{max} of 4.3 day^{-1} shown in Figure 2.3. This calculation assumes that the electron transport rate is constant throughout the cell cycle and ignores losses during growth. To achieve any particular (net) specific growth rate, the maximal rate of electron transport must thus be faster than the average rate used in the calculation. A short-term estimate of μ_{max} may be calculated from the maximum photosynthetic rate, P_{max} from Figure 2.2, again by converting oxygen to carbon using an assumed photosynthetic quotient and carbon to biomass ratio (2 g AFDW per g C). From the 160 nmole O_2 per mg AFDW per minute shown in Figure 2.2, μ_{max} is calculated to be 0.20 hour^{-1}. This estimate is close to that obtained by assuming an electron transit time of 5 ms. The conclusion, based on measured specific growth rate and maximum photosynthetic rate, is that the electron transport rate must be at least as fast as 200 s^{-1}, and that carbon fixation can keep up with such a rate. Faster maximum electron transit times must be possible for strains that have faster measured average reproductive rates, such as a 3-hour doubling time (which would be an average turnover time of 4.5 ms) for *Chlorella* (Pirt et al. 1980), 2.6 hours for *Synechococcus* sp. PCC7002 (Ludwig and Bryant 2012), and 2 hours for *Synechococcus* sp. UTEX 2973 (Yu et al. 2013).

The latter growth rate requires an average electron transport time of only about 3.0 ms, based on inverting the earlier calculation.

How much Rubisco is required to support a given turnover time of the electron transport system? Assuming R is the biomass fraction of Rubisco, X is the rate of turnover of Rubisco (in moles CO_2 per mole Rubisco), and the biomass is 50% carbon, μ_{max} equals $(R/550,000) \cdot (X) \cdot (86,400\ s \cdot day^{-1})/(0.042\ mol\ C/g\ cell\ mass)$. With values of X and R equal to 12 and 0.125, μ_{max} is about 5.6 day^{-1}, or 0.23 $hour^{-1}$, or a doubling time of about 3.1 hours, or enough to support a 5 ms turnover time. Clearly, when carbon fixation rates are discussed, X and R always occur as a product. The values chosen are within norms, although X has been asserted to be twice as great and thus R half as great (Losh et al. 2013), but in this work, only growth rates up to 1.2 day^{-1} were considered. Our results (data not shown) indicate that even cells growing at less than 1.2 day^{-1} are almost immediately able to fix carbon at three to five times this rate when switched to higher light.

This simplified photosynthesis model has electrons from n PSUs delivering electrons to m molecules of Rubisco. As an aside, with the earlier assumptions of the chlorophyll and Rubisco content, m/n can be estimated. Again, let C be the mass fraction of chlorophyll per gram of the biomass and R be the Rubisco fraction. The ratio of chlorophyll to Rubisco, mol to mol, is $(C/900)/(R/550,000)$ where approximate molecular weights of chlorophyll and Rubisco have been inserted. Then m/n is the number of chlorophyll in the PSU divided by this number, or $600/((C/900)/(R/550,000))$. If $C = 0.03$, again typical of low-light acclimated chlorophytes, and $R = 0.125$, then m/n is about 4. Clearly, this decreases as R decreases due to a higher Rubisco turnover rate. This analysis does not imply physical proximity of Rubisco molecules to PSUs.

In flashing-light experiments (Phillips and Myers 1954) in which specific growth rates were measured (as opposed to short-term photosynthetic rate measurements), a minimum dark time of 6–10 ms was required for saturation of flash yield at the shortest flash duration (single turnover flashes) as shown in their figure 8, though this is not emphasized in their paper. Radmer and Kok (1977) specify a minimum dark time for PSU turnover of greater than 5 ms. In studies on photoacclimation, measurements of the turnover time of the electron transport system were determined to be less than 5.6 ms (Myers and Graham 1971), less than 3.5 ms (Sukenik et al. 1987), and less than 3.0 ms (Dubinski et al. 1986). In one of these studies (Sukenik et al. 1987), PSU turnover time was proportional to PSU content, but uncorrelated to Rubisco content (which was a nearly constant fraction of cell mass), making it likely that carbon fixation, not electron transport through PSUs, limits photosynthesis in saturating light. Neither of these studies ascertained a minimum electron transport time because light acclimation was only taken to a certain point. An estimate of only 1.6 ms (Sukenik et al. 1987) was based on calculations of diffusion of quinones through a lipid membrane, but this was not experimentally validated.

It is clear from this simple model of photosynthesis that photosynthetic rate may be variable, changing not only due to changes in PSU numbers, Rubisco levels, and other adaptation and acclimation factors, but also, and on a relatively short time scale, to changes in response to the regulation of the enzymes in the metabolic pathway. Thus, the P–I curve described earlier may differ (in either or both the initial slope and the maximum rate) at each growth rate or might change depending on the time a cell spends at each light level, that is, due to the effect of mixing in the water column (e.g., moving from high- to low-light intensities). Only light absorption responds instantaneously to the light level as cells move about in the water column. However, most models of photosynthetic production assume that the photosynthetic rate also changes instantaneously with changes in light intensity, as depicted by the P–I curve. This assumption is valid only if cells are moving slowly (on the order of many seconds) between different light intensities.

2.4   IMPROVING THE EFFICIENCY OF PHOTOSYNTHESIS

Since the saturating irradiance is much lower than sunlight throughout most of the day, algae and plants typically use only a small fraction of the absorbed photons for CO_2 fixation. The remainder is lost, wasted mainly as heat with some fluorescence. In this manner, photosaturation decreases the efficiency and productivity of photosynthesis in algae and plants. In addition, excess photon absorption actually damages the photosynthetic apparatus, as evidenced by photoinhibition observed at high irradiances. Thus, increasing the saturating irradiance of photosynthesis is a plausible way to increase photosynthetic efficiency.

A strategy for increasing the saturating irradiance is to decrease the size of the light-harvesting antennae, that is, the number of pigment molecules. This would decrease light absorption per photosynthetic unit, bringing the rate of light absorption per PSU more in line with the rate of electron transport, thus increasing the saturating irradiance and overcoming the wasteful consequence of the light saturation effect. An alternative way of increasing the saturating irradiance would be to increase the rate of electron transport through the PSU by ultimately alleviating the limitation set by rates of carbon fixation. This is problematic due to the likely limited degree to which the rates of the enzyme reactions and the amounts of enzymes could be increased. As shown in the discussion earlier on photosynthetic electron transport rates, the maximum photosynthetic rates achieved by some organisms are already high and could not likely be increased much. For example, increasing the turnover rate of Rubisco or increasing the amount of Rubisco in the cell would be very challenging. Rubisco content is already quite high, estimated in the discussion earlier to be at least 20% of the cellular protein in algae, a likely maximum considering other metabolic requirements. Another enzyme in the carbon fixation cycle, sedoheptulose-1–7 bisphosphatase (Rosenthal et al. 2011), is reported to actually limit carbon fixation under conditions of somewhat elevated CO_2. However, flux analysis typically reveals that metabolism is coordinated so that no single enzyme is greatly more rate limiting than others in the pathway. Given the large amount of Rubisco, it seems plausible that it would be the rate-limiting step.

In conclusion, decreasing light absorption through reduction in the amount of the chlorophylls and other major light-harvesting pigments that form the antenna complexes is the most plausible way to increase, possibly double, photosynthetic efficiency, thus productivity, by algal cultures.

During photoacclimation, the increasing irradiance to which cells are exposed causes a decrease in the content of photosynthetically active pigment (although not necessarily of photo protective pigments). This decrease may manifest itself as a reduction in the antenna size or as a decrease in the number of PSUs, or both, depending on strain and conditions (Myers and Graham 1971; Sukenik et al. 1987). Only so much can be attained by decreasing the number of PSUs. Once the PSUs are turning over at maximal rate, it is futile to further reduce their number. Past this limit, the reduction in PSU number will actually cause the maximum rate of photosynthesis to decline, which will decrease the saturating irradiance.

The increase in saturating irradiance of a suspension of cells due to decreasing pigment in the cells is inversely proportional to the decrease in the rate of light absorption. However, due to the phenomenon of pigment packaging (see Section 2.6.3), the decrease in absorption per unit biomass is less than proportional to the decrease in the chlorophyll content of the biomass. For example, a 75% decrease in pigment may lead only to about a 60% decrease in absorption by the suspension. Presumably, this leads to a $1/(1 - 0.6) = 2.5$, rather than a fourfold increase in I_s. As will be shown in the following text (Section 2.6.1), any increase in photosynthetic efficiency brought about by this increase in I_s is also less than linear. Thus, the overall increase in efficiency from a fourfold decrease in antenna pigment content may only be about twofold. This approach to increasing productivity by algal cultures has been studied for almost 20 years (Melis et al. 1999; Nakajima and Ueda 1999, 2000), but has been difficult to realize in practice.

2.5 LIGHT AS A LIMITING NUTRIENT

In order to model growth and productivity, the μ–S response curve must be known, where S is the concentration, or availability, of the growth-limiting nutrient. Models for dependence of specific growth rate on dissolved nutrients are not valid for photosynthetic organisms under light limitation because light is different from most nutrients in several important ways. First, light does not enter the culture vessel with the culture medium, so its availability is not increased by increasing the dilution rate of a continuous culture (Weissman and Benemann 1979). Also, light cannot be stored. Its use must be essentially immediate at whatever rate and efficiency the algal cells can achieve under the prevailing conditions. In addition, there is always some structure implied in any model in which light is the limiting nutrient with specialized light-harvesting and charge separation apparatuses involved in the first step and then utilization of this charge separation by biochemical reaction as a necessary subsequent step. The balance between the rates of the two processes must be considered in any model of productivity and efficiency. This balance is influenced by the variable adaptation of the photosynthetic apparatus to light, as mentioned earlier. Finally, in dense algal cultures, the light field is spatially heterogeneous, requiring integration over the entire depth of the culture, to determine culture growth rate and productivity. To this spatial heterogeneity must be added a temporal heterogeneity, as the individual algal cells move through the light field, due to any mixing of the culture fluid, resulting in possible interactions between the time constants for mixing and the dynamics of the photosynthetic response.

A mechanistically based model of algal growth must use the rate of photosynthesis as a starting point from which specific growth rate is derived. Essentially, all models are of the form of the photosynthetic rate being proportional to an absorption term times a quantum efficiency term times a saturation efficiency factor times irradiance. The absorption terms and the saturation efficiency factor are functions of light and of time. Each depends on the physiological state, as determined by light history, of the individual algal cells.

In all but the most dilute cultures, there is a gradient of light caused by the absorption of light by the cells. This means that, even in continuously diluted cultures (i.e., macroscopic steady states), a true steady state is never actually attained at the cellular level due to the movement of cells through the light gradient, resulting in cells that have been exposed to different light levels and light histories to which they have adapted. Thus, they differ, especially in the balance between the absorption of light and the biochemical utilization of the charge separation.

The light gradient is determined by the extinction coefficient of the suspension, which in turn is determined by the optical cross section per unit biomass and the biomass concentration. The light intensity and periodicity to which the algal cells are exposed regulate the cross section, on both a short- and a long-term basis, and the biochemical reactions of photosynthesis. The rate of these reactions and their coordination with the other metabolic reactions will, in turn, regulate the efficiency with which the absorbed light is actually utilized in photosynthesis. Thus, simplifying assumptions must be made in any equations that attempt to relate specific growth rate to light intensity, biomass density, light conversion efficiency, and the resulting productivity.

2.6   ADVANCES IN UNDERSTANDING LIGHT AS A LIMITING NUTRIENT

2.6.1   MEASUREMENT OF LOW-LIGHT PRODUCTIVITY, LIGHT SATURATION, AND MAINTENANCE

Even before the publication of the first description of the chemostat (Novick and Szilard 1950), Myers and Clark (1946) had already patented a continuous dilution culture system based on keeping the optical density in the reactor constant, a turbidostat. Myers et al. (Cook 1951; Myers and Graham 1959) studied the productivity of steady-state cultures in terms of density and/or specific growth rate, in essentially similar experiments to those shown in Figure 2.1. At any point on the figure, biomass productivity is equal to the product of specific growth rate and biomass density. They interpreted the results as follows. At low density (thus high, near maximum, specific growth rate), not all of the photons are absorbed by the water column, resulting in low productivity and efficiency per unit of incident light. As the dilution rate is decreased (specific growth rate is decreased), the biomass density increases, decreasing the average light level in the culture, and a new steady state is attained at an average culture specific growth rate equal to the dilution rate. For biomass productivity to increase, the biomass density must increase faster than the specific growth rate decreases. The steepness of this increase in productivity may be influenced by a changing light saturation parameter caused by a changing pigment content of the biomass (see photoacclimation, Section 2.6.2). At some point (after all of the light is absorbed), the increase in productivity slows, a maximum in productivity is reached, and then it declines as higher and higher cell densities (lower specific growth rates) require more of the light energy to be used for cell maintenance. Specific growth rates decline eventually to zero (as does biomass productivity) when density is so high that light absorption is balanced with respiration (cell maintenance). There is often, but not always, a relatively broad maximum of productivity versus specific growth rate (and thus cell density) when respiration rates of the algae are low, as is evident also from the low compensation point (see Figures 2.2 and 2.3). Although it was not discussed in these early papers, for many algae, there is also a depressive effect on efficiency (productivity) at high-light intensities—photoinhibition. This will be greater at high-light and low biomass densities (more time at photoinhibitory light), causing the productivity of dilute (but still not completely photosaturated) cultures to be lower than the maximum productivity observed at higher culture density.

Much of the early work in photosynthesis with microalgae, mainly *Chlorella*, addressed the question of quantum efficiency—how many photons are needed to fix one molecule of CO_2. This had been a hotly contested subject since the 1920s, when Otto Warburg first studied this problem and claimed that only four to five photons were required in this process (see review by Nickelsen 2007). Later work by the pioneers in this field (Emerson and Arnold 1932; Van Oorschot 1955; Kok 1960; Myers 1980) convincingly demonstrated a minimum of nine to ten photons/ CO_2 fixed, which agrees with the Z scheme of photosynthesis. Low irradiance was typically used in all these studies, resulting in a roughly 20%–25% maximum energy efficiency in converting "photosynthetically active radiation" (PAR), essentially the visible spectrum of light, energy into chemical energy. However, this controversy continues, with subsequent claims being published of only six quanta (Pirt et al. 1980; Pirt 1986) or fewer (Greenbaum et al. 1995) being required, suggesting a photosynthetic efficiency of up to 50% of light to chemical energy conversion.

Another motivation for studying algal productivity was for industrial purposes, especially food production, which became a high visibility project during the late 1940s and early 1950s (Burlew 1953). Of course, this necessitated using high, that is, sunlight, intensities, which, however, greatly reduced photosynthetic efficiencies, compared to the lower irradiances used in the laboratory (Cook 1951). An explanation for this result is that the "light saturation effect" was already evident in the experiments of Emerson and Arnold previously described. More detailed studies characterized the so-called flashing-light effect (Phillips and Myers 1954; Kok 1956b) in which short flashes of a few milliseconds of high-intensity light, followed by about fivefold longer dark periods, resulted in photon utilization efficiencies similar to those observed at low-intensity continuous light. The explanation for this phenomenon is that during the short flashes of high light, enough photons are absorbed to saturate, but not over saturate, the photosynthetic reaction centers. The dark period is required to allow sufficient time for the dark reactions to proceed. Without a dark period, the photons continue to be absorbed, but without any means to transform them into chemical energy, they are lost as heat or fluorescence and wasted. Kok (1956a) presented a model for the damaging and inhibitory effects of supersaturating light on algae, the "photoinhibition effect." Light saturation and photoinhibition are the major, fundamental, limitations to photosynthesis and productivity by algal mass cultures.

Another major insight developed during the early work on algal mass cultures was the use of the saturating irradiance as the basis for a quantitative relationship between light intensity and productivity (i.e., efficiency) in a dense algal culture. This is given by the Bush equation (Burlew 1953):

$$F = (I_s/I_o) (\ln [I_o/I_s] + 1) \tag{2.1}$$

where F is the Bush efficiency, that is, the efficiency of photosynthesis as a function of incident light intensity (I_o), relative to the maximum efficiency at or below the saturating light intensity (I_s). The equation is graphed in Figure 2.5, where it is assumed that all of the incident light is absorbed.

This equation appears as a multiplicative factor in the column integration of the simplest P–I curve (first-order rise to P_{max}) with light attenuation using the Beer–Lambert's law (Shelef et al. 1969). This simple model is still useful today. When combined with a biomass-specific respiratory term and subtracting out transmitted light, it reflects the maximum in productivity relationship described earlier:

Productivity = (Maximum energy efficiency) · (Bush efficiency) · (I_o) − Respiration

This result could be used to form an integrated output for productivity over time, for example, a day. The model is limited in many ways including the use of a constant value for the saturating irradiance (lack of adaptation) and for respiration, as well

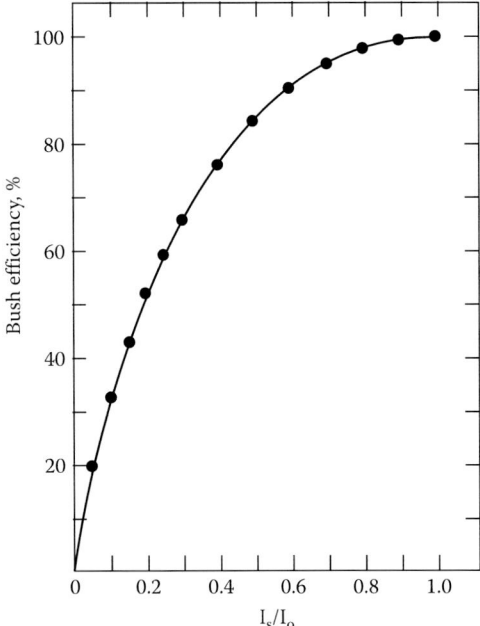

FIGURE 2.5 Bush efficiency: photosynthetic efficiency as a function of the ratio of saturating irradiance to incident irradiance.

as the absence of a photoinhibitory term. Consequently, the results in Figure 2.4 can only be approximately predicted with this simple model but still within about 30% for well-mixed systems such as small ponds or photobioreactors for which temperature does not vary much. Estimates for two of the three parameters of a P–I curve (maximum quantum efficiency, saturating irradiance, P_{max}) are usually required as input to a simple column integration, but maximum efficiency is assumed in this equation.

A major focus of much research in this field was, and continues to be, finding some way around the constraint on productivity imposed by light saturation. The flashing-light effect discussed earlier is one possible means of increasing algal productivity in mass culture. However, it was quickly realized that this would require very intense mixing, even an organized type of mixing, to obtain any significant improvement in productivity. Such intense mixing would require an enormous amount of energy, which is impractical. Organized mixing, in which all cells experience the needed light regime (as noted earlier, a few milliseconds in the light, followed by 5- to 10-fold longer, but not much more, in the dark), is not possible outside of the laboratory.

Another potential limitation on productivity is respiration, more generally, "cell maintenance" (Pirt 1965). Most data on respiration by algal cultures are obtained by removing samples from growing cultures and then measuring respiration in an O_2 electrode in the dark. However, respiration rates from these perturbed samples are not likely to be representative of maintenance requirement of cells growing in the light.

A simple model for cell maintenance developed for algae (Gons and Mur 1975) is similar to the model of variable yield for energy substrates (Pirt 1986). Absorbed light is assumed to be converted, at constant overall conversion efficiency, c, into chemical energy, and then used for either growth or maintenance. Growth and maintenance coefficients, μ and m, are defined as biomass specific rates. Dense culture data for various specific growth rates are needed to determine the overall photosynthetic conversion efficiency and the biomass specific maintenance coefficient. Figures 2.6 through 2.8 show several examples of this type of modeling. Figure 2.6 represents data from indoor, dense cultures under 200 $\mu E \cdot m^{-2} \cdot s^{-1}$ continuous light. Given the low irradiance, there are no effects from photoinhibition. Also, there is no dark period to increase maintenance. The data in Figures 2.7 and 2.8 are from the same cultures as in Figure 2.1: outdoor continuous cultivation for which the daytime maximum temperatures were 35°C–38°C, while the overnight temperatures were from 24°C to 26°C. Here, maintenance losses include the losses overnight in addition to the maintenance during the day, including any repair due to photoinhibition. Figure 2.8 shows how maintenance energy requirements and thus net productivity vary with dilution rate. Table 2.1 summarizes the overall photosynthetic efficiency c (for converting light energy to biomass growth and maintenance energy) and the maintenance coefficients m measured from these experiments and for two more species grown indoors. Although maintenance coefficients can be quite low, indeed remarkably low, compared to the maximum specific growth rate μ_{max}, mass cultures are operated at specific growth rates substantially lower than the maximum. Table 2.1 and Figure 2.8 demonstrate that at low dilution rate (high density), even

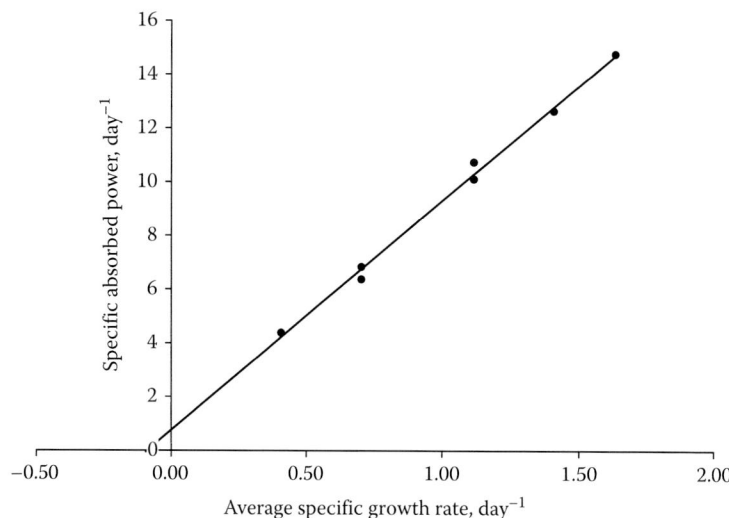

FIGURE 2.6   Specific light absorption versus specific growth rate for cultures of *Chaetoceros muelleri* grown in the laboratory at 200 $\mu E \cdot m^{-2} \cdot s^{-1}$. (Weissman, J. C. and Goebel, R. P., U.S. DOE Aquatic Species Program, 1986, unpublished.)

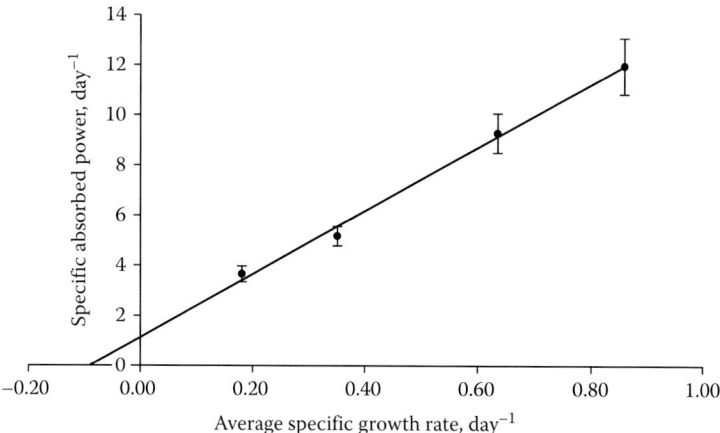

FIGURE 2.7   Specific light absorption versus specific growth rate for the continuously diluted 1.4 m² outdoor pond cultures of *Chaetoceros muelleri* shown in Figure 2.1. (Weissman, J. C. and Goebel, R. P., U.S. DOE Aquatic Species Program, 1986, unpublished.)

algae with low maintenance coefficients start to lose productivity. Losses for algae with maintenance coefficients similar to the *Spirulina* and the *Chlorella* species shown in Table 2.1 would be large. Thus, the potential for losses is substantial unless proper screening is used to provide strains with low maintenance. As a result of such screening, maintenance should be overall a lesser limitation in the productivity of algal mass cultures compared to light saturation and photoinhibition.

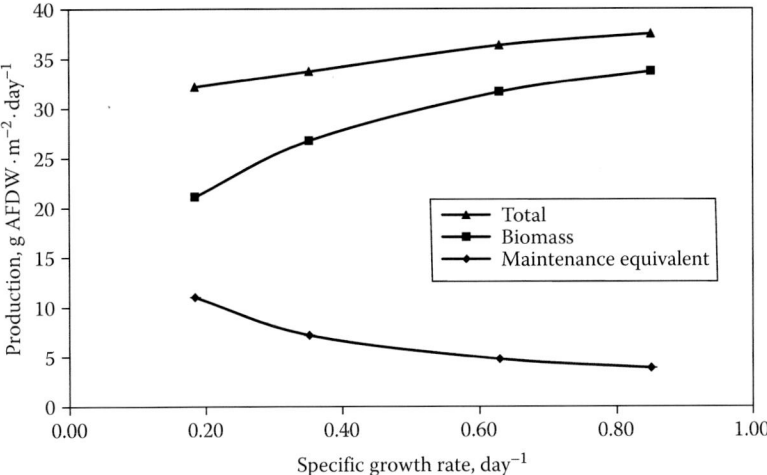

FIGURE 2.8 For cultures in Figure 2.7, biomass productivity (μX), maintenance energy (mX), and total conversion of photosynthetic energy to chemical energy (= $\mu X + mX$) as a function of specific growth rate. X is biomass density in g $AFDW \cdot L^{-1}$. Maintenance energy is shown as biomass productivity with energy equivalent to that used for maintenance. (Weissman, J. C. and Goebel, R. P., U.S. DOE Aquatic Species Program, 1986, unpublished.)

TABLE 2.1
Parameters from the Maintenance Model

Alga	Maintenance Coefficient m, day⁻¹	Photosynthetic Efficiency c, %PAR	Maximum Specific Growth Rate μ_{max}, day⁻¹	Ratio m/μ_{max}
C. muelleri 6[a]	0.09	7.9	3.2	0.03
C. muelleri 14[a]	0.09	11.6	4.4	0.02
C. pyrenoidosa[b]	0.27	13	2.3	0.12
Spirulina[c]	0.12	19	1.2	0.10

[a] Weissman, J. C. and Goebel, R. P., U.S. DOE Aquatic Species Program, 1986, unpublished.
[b] From data of Dabes et al. (1970).
[c] Weissman and Benemann (1979).

2.6.2 PHYSIOLOGICAL ACCLIMATION TO LIGHT

Myers and Graham (1959) were among the first to propose that the steady states of dense continuous algal cultures are different physiological light adaptations, analogous to sun/shade plants, with decreasing chlorophyll content with higher irradiance. In a seminal paper, Myers and Graham (1971) demonstrated that points on the μ–I curve (dilute, acclimated steady states) were different physiological states not only with different chlorophyll content in the cells but also with distinct P–I curves. Faster-growing cells (acclimated to higher irradiance) had a higher I_s and P_{max}.

However, for the most part, the maximum photosynthetic rate (Q_{max}) per cell mass was not very different for the various physiological states. Most importantly, they were able to apportion the effect of the adaptive changes of chlorophyll content on P_{max}/chlorophyll both to a decrease in the PSU size and to a decrease in PSU turnover time for cells grown in higher light. They measured both the Emerson–Arnold number (molecules of chlorophyll per oxygen evolved in oxygen flash experiments) and the P–I curve for each acclimation state. Dividing the maximum continuous rate of photosynthesis, μmoles oxygen per mg chlorophyll per hour, by the Emerson–Arnold number (number of chlorophyll per oxygen) yields the PSU turnover rate. This demonstrated that a reduction in chlorophyll as an acclimation to light was attained by both varying the antenna size and the number of PSUs. This implied that the rate-limiting step in photosynthesis was beyond the PSU itself; otherwise, P_{max} would decrease when the number of PSUs decreased. At this time, the authors were not aware of the nonlinear relationship between chlorophyll content of the algal cells and absorption of light due to the packaging effect. The data of Myers and Graham on respiration (measured as oxygen consumption in the dark) are often used to form a growth rate–based model of cell maintenance as an alternative to the mass-based model described earlier. In such a growth rate model, respiration is an adaptive parameter equal to a constant plus another constant times the specific growth rate of the culture. It is difficult to know how well dark respiration rates of perturbed samples from growing cultures, as measured by Myers and Graham and most others, describe the maintenance of cells growing in the light.

In addition to the studies of acclimation performed by the Falkowski laboratory discussed in Section 2.3.2, Falkowski and Owens (1980) reported these two different strategies of light acclimation in two different algae: in *Skeletonema costatum*, the size of the PSUs decreased due to acclimation to high light, while in *Dunaliella tertiolecta*, the number of PSUs decreased. Falkowski et al. (1985) calculated quantum requirements for dilute, growing culture of three algal strains, obtaining 9–11 quanta for lowest values for *Isochrysis galbana* and *Thalassiosira weissflogii*. This again emphasizes that maintenance must be very low, possibly requiring only one quanta per oxygen (Myers 1980).

Many more studies have shown a decrease in pigment content with high-light acclimation (MacIntyre et al. 2002). An example of this, as well as other important physiological aspects of light acclimation, is shown in Figures 2.9 through 2.11 for the same experiment shown in Figure 2.3. The lowering of the pigment content is accompanied by a proportionate decrease in protein, as shown in Figure 2.10. In this case, nearly half of the protein in the cell is associated with the light absorption apparatus. As evidenced by the high specific growth rates achieved, the proteins associated with carbon fixation and reproduction are not reduced. As calculated earlier, another 20% of the cell protein may be Rubisco. Removing protein as a sink for reductant, such as during N-limitation, forces electrons onto other electron-accepting carbon compounds. Figure 2.11 shows the increase in non-nitrogen-containing, reductant-accepting compounds—carbohydrates in this strain. It is unclear how much of this requires changes in genetic expression or is based on metabolic regulation. Except for their fast rate of growth, the physiological state of cells exposed to supersaturating irradiance, where light-absorbing pigments are reduced giving the cultures a

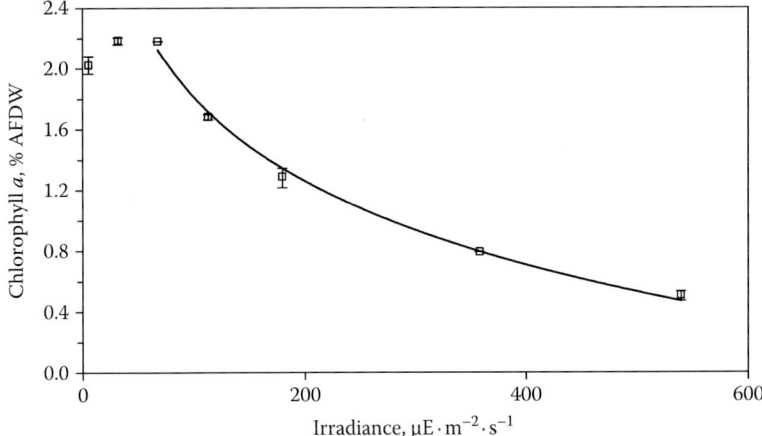

FIGURE 2.9   Chlorophyll *a* content of the optically thin cultures of *Chaetoceros muelleri* as a function of irradiance (the same cultures as in Figure 2.3). (Weissman, J. C. and Goebel, R. P., U.S. DOE Aquatic Species Program, 1986, unpublished.)

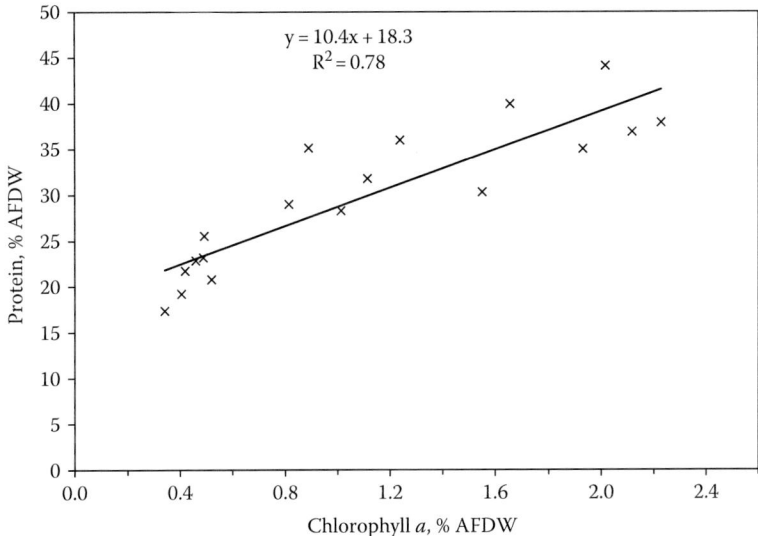

FIGURE 2.10   Protein content of the biomass versus the chlorophyll *a* content of the biomass (the same cultures as in Figures 2.3 and 2.9). (Weissman, J. C. and Goebel, R. P., U.S. DOE Aquatic Species Program, 1986, unpublished.)

pale yellow-green color, could be mistaken for a nitrogen-deficient state even though nitrogen is plentiful. The other major difference is that in the high-light adapted state, the cells have high photosynthetic capacity (high photosynthetic rates per chlorophyll as well as per unit biomass), while in nitrogen-deficient states, photosynthesis is shutting down and photosynthetic capacity decreases. This is a major limitation

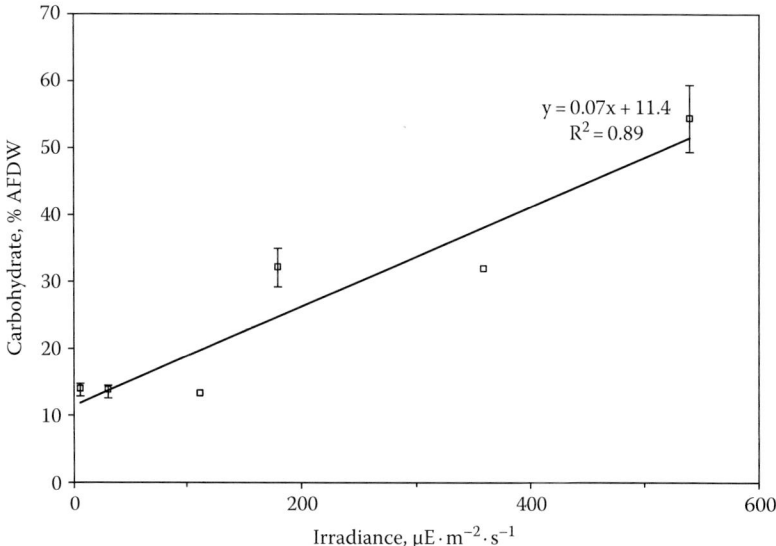

FIGURE 2.11   Carbohydrate content of optically thin cultures of *Chaetoceros muelleri*, the same cultures as in previous figures. (Weissman, J. C. and Goebel, R. P., U.S. DOE Aquatic Species Program, 1986, unpublished.)

when N limitation is used to attempt to increase the oil or starch content of algal cultures. Nevertheless, significant amounts of storage compounds may accumulate during the initial day(s) of nitrogen deficiency prior to the decrease in photosynthetic capacity or in semicontinuously diluted, nitrogen-deficient cultures (Weissman and Goebel 1987).

2.6.3   PIGMENT PACKAGING

Absorption cross sections for cells grown at various irradiances measured by, for instance, Falkowski et al. (1985), reveal the "packaging effect" (Kirk 2011): the decreasing absorption cross section of algal suspensions per unit pigment as cellular pigment content increases. The gross packaging effect is due to confinement of pigment within discrete particles (cells), which differs from the absorption that would be obtained if the pigments were uniformly distributed in solution.

   Given two equal macroscopic volumes equally illuminated and containing equal numbers of pigments, with the pigment uniformly and randomly distributed in one, and with pigment constrained to discrete particles in the other, the total absorption of light by the second volume will be less than the first. An easy way to see that there must be a difference between the two scenarios is to consider cells that are so highly pigmented that all light impinging on the cell is absorbed. Then, the attenuation of light within a suspension is determined solely by the concentration of cells. The key is that no matter how much the pigment is increased in the cells, the fraction of void spaces in the solution where there are no cells, and hence no absorption, remains constant so that that a certain amount of light must be transmitted. For a fixed cell concentration, the absorption in a suspension is constrained by the geometric cross section of the cells as an upper limit. From the perspective of pigment light absorption, sufficiently enhanced attenuation of light due to the concentrating effect of particle confinement reduces the average rate of light absorption in the particles relative to a random uniform distribution in solution. Because of the loose analogy with the action of a mechanical sieve, the approach to this limit has been called the "sieve" effect.

From the perspective of photosynthetic activity, the consequence of confining the pigment within cells, unlike a solution, which can be made very dilute, is that not all of the pigment molecules "see" the same irradiance as the light incident on the cell wall. Due to attenuation of light as it traverses the cell, most pigment molecules experience lower-light levels and hence absorb less light energy as a function of time leading to decreased absorption on a per pigment basis. The distribution of pigment within the cell is not uniform, but since cellular structures, especially at the level of grana, are smaller than the wavelength of visible light, the light distribution in the cell appears closer to the case where the pigment distribution is uniform due to diffraction of light around the structures.

Consequently, to a good approximation, light absorption of Beer's law holds within a cell, so this approximation will be used to describe attenuation in the following discussion. However, small deviations can be expected due to chloroplast structure and may be accentuated in extreme cases of irregular organelle geometry. Comparative studies (Berner et al. 1989) clearly show that the influence of intracellular structure on light attenuation is hindered by cell size changes that often accompany changes in chloroplast structure due, for example, to light acclimation.

Other simplifications are possible. For example, the precise shape of a cell, for example, spheroidal unicellular alga, is not very influential in determining the total cell absorption. Exact calculations of cellular absorption cross section using Mie scattering theory for spherical particles have been used widely, especially in the oceanographic literature (van de Hulst 1981; Morel and Bricaud 1986a,b). However, the absorption of a spherical cell can be approximated with a high degree of accuracy by a "rectangular" cell with a depth dimension equal to the mean chord length of the sphere it is meant to represent.

In this approximate sense, if a cell absorbs 10% of the light incident on it with a given amount of cellular chlorophyll (ignoring other pigments), then with four times as much cellular chlorophyll, it will absorb about 30% of the light. Simply due to nonlinear light attenuation (Beer's law) the light absorbed by the cell per amount of chlorophyll is 3/4 the value of the less pigmented cell. Even if Beer's law is not strictly valid, the absorption of light through the cell is well less than linear with respect to pigment content unless the cell is very small and/or the pigment content is dilute. Thus, in a given cell, absorption increases by diminishing returns: the more pigment the more absorption of light across the cell, but less absorption per amount of pigment. However, there is no packaging effect if absorption within the cell is linear with pigment content (low-light absorption across the cell), and thus no sieve effect, since the light absorption averaged over cells and void spaces is then also linear with respect to cellular pigment content.

To put this in practical terms, it is possible to express the extinction of light by the suspension as the product of an (optical) absorption cross section of the suspension ($m^2 \cdot g^{-1}$ chlorophyll) times the cellular pigment content times the amount of cells. For a constant size and concentration of cells, when pigment content of the cells is decreased, the total extinction coefficient of the suspension is lowered, but the optical cross section per pigment ($m^2 \cdot g^{-1}$ chlorophyll) increases due to the nonlinear relationship between cellular cross section and pigment content. The latter relationship is often found experimentally, as algae acclimate to high-light conditions.

Pigment packaging is very important when considering the relationship of light absorption to biomass accumulation. For example, as discussed earlier, one approach to increasing photosynthetic efficiency of algae for biofuels applications involves reducing the cellular pigment concentration in an attempt to alleviate photosaturation in natural high-light conditions. Pigment packaging can have a significant impact on the effectiveness of this approach. As an example, consider a reduced pigment mutant that has 25% as much pigment as its wild type parent. Due to pigment packaging, this reduction in cellular pigmentation leads to a less than linear change in the cellular absorption. For example, data from our laboratory (Weissman et al. 2013) show that such a 75% reduction in pigment in high-light acclimated cells versus low-light adapted cells results in only a 60% reduction in absorption of the suspensions of the high-light acclimated cells. Thus, a pigment reduction strategy to increase the saturating irradiance, and hence photosynthetic efficiency at high light, is affected by this phenomenon as was also discussed in Section 2.4. Pigment packaging is also an impediment to estimating the biomass in a natural system since remote sensing measures absorption or a metric based on absorption. But due to pigment packaging, absorption is not a true measure of pigment concentration, which is often used as an estimator of biomass.

2.6.4 TIME IN THE LIGHT

It has been known for a long time that the rate of photosynthesis, in an actual culture or as measured by short-term photosynthesis experiments, is determined by both the instantaneous light absorption, which is an adaptive feature (Myers 1970), and the prevailing rate of biochemical reactions. However, it has been difficult to incorporate the dynamics of photosynthesis in a light gradient into photosynthetic models. The primary method for doing so is to model the hydrodynamics in the systems used to grow algae and coordinate this with P–I relationships.

Consider two contrasting, simplified cases based on dense cultures. In the first, movement in and out of the light in a dense culture is so fast that an average metabolic state is maintained. In this case, P–I curves or μ–I curves are not relevant though a light level could be defined, which is assigned to the measured μ. In the second case, the movement is so slow that the photosynthetic rate adjusts to the light levels as cells move about the culture depth. Even in this case, modeling via P–I curves would be limited by dynamic lags and hysteresis, for example, individual cell history, in light acclimation and photoinhibition, as the individual cells move in and out of the light, over a period of seconds to minutes, even hours. In both cases, acclimated μ versus I or snapshot P–I curves do not represent the actual photosynthetic capacity of the cells *in situ*. Estimating the dynamics for changes in photosynthetic parameters in response to changes in irradiance has been attempted (Geider et al. 1996) but responses are still poorly understood. Marra (1980) summarized experiments on the response to movement into high light for three diatoms and the haptophyte *I. galbana*. It took a couple of minutes for the P_{max} to fall to half of its initial value in high light versus remaining essentially constant in subsaturating light. Han et al. (2000) modeled photoinhibition using rate constants for damage and repair to the D1 protein of photosystem II. Damage was related to irradiance and the concentration of D1. Using relative values of D1 from 1 (at night) to 0.4 (morning) and then 0.2 (afternoon) and various damage rate constants and repair rate constants on the order of hours, they were able to simulate the experimental results of Marra (1980) for photosynthesis over the day. This is just an example of how incorporating physiology can lead to semiquantitative models. For large ponds with slow movement of the algae in the vertical direction (e.g., $0.3 \ cm \cdot s^{-1}$ or less), time spent in high light could have a large influence on biomass productivity.

Other processes could also have a great impact when cells move in and out of the light. Antenna non-photochemical quenching (NPQ), turns on in a matter of seconds to minutes at high light, but may turn off more slowly. In reactors for which the time constants of mixing (in and out of high-light zones) are on the order of seconds to tens of seconds, antenna NPQ may be active even when the cells are in low, subsaturating, light. This would reduce photosynthetic efficiency and hence productivity. Worse yet, the level of antenna NPQ might ratchet up to a nearly constant high level if it increases faster than it decreases.

Similar issues may pertain to dark respiration. Many algal strains seem to be able to maintain themselves with very low expenditure of energy in the light (low compensation point, when measured in short-term experiments), but extended time in the dark (minutes) may lead to increased rate of breakdown of previously formed cell mass, which is respiration.

Regarding CO_2 fixation, which typically limits the rate of photosynthesis, the time constants are not well known. Estimates based on a content of Rubisco and turnover times require assumptions about the amount and activation state of Rubisco and the availability of its substrates. Assuming a maximum rate of turnover of Rubisco of 24 CO_2/sec, requiring 48 NADPH/sec, and with a steady-state pool of 100 NADPH per Rubisco present in the light, then, allowing for a required NADPH/NADP ratio of >1, it would take about 1 second to use up this substrate in the dark, absent continuous regeneration. The mixing of an algal culture will thus interact with photosynthesis depending on the time constants for mixing. However, in large ponds with slow mixing (20–30 $cm \cdot s^{-1}$), typical for algal mass cultures, the movement of algal cells throughout the pond depth (typically 30 cm) would probably not benefit much from this capacitance because cells linger at one light level for many seconds, even minutes. In highly mixed laboratory cultures, commercially closed photobioreactors, or even small experimental ponds, cells would likely stay at a particular irradiance for a second or less and thus may exhibit a pronounced lag in changing photosynthetic rates. If mixed fast enough and randomly, they could even adjust to a constant or highly damped rate. Thus, projecting results from laboratory or small-scale cultures to large-scale algal ponds may be affected by the differences in the interactions of photosynthetic time constants and mixing. Indeed, algal cultures with an increased I_s such as the ones obtained with reduced antenna mutants (see discussion in the following texts) may not manifest the full benefits of such altered I_s in laboratory or small-scale cultures, as compared to large ponds. Of course, the latter may exhibit other limitations due to other processes that reduce productivity.

2.6.5   MODELING COLUMN-INTEGRATED PHOTOSYNTHESIS

As described earlier, the productivity of a water column containing algae can be estimated by integrating the photosynthetic rate at each depth over the total depth (given a model for the light field), something of interest to oceanographers and limnologists as well as in algal mass cultures. The key is to know the actual photosynthetic rates and the extinction constant for the water column to allow calculation of daily column integrated photosynthesis with the least number of measurements. Ryther (1956) was one of the first to measure the rates of photosynthesis of marine phytoplankton as a function of irradiance, which, together with dark respiration, the extinction coefficient, and diurnal surface radiation values (sunlight), he used to calculate column photosynthesis (g $C \cdot m^{-2} \cdot day^{-1}$). Subsequently, he measured the acclimation of marine algae to light as a change in the saturating irradiance and photoinhibition, and thus the changing shape of the P–I curve in water columns (Ryther and Menzel 1959). Ryther and Yentsch (1957) further simplified column integration in general by noting that in many cases the saturated rate of photosynthesis was a constant times the chlorophyll concentration. Such an approximation is also used in remote sensing, where areal chlorophyll concentrations and "regional" P–I curves are used. However, these are not very accurate, perhaps only ±50%, as this ignores the actual variations of P_{max} with chlorophyll content of the algae and the variable relationship between chlorophyll concentration and absorption due to pigment packaging, which is nonlinear.

Numerous models were subsequently put forth for the P–I and μ–I curve applicable to predict productivity in both mixed and stagnant water columns. Without photoinhibition and dark respiration, only two parameters are needed to formulate all of the uncomplicated mathematical models put forth (Jassby and Platt 1976): the slope and maximum rate of the P–I curve. If photoinhibition is important, a third parameter, at least, is required. Depending on the accuracy of prediction desired, additional measurements are needed to establish these parameters. The various models of the P–I curve used differ mostly in shape parameters, that is, how fast they pull away from the initial slope. The simple rectangular hyperbola pulls away faster than the exponential model, with the hyperbolic tangent (Jassby and Platt 1976) pulling away even more slowly, predicting a longer, steeper initial rise of photosynthesis with increasing irradiance.

Another major goal of algal physiologists and oceanographers has been to formulate generalizable models of algal growth as a function of nutrient status. Growth at low nutrient levels is an important feature of natural systems. In algal mass cultures, nutrients can be supplied *ad libitum*, and nutrient limitation is something that can be manipulated as desired. Attempts were also made to merge models for light limitation and dissolved nutrients limitation, given that the nutrient status often affects the chlorophyll to carbon ratio of the cells. Models became more detailed with an increasing number of equations and constants that needed to be determined (e.g., Laws and Bannister 1980; Kiefer and Mitchell 1983). Many attempted to devise simplifying generalizations, for example, the quantum efficiency is unaffected by the degree of nutrient limitation (Kiefer and Mitchell 1983) or that C/N ratio (or the physiological state in general) is uniquely related to relative growth rate regardless of irradiance (Goldman 1980). These assumptions, and others, limit the applicability of the models, especially in high light.

Cullen (1990) used the data of Sakshaug et al. (1989) to fit an exponential curve to the μ–I data. The goal was to produce a model requiring only a few parameters, which could describe primary production in the ocean. It is based on data from growth of dilute algal suspensions that were acclimated to the light levels to which they are exposed. The objective was to use this model for both nutrient- and light-limited cultures because the cross section and chlorophyll content parameters could incorporate the effects of nutrient limitation. A characteristic of this model is that certain parameters are always combined as (multiplicative) products. One combination, the chlorophyll-based optical cross section times the chlorophyll per unit cell carbon, yields the carbon-based cross section, which could be more fundamental than either alone, although chlorophyll is a more accessible measurement than algal carbon, in general. Another combination, the PSII cross section times the turnover time for electron transport, is assumed to be constant. Although this combination has no fundamental meaning since they are physiologically distinct, each would be constant for nutrient-saturated cells adapted to a constant irradiance. It is not at all clear that the product would be constant for different levels of nutrient limitation, even at a single irradiance. One might expect the cross section to decrease with increased nitrogen limitation and the amount of cellular enzymes to also decrease. The latter would increase the turnover time but not necessarily inversely to the decrease in cross section. Cullen goes on to graph the model-fitted μ–I curve and three P–I curves for a different alga acclimated to three light levels (20, 100, 2200). The photosynthetic rates from the three P–I curves read at the irradiance of growth fall very closely on the model. Two of the three irradiances to which the cultures were acclimated are below saturation. This is why a μ–I curve can sometimes be used: at subsaturating irradiances the photophysiology can be nearly the same for the acclimated alga and the differences in slope at subsaturating irradiance is hard to detect. In the end, Cullen concludes that the old models of Ryther are as useful in representing most data as this more advanced one.

Volume integration of various photosynthetic models is reviewed in Béchet et al. (2013). These examples underscore a major issue with photosynthetic modeling: the models become too complex (many variables and parameters) to be practical, while still not being adequately based on fundamental photophysiology. It would seem that complex modeling would need to be justified either when a particular alga is of sufficient interest, for example, it is expected to be used in large-scale production to justify a very detailed physiological description or that the modeling effort (based on photophysiology) is used to devise experiments to further our understanding.

2.7 PRODUCTIVITY OF OUTDOOR ALGAL MASS CULTURES

Outdoor algal cultivation began in the early 1950s, with a project for cultivation of *Chlorella* for human food production using large (approximately 100 m²) plastic bag cultures on a rooftop at MIT (Burlew 1953). Work in Japan in the 1950s, developed *Chlorella* cultivation using circular ponds, a technology that became the first commercial production of microalgae starting in the 1960s. Mass culture of *Scenedesmus*, another green alga, was initiated in Germany in the 1960s, using both small circular and also paddle wheel mixed raceway ponds, and in the 1970s, Germany carried out several projects using raceway cultures in India, Peru, Egypt, and Israel. During the 1970s, commercial *Spirulina* cultivation was initiated in Mexico and Thailand, the latter using paddle wheel raceway ponds. A very shallow culture system was developed in Czechoslovakia (Masojidek et al. 2011), and large-scale tubular photobioreactors were operated in several Soviet Republics for animal feed production. At the University of California, Berkeley, shallow raceway ponds were developed during the 1950s for wastewater treatment, and the first large (0.1 ha) experimental paddle wheel mixed ponds operated in the 1970s (Benemann et al. 1980). During the 1980s, a concerted effort was the U.S. DOE–sponsored Aquatic Species Program (ASP), mostly from 1980 to 1992, with some work continuing until the mid-1990s (see Sheehan et al. 1998, for a review). During the 1990s in Japan, a very large effort was carried out to develop photobioreactors for algal CO_2 capture and utilization as part of the Research Institute of Innovative Technology for the Earth (RITE) project. A large R&D effort for production of algal biofuels, and other products, has been ongoing in the United States and worldwide for the past decade, with billions of dollars invested. The ASP results still represent the most relevant information on the current state of the art in this field.

Initial work by the ASP used culture collection strains, but these generally performed poorly outdoors, being relatively slow growing and rapidly replaced by invading species. The ASP embarked on a strain isolation, characterization, and screening effort, isolating algal strains from different habitats, in particular from brackish and saline puddles, coastal pools, and saline lakes, in hot, arid, sunny locations in the southwestern United States as well as coastal regions such as Hawaii. Initial enrichments used temperature–light–salinity gradient culture platforms. These strains were used in outdoor cultivation experiments in California, New Mexico, Hawaii, and, to a more limited extent, Israel. One project (Weissman and Goebel 1987) imposed further screening criteria on strains grown meticulously to establish specific growth rates and biomass productivity. These screening experiments emphasized high light, high temperature, high dissolved oxygen, and eventually high pH (low free CO_2). Each ASP project used their own strains, but with one or two strains used in common. Culture systems were similar, of the raceway type, usually small 1–10 m². The maximum productivities reported for the various locations were similar (Table 2.2) at about 30–40 $g \cdot m^{-2} \cdot day^{-1}$, but over different periods: year-round in Hawaii, 3 months in CA, 1–2 months in Israel, and 2 months in New Mexico. The daytime pond temperature was the main determinant in the duration of the highest productivity periods in these different climates. In Hawaii,

TABLE 2.2
Summary of Outdoor U.S. DOE Aquatic Species Program Productivity Results

Principal Investigator and Location	Mean Productivity, $g \cdot m^{-2} \cdot day^{-1}$ (max)	Mean %PAR (max)	Time in Culture, Days	Scale, m^2	Alga
Laws, Hawaii, 1984–1985	37 (65)	9 (15+)	78	48	*Tetraselmis* sp.
Laws, Hawaii, 1985–1986	30 (35)	9.6	400	48	*Cyclotella* sp.
Laws, Hawaii, 1990	15	3.5	Variable	24	*Tetraselmis* sp.
Ben Amotz, Israel, 1986	27 (40)	9.5	90	0.35	*Chaetoceros* sp.
Richmond, Israel, 1985–1986	25		60	2.5	*Nannochloropsis* sp.
	28		14	2.5	*Isochrysis* sp.
Weissman, CA, 1986	32 (40)	7 (10)	125	1.4	*Chaetoceros muelleri*
	32 (40)	6.8 (9)	43	1.4	*Cyclotella* sp.
Weissman, CA, 1983	16	3.6	240	100	*Scenedesmus* sp.
Weissman, NM, 1988	30 (50)	6.8	50	3	*Amphora* sp.
	18.4	4.8	180	3	*Tetraselmis* sp.
	20	7	40	3	*Cyclotella* sp.
Weissman, NM, 1989	20 (24)	5	100	1000	*Cyclotella* sp. *Monoraphydium* sp.
Weissman, NM, 1988–1990	10 (24)	3	730	1000	*Cyclotella* sp. *Monoraphydium* sp.

in 10 m^2 raceways, multiple short air foils (flow deflectors with geometric cross section similar to an airplane wing) were installed across the channels, in rows about 6 ft apart, to induce regular vertical mixing (Laws et al. 1986). Very high production rates, 65 $g \cdot m^{-2} \cdot day^{-1}$, were reported for some days (averages remained 45 $g \cdot m^{-2} \cdot day^{-1}$ over 3 days), but these were not reproduced in subsequent work (Laws and Berning 1991). Still, the highest annual productivities were achieved in the small-scale Hawaiian systems.

The ASP work culminated in the operation of two 0.1 ha (1000 m^2) demonstration ponds in Roswell, New Mexico, and, as anticipated from other work, these larger ponds were less productive than the smaller test ponds at this location (Table 2.2). Several factors could cause this difference including greater sedimentation in the large ponds due to less vertical mixing, which was clearly evident and which would also expose cells to longer periods of high light and darkness. Another factor could be large variations in CO_2 concentrations around the ponds, possibly in conjunction with high O_2 concentrations. The New Mexico results showed that year-round, stable (no restarts) cultivation was possible, that CO_2 utilization efficiencies (carbon into the algae divided by carbon inputs to the pond) could be greater than 90% as long as algae productivities were greater than $10 \, g \cdot m^{-2} \cdot day^{-1}$ (CO_2 outgassing to the atmosphere is the main loss factor at low productivity) and that water loss is mainly due to evaporation, and similar between lined and unlined ponds (which exhibit little percolation due to self-sealing). Summer and winter strains succeeded each other in the large ponds. There is no evidence that any inoculated strain became dominant in the large ponds. Although a *Cyclotella* sp. was inoculated, a different, larger species of this alga invaded and became virtually unialgal as the dominant summer strain. In Hawaii, both a native *Tetraselmis* sp. and an inoculated *Cyclotella cryptica* dominated small ponds for over a year, though for the latter it was not clear how often it had to be reinoculated.

The ASP was focused on the production of microalgal oil in southwestern United States where favorable land, CO_2, and ample subsurface brackish water resources were thought to be available. However, except for land, this is not correct: temperatures are not favorable, brackish well water is limited and difficult to use, and CO_2 sources are also limited. At the Roswell site, it had to be purchased from vendors and trucked in as liquefied CO_2, at high cost. This project did not address other critical issues such as harvesting of the biomass, producing biomass with a high oil content, oil extraction, and water and nutrient recycling. The greatest obstacles to economical mass algal production remain today: high capital costs, especially for the ponds (and much higher for closed photobioreactors); relatively low annual biomass productivities at large scale; cell harvesting, and availability of an economic source of CO_2.

2.8 FACTORS OTHER THAN LIGHT AFFECTING BIOMASS PRODUCTIVITY OF DENSE CULTURES

2.8.1 TEMPERATURE

Temperature has the potential to significantly constrain biomass productivity since enzymatic reactions are temperature-dependent. It is still a question whether within an organism's temperature optimum it is as efficient as an organism that grows at a higher temperature. The problem here is in the meaning of the term efficient. Cleary, if this were literally true, the biomass productivity of an organism, which has a temperature optimum of, say, 10°C, would be as high as one that grows well at 25°C. We know of no example of this. During the last 70 years of research in which biomass productivity has been measured, the highest productivities come from the organisms with temperature optima between 25°C and 35°C, and possibly higher. However, there is a high correlation between temperature and sunlight input. Less certain is the case of maximum quantum efficiency in low light. Since temperature influences several physiological processes, its effects may be difficult to predict. For example, with a respiration rate of 10% of P_{max} at 30°C, even with a twofold decrease in the maximum rate of photosynthesis due to lowered temperature, a concomitant decrease in maintenance rate could help result in only a 25% decrease in productivity.

In most of the ASP outdoor work, warm water strains were tested. Some were tested at lower temperatures. The data in Figure 2.12 show the effects of temperature

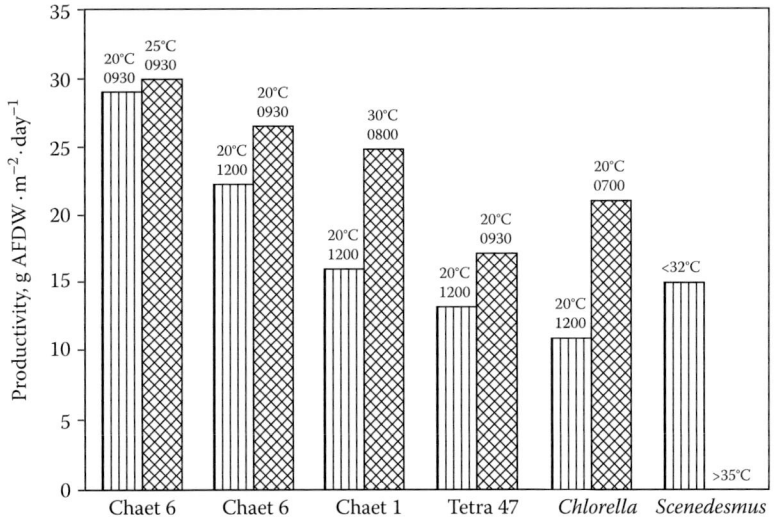

FIGURE 2.12 Productivity of outdoor grown, semicontinuously diluted, 1.4 m^2 cultures of several algae under different condition of temperatures. Chaet, *Chaetoceros muelleri*, and Tetra, *Tetraselmis suecica*, Aquatic Species Program species. *Chlorella* sp. and *Scenedesmus* sp., indigenous strains from Vacaville, CA. Numbers above the bars represent the temperature reached at the time of day indicated. (Weissman, J. C. and Goebel, R. P., U.S. DOE Aquatic Species Program, 1985 and 1986).

on some common strains. One case shows the phenomenon of fast culture death when temperature exceeds the maximum. Also shown are cases of great concern to temperate and even subtropical growth areas: the lowering of daily productivity by cool morning temperatures. Temperature lowers productivity by decreasing the maximum rates of photosynthesis, but also by making some warm water species more vulnerable to damage from photoinhibition by keeping reaction centers closed and accumulating reactive oxygen species (Vonshak et al. 2001).

2.8.2 VARIABLE CONCENTRATIONS OF CARBON DIOXIDE AND PH

In any large-scale algal production system, the introduction of CO_2 will occur at discrete injection stations. The injected CO_2 will be stored predominantly as part of the carbonate buffer system. Thus, the amount of storage will depend on the volume (depth) of the pond and the alkalinity of the system's (pond) water. As the suspension travels from one injection station to another, the carbon will be both taken up by the growing algal cells and be lost to the atmosphere through the surface. These processes compete for the CO_2. High carbon utilization efficiencies, greater than 90%, may be achieved. This requires efficient injection, productive algal growth, and pH ranges in between injection stations that keep the dissolved CO_2 concentrations low for most of the transit time. These pH ranges have been established for a range of water alkalinities (Weissman and Goebel 1987), but generally pH starting at about 7 after injection and reaching about 9 prior to the next injection allow

TABLE 2.3

Productivity in 3 m² Outdoor Ponds of *Cyclotella* sp. at High-and Low-Dissolved CO$_2$

pH	7.2	8.3
CO$_2$, µM	300	20
Productivity, g·m^{-2}·day^{-1}	19.4 ± 1.1	18.5 ± 1.2

Source: Weissman, J. C. et al., U.S. DOE Aquatic Species Program, 1988, unpublished.
Note: Grown on saline groundwater of 18 ppt.

very high overall carbon-utilization efficiencies. In horizontal closed systems, the CO$_2$ will still be lost at gas transfer stations, which are much more frequent than in ponds due to the lower system water volume (Weissman et al. 1988). Carbon dioxide loss from vertical systems in which CO$_2$ is injected from the bottom is even worse because the height of the water column is usually not enough for all of the CO$_2$ to be absorbed.

There are very few tests of the potential reductions in the productivity of algae due to the periods of low CO$_2$ concentrations (at pH 8 and above in seawater) that occur in large systems according to the discussion earlier. Table 2.3 shows data from small pond cultures (3 m²) with a native *Cyclotella* sp. from the ASP test facility in Roswell, New Mexico, operated at constant conditions of pH and CO$_2$. For constant conditions, there was no significant difference between the high CO$_2$ and low CO$_2$ cases, by paired t-tests. Similar results were found with an *Amphora* sp., and only small (10%) productivity decreases were measured in the same system for a *Chaetoceros* and a *Tetraselmis* sp.

More detailed and better controlled indoor experiments conducted with the diatom *C. muelleri* are summarized in Figures 2.13 and 2.14. The specific growth rate of optically thin, semicontinuously diluted cultures illuminated at 200 µE·m^{-2}·s^{-1} incident irradiance (which is saturating for growth rate; see Figure 2.3) are given in Figure 2.13 for many combinations of CO$_2$ concentration and total dissolved inorganic carbon (DIC), which, together with the media composition, determine the pH. There is a steep increase in specific growth rate around 100 µM DIC. For the most part, bicarbonate fully substitutes for dissolved CO$_2$ in supporting maximum specific growth rate. The same is not true of dense culture productivity, as shown in Figure 2.14. Experimentally, it is much more difficult to maintain very low concentrations of CO$_2$ in dense cultures, which have significant carbon uptake rates. In these experiments, the 1 L Roux flasks were sparged vigorously with the gas. The effluent was connected to an infrared gas analyzer to determine the concentration of CO$_2$. Based on the infrared gas analyzer readings and a model for transfer of CO$_2$ from the gas phase to the liquid phase, a computer-generated control algorithm kept the CO$_2$ in the liquid phase to within 10% of the set point at 10 µM CO$_2$ and above. At a set point of 1 µM, the control maintained CO$_2$ concentration from 1 to 2 µM. The first column in the figure is the control operated under conditions considered optimal. The second column shows that productivity is reduced to almost 75% when

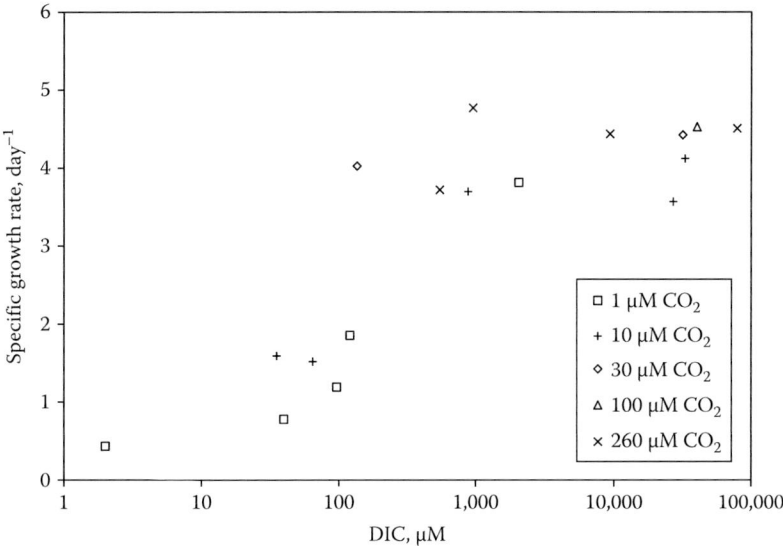

FIGURE 2.13 Specific growth rate of optically thin cultures of *Chaetoceros muelleri* grown at 200 $\mu E \cdot m^{-2} \cdot s^{-1}$ as a function of total dissolved inorganic carbon (DIC) in the medium. (Weissman, J. C. and Goebel, R. P., U.S. DOE Aquatic Species Program, 1986, unpublished.)

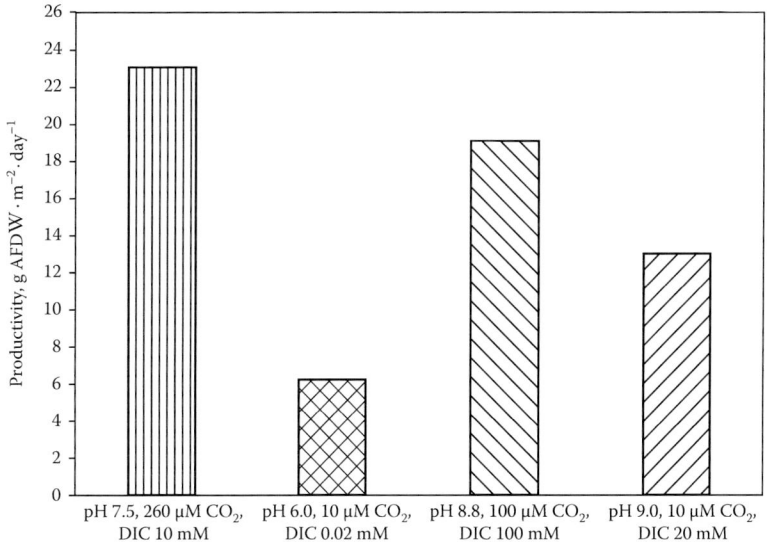

FIGURE 2.14 Productivity of dense cultures of *Chaetoceros muelleri* grown at 200 $\mu E \cdot m^{-2} \cdot s^{-1}$ with carbon dioxide concentration controlled at the indicated values, with the medium DIC as shown and pH as measured. (Weissman, J. C. and Goebel, R. P., U.S. DOE Aquatic Species Program, 1986, unpublished.)

both the CO_2 and the total DIC concentrations are low. The third column is a high pH and high alkalinity (low divalent ion) control for the last column, which has CO_2 as low as for column 2, but with 20 mM DIC. Relative to the high alkalinity control, the productivity is reduced by about 30% at 10 μM CO_2 when DIC is increased greatly. Still, this is a considerable reduction in efficiency presumably due to the ATP required to concentrate carbon. With varying CO_2 and pH, the full penalty may not be assessed.

2.9 CONCLUSIONS

Photosynthesis itself is an inherently complicated process. In trying to predict growth and productivity of mass algal cultures or primary productivity of aquatic ecosystems, the basic processes of photosynthesis must be integrated into cell growth amidst varying environmental inputs. It is advantageous to understand the fundamentals of the problem in their simplest form prior to building complexity. The history of growth kinetic analysis has followed this course, starting by model-ing the input–output relationships with a few key parameters and analyzing how the model performs. Performance is based on the level of accuracy that is required and attained for a specific objective. Several examples have been provided to show that productivity can be roughly estimated with only these key parameters. The elemen-tary description of photosynthesis and the development of simple models like the Bush equation and one-parameter maintenance models, as well as the early oceano-graphic models, actually predict productivity in terms of light input within ±30%. Some of the major improvements to models have come in more refined estimates of the parameters, including the absorption of light. However, light absorption is not an accurate surrogate for biomass.

The basis of much of growth kinetics and modeling of algal productivity is the photosynthesis–irradiance curve. The basic parameters of this relationship should be based on the underlying photosynthetic mechanism and not just fitting curves. The complexity of photosynthesis has made it difficult to easily model the P–I curve, let alone the extension of this paradigm to include the dynamics of the pho-tosynthetic response and growth to varying inputs. This is particularly challenging for the modeling of mass algal cultures due to the movement of cells through a light gradient. In addition, the adaptability that organisms must possess to survive makes it difficult to formulate models that can be generalized, especially beyond the parameters measured for the species and conditions studied. Thus, to go beyond the simplest models with just a few parameters requires substantial effort. This explains the wealth of parameters and models based on them that become hard to apply. Yet, these efforts slowly expand the understanding of the photophysiology that drives progress.

Basic data from the growth and productivity of an algal species were presented to set the stage for discussing how the fundamentals of photosynthesis and the physio-logical parameters of each species determine these results. Early researchers focused on the photosynthetic process and the losses along the way to biomass production. The greatest losses are due to the mechanism of photosynthesis itself manifested as the maximal efficiency measured in low irradiance, the saturation of photosynthesis with irradiance, photoinhibition, and maintenance requirement, usually in that order. Of these, perhaps photosaturation is the least amenable to alleviation through strain screening because species selection is based on the ability to scavenge nutrients better than the competition, requiring high pigmentation. A fifth element important to productivity, the interplay between the dynamics of photosynthetic processes and the time constants of mixing was investigated early on as the flashing-light effect. The practicality of this has been viewed as minimal for a long time, but as the dynamics of other processes (e.g., pools of intermediates that serve as capacitances between absorption and carbon fixation, the activation and inactivation of Rubisco, the initiation and termination of NPQ, and the damage and repair of photosynthetic proteins as a form of photoinhibition) are considered, new possibilities arise for optimizing productivity in a given system. Photoinhibition and photo dynamics are presently the least understood.

Modeling productivity involves the consideration of the photosynthetic response as a function of light and averaging it over the volume of cultivation. This is true in mixed or stagnant systems, dilute or dense. The response curve that is utilized depends on the mixing; it may be the P–I curve of a particular state of adaptation (light acclimation) or the sequential μ–I response curve of a series of light acclimation states. Initial work focused on simplifications that allowed volume integration based on only a few parameters, some of which were average values typical of many algal species, like the Bush equation or the work of Ryther and colleagues. The photosynthetic response and the integrated productivity are often conveniently written as a product of light absorption parameters (biomass-specific absorption), a maximum efficiency factor, and a factor for the effect of irradiance on photosynthetic rate or efficiency.

A simplified model of photosynthesis was discussed to highlight the importance of the concept of the saturating irradiance on photosynthetic productivity and modeling it. This parameter defines the balance between light absorption and the utilization of the subsequent stable charge separation in biochemical reactions. The latter is described by the rate of electron transport through the photosynthetic units and the rate of carbon fixation. It was shown that typical low-light adapted algae absorb light more than 10–20 times faster than they can use it, resulting in saturating irradiance of 100–200 $\mu E \cdot m^{-2} \cdot s^{-1}$, for most algae. Increasing the saturating irradiance thus became the focal point for increasing photosynthetic efficiency by alleviating one of the greatest losses in photosynthetic production. This can be accomplished by one of two ways: decreasing absorption or increasing the maximum rate of photosynthesis. The former has a nonlinear (diminishing return) effect on saturating irradiance, the latter a linear one. The Bush equation (and other light utilization factors), which is an expression of how photosynthetic efficiency changes as a function of the ratio of saturating irradiance to incident irradiance, is a curve of diminishing return. In addition, due to the effects of pigment packaging, the decrease in absorption by a cell due to decrease in pigment content is also a curve of diminishing return. Still, pigment content can be drastically reduced, as demonstrated by light acclimation experiments, such that, say, a fourfold decrease in pigment could lead to a twofold to threefold increase in saturating irradiance and thus a possible doubling of photosynthetic efficiency. On the other hand, it is essentially impossible to attain this level of increase by increasing the rate of photosynthesis. The rate of carbon fixation is already high (at high CO_2 levels) as evidenced by the large amount of Rubisco in the cell (up to 20% of the cell must be Rubisco to account for measured maximum specific growth rates). Even increasing this by 20%–30% seems unlikely. In addition, carbon fixation rates are not that much slower than the maximal rate of turnover of the photosynthetic units. Therefore, to increase the maximal rate of photosynthesis, not only would Rubisco and many other enzymes of carbon fixation have to be increased, but so would electron transport rates of the PSUs.

A simple model of maintenance losses, including the ones during growth in the daytime and during dark periods, was described and used to demonstrate that it is possible to find algae with sufficiently low maintenance to be good candidates for biomass production. Photoinhibition, on the other hand, is more difficult to estimate in dense cultures but is very evident in the upper layers of natural systems.

Open pond biomass productivity data from the Aquatic Species Program sponsored by the U.S. DOE in the 1980s and 1990s are presented. It is clear that consistently high productivities can be maintained with algae grown under favorable conditions of nutrient sufficiency (including elevated levels of CO_2) in warm climates. The duration of cultivation at high productivity was dependent on the duration of warm weather: year-round in Hawaii and 4 months in California. Small ponds in New Mexico were just as productive during the summer. Summer productivities in much larger ponds were only 50%–75% as high as in small ponds, and year-round productivity in this climate was only 25% as high as that in Hawaii.

Along with light, and no less impactful on productivity, are temperature and concentrations of CO_2. Clearly, for a particular alga, there is a sustained temperature that is lethal. But in order to attain the highest productivity, the temperature must get to the optimal range in the morning. So, cold nights may reduce losses, but if the temperature does not quickly rise, productivity potential will be lost. The optimization of CO_2 level in large, open ponds is a multivariable issue. Although algae can use CO_2 at very low concentration, below that in equilibrium with air, the carbon concentrating mechanisms require ATP. Providing CO_2 at consistently higher concentrations is untenable due to outgassing losses. In any case, in a large system with discreet injection stations, CO_2 and pH will vary during the transit time from one to the next station. Results reported here indicate that, in dense, light-limited, culture, the use of carbon at low concentration could reduce biomass productivity by at least 25%.

In summary, understanding and predicting the growth of algal cultures requires fundamental knowledge of both photosynthesis and the kinetics of growth with light as a limiting input. Although approximate estimates of primary productivity are attainable with simple conceptual models, the inherent complexity of photoautotrophic growth makes it very difficult to accurately predict growth and productivity, especially in a dynamic environment. Light is an unusual nutrient, not only in that it cannot be stored and subsequently used, but also, in the natural environment, it varies over a large range. Algae have evolved to scavenge it at low irradiance by developing large pigment antennae. At low irradiance, many algal species use light as efficiently as the basic photosynthetic mechanism allows. However, unlike many enzyme-based nutrient uptake systems, absorption of light does not saturate. Downstream metabolic limitations, primarily in carbon fixation and reduction, do saturate at rates corresponding to absorption at relatively low irradiance. This prevents efficient use of light at high irradiance. Algae protect themselves from high irradiance by dissipating absorption, but they cannot adapt quickly or to a great enough extent to use it efficiently nor to completely prevent deleterious results. Hence, discussions of algal growth and productivity are often dominated by light saturation and inhibition.

REFERENCES

Béchet, Q., A. Shilton, and B. Guieysse. 2013. Modeling the effects of light and temperature on algae growth: State of the art and critical assessment for productivity during outdoor cultivation. *Biotechnol. Adv.* 31:1648–1663.

Benemann, J. R., B. L. Koopman, J. C. Weissman, D. M. Eisenberg, and P. Goebel. 1980. Development of microalgae harvesting and high rate pond technologies in California. In *Algae Biomass: Production and Use*, eds. G. Shelef and M. Soeder, pp. 457–496. Amsterdam, the Netherlands: Elsevier.

Berner, T., Z. Dubinsky, K. Wyman, and P. G. Falkowski. 1989. Photoadaptation and the "package" effect in *Dunaliella tertiolecta* (Chlorophyceae). *J. Phycol.* 25:70–78.

Burlew, J. S. 1953. The current status of large-scale cultivation of algae I. In *Algal Culture: From Laboratory to Pilot Plant*, ed. J. S. Burlew, Vol. 600, pp. 1–23. Washington, DC: Carnegie Institute of Washington.

Cook, P. 1951. Chemical and engineering problems in large-scale culture of algae. *Ind. Eng. Chem.* 43:2385–2389.

Cullen, J. J. 1990. On models of growth and photosynthesis in phytoplankton. *Deep-Sea Res.* 37:667–683.

Dabes, J. N., C. R. Wilke, and K. H. Sauer. 1970. Behavior of *Chlorella pyrenoidosa* in steady state continuous culture. Report to DOE, #UCLBL-19958, Oak Ridge, TN.

Dubinski, Z., P. G. Falkowski, and K. Wyman. 1986. Light harvesting and utilization by phytoplankton. *Plant Cell Physiol.* 27:1335–1349.

Emerson, R. and W. Arnold. 1932. The photochemical reaction in photosynthesis. *J. Gen. Physiol.* 16:191–205.

Falkowski, P. G., Z. Dubinsky, and K. Wyman. 1985. Growth-irradiance relationships in phytoplankton. *Limnol. Oceanogr.* 30:311–321.

Falkowski, P. G. and T. G. Owens. 1980. Light-shade adaptation. *Plant Physiol.* 66:592–595.

Geider, R. J., H. L. MacIntyre, and T. M. Kana. 1996. A dynamic model of photoadaptation in phytoplankton. *Limnol. Oceanogr.* 41:1–15.

Goldman, J. C. 1980. Physiological processes, nutrient availability, and the concept of relative growth rate in marine phytoplankton ecology. In *Primary Productivity in the Sea*, ed. P. G. Falkowski, pp. 170–194. New York: Plenum.

Gons, H. G. and L. R. Mur. 1975. An energy balance for algal populations in light-limiting conditions. *Verh. Internat. Verein. Limnol.* 19:2729–2733.

Greenbaum, E., J. W. Lee, C. V. Tevault, S. L. Blankinship, and L. J. Mets. 1995. CO_2 fixation and photoevolution of H_2 and O_2 in a mutant of *Chlamydomonas* lacking Photosystem I. *Nature* 376:438–441.

Han, B.-P., M. Virtanen, J. Koponen, and M. Straskraba. 2000. Effect of photoinhibition on algal photosynthesis. *J. Plankton Res.* 22:865–885.

Hill, R. and F. Bendall. 1960. Function of two cytochrome components in chloroplasts: A working hypothesis. *Nature* 186:136–137.

Jassby, A. D. and T. Platt. 1976. Mathematical formulation of the relationship between photosynthesis and light for phytoplankton. *Limnol. Oceanogr.* 21:540–547.

Kiefer, D. A. and B. G. Mitchell. 1983. A simple steady state description of phytoplankton growth based on absorption cross section and quantum efficiency. *Limnol. Oceanogr.* 28:770–776.

Kirk, J. T. O. 2011. *Light and Photosynthesis in Aquatic Ecosystems*, 3rd edn. Cambridge, UK: Cambridge University Press.

Kok, B. 1956a. On the photoinhibition of photosynthesis by intense light. *Biochim. Biophys. Acta* 21:234–244.

Kok, B. 1956b. Photosynthesis in flashing light. *Biochim. Biophys. Acta* 21:245–258.

Kok, B. 1960. Efficiency of photosynthesis. In *Encyclopedia of Plant Physiology*, ed. W. Ruhland, Vol. 5, pp. 566–633. Berlin, Germany: Springer.

Laws, E. A. and T. T. Bannister. 1980. Nutrient and light-limited growth of *Thalassiosira fluviatilis* in continuous culture, with implications for phytoplankton growth in the ocean. *Limnol. Oceanogr.* 25:457–473.

Laws, E. A. and J. L. Berning. 1991. A study of the energetic and economics of microalgal mass culture with the marine chlorophytes *Tetraselmis suecica*: Implications for use of power plant stack gases. *Biotechnol. Bioeng.* 37:936–947.

Laws, E. A., S. Taguchi, J. Hirata, and L. Pang. 1986. High algal production rates achieved in a shallow outdoor flume. *Biotechnol. Bioeng.* 28:191–197.

Losh, J. L., J. N. Young, and F. M. M. Morel. 2013. Rubisco is a small fraction of total protein in marine phytoplankton. *New Phytol.* 198:52–58.

Ludwig, M. and D. A. Bryant. 2012. *Synechococcus* sp. strain PCC7002 transcriptome: Acclimation to temperature, salinity, oxidative stress, and mixotrophic growth conditions. *Front. Microbiol.* 3:1–14.

MacIntyre, H. L., T. M. Kana, T. Anning, and R. J. Geider. 2002. Photoacclimation of photosynthesis irradiance response curves and photosynthetic pigments in microalgae and cyanobacteria. *J. Phycol.* 38:17–38.

Marra, J. 1980. Vertical mixing and primary production. In *Primary Productivity in the Sea*, ed. P. G. Falkowski, pp. 1–6. New York: Plenum.

Masojidek, J., J. Kopecky, L. Giannelli, and G. Torzillo. 2011. Productivity correlated to photobiochemical performance of *Chlorella* mass cultures grown outdoors in thin-layer cascades. *J. Ind. Microbiol. Biotechnol.* 38:307–317.

Melis, A., J. Niedhardt, and J. R. Benemann. 1999. *Dunaliella salina* (Chlorophyta) with small chlorophyll antenna sizes exhibit higher photosynthetic productivities and photon use efficiencies than normally pigmented cells. *J. Appl. Phycol.* 10:515–525.

Mitchell, P. 1975. The protonmotive Q cycle: A general formulation. *FEBS Lett.* 59:137–139.

Morel, A. and A. Bricaud. 1986a. Inherent optical properties of algal cells including picoplankton: Theoretical and experimental results. *Can. Bull. Fish. Aquat. Sci.* 214:521–559.

Morel, A. and A. Bricaud. 1986b. Light attenuation and scattering by phytoplanktonic cells: A theoretical modeling. *Appl. Optics* 24:571–580.

Myers, J. 1970. Genetic and adaptive physiological characteristics observed in the Chlorellas. In *Prediction and Measurement of Photosynthetic Productivity*, ed. I. Malik, pp. 447–454. Wageningen, the Netherlands: Center for Agricultural Publishing and Documentation.

Myers, J. 1980. On the algae: Thoughts about physiology and measurements of efficiency. In *Primary Productivity in the Sea*, ed. P. G. Falkowski, pp. 1–6. New York: Plenum.

Myers, J. and L. B. Clark. 1946. Culture conditions and the development of the photosynthetic mechanism. II. An apparatus for the continuous culture of *Chlorella*. *J. Gen. Physiol.* 28:103–112.

Myers, J. and J.-R. Graham. 1959. On the mass culture of algae. II. Yields as a function of concentration under continuous sunlight irradiance. *Plant Physiol.* 34:345–355.

Myers, J. and J.-R. Graham. 1971. The photosynthetic unit in *Chlorella* measured by repetitive short flashes. *Plant Physiol.* 48:282–286.

Nakajima, Y. and R. Ueda. 1999. Improvement of microalgal photosynthetic productivity by reducing the content of light harvesting pigment. *J. Appl. Phycol.* 11:195–201.

Nakajima, Y. and R. Ueda. 2000. The effect of reducing light-harvesting pigment on marine microalgal productivity. *J. Appl. Phycol.* 12:285–290.

Nickelsen, K. 2007. Otto Warburg's first approach to photosynthesis. *Photosynth. Res.* 92:109–120.

Novick, A. and L. Szilard. 1950. Experiments with the chemostat on spontaneous mutations of bacteria. *Proc. Natl. Acad. Sci. USA* 36:708–719.

Phillips, J. N. and J. Myers. 1954. Growth rate of *Chlorella* in flashing light. *Plant Physiol.* 29:152–161.

Pirt, S. J. 1965. The maintenance energy of bacteria in growing cultures. *Proc. R. Soc. Lond. B Biol. Sci.* 163:224–231.

Pirt, S. J. 1986. The thermodynamic efficiency (quantum demand) and dynamics of photosynthetic growth. *New Phytol.* 102:3–37.

Pirt, S. J., Y.-K. Lee, A. Richmond, and M. Pirt. 1980. The photosynthetic efficiency of *Chlorella* biomass growth with reference to solar energy utilization. *J. Chem. Technol. Biotechnol.* 30:25–34.

Radmer, R. and B. Kok. 1977. Photosynthesis: Limited yields, unlimited dreams. *Bioscience* 27:599–605.

Rosenthal, D. M., A. M. Locke, M. Khozaei, C. A. Raines, S. P. Long, and D. R. Ort. 2011. Over-expressing the C3 photosynthesis cycle enzyme sedoheptulose-1–7 bisphosphatase improves photosynthetic carbon gain and yield under fully open air CO_2 fumigation (FACE). *BMC Plant Biol.* 11:123.

Ryther, J. H. 1956. Photosynthesis in the ocean as a function of light intensity. *Limnol. Oceanogr.* 1:61–70.

Ryther, J. H. and D. W. Menzel. 1959. Light adaptation by marine phytoplankton. *Limnol. Oceanogr.* 4:492–497.

Ryther, J. H. and C. S. Yentsch. 1957. The estimation of phytoplankton production in the ocean from chlorophyll and light data. *Limnol. Oceanogr.* 2:281–286.

Sakshaug, E. D., A. Kiefer, and K. Anderson. 1989. A steady state description of growth and light absorption in the marine planktonic diatom *Skeletonema costatum*. *Limnol. Oceanogr.* 34:198–205.

Sheehan, J., T. Dunahay, J. Benemann, and P. Roessler. 1998. A look back at the U.S. department of energy's aquatic species program: Biodiesel from algae; close-out report, Golden, CO. http://www.osti.gov/servlets/purl/15003040-tW7nZs/native/ (accessed March 5, 2015).

Shelef, G., W. J. Oswald, and C. G. Golueke. 1969. The continuous culture of algae biomass on waste. In *Continuous Cultivation of Microorganisms*, ed. I. Malek, pp. 601–629. Prague, Czech Republic: Academy.

Sukenik, A., J. Bennett, and P. Falkowski. 1987. Light-saturated photosynthesis—Limitation by electron transport or carbon fixation? *Biochim. Biophys. Acta* 891:205–215.

Van de Hulst, H. C. 1981. *Light Scattering by Small Particles*. New York: Dover.

Van Oorschot, J. L. P. 1955. Conversion of light energy in algal culture. *Mededel. Landbouwhogeschool te. Wageningen* 55:225–276.

Vonshak, A., G. Torzillo, J. Masojidek, and S. Boussiba. 2001. Suboptimal morning temperature induces photoinhibition in dense outdoor cultures of the alga *Monodus subterraneus* (Eustigmatophyta). *Plant Cell Environ.* 24:1113–1118.

Weissman, J. C. and J. R. Benemann. 1979. Biomass recycling and species competition in continuous cultures. *Biotechnol. Bioeng.* 21:627–648.

Weissman, J. C. and R. P. Goebel. 1987. Factors affecting the photosynthetic yield of microalgae. 1986 Aquatic species program annual report, SERI/SP-231-3071, Golden, CO, February 1987, pp. 139–168. www.nrel.gov/docs/legosti/old/3071.pdf (accessed March 5, 2015).

Weissman, J. C., R. P. Goebel, and J. R. Benemann. 1988. Photo bioreactor design: Mixing, carbon utilization, and oxygen accumulation. *Biotechnol. Bioeng.* 31:336–344.

Weissman, J. C., M. Likhogrud, K. McNeely, R. Prince, and R. Nielsen. 2013. Quantum requirement for growth of dilute algal cultures. Paper presented at *The Third International Conference on Algal Biomass, Biofuels and Bioproducts*, Toronto, Ontario, Canada.

Yu, J., M. Liberton, P. Clifton, R. Head, J. Brand, and H. B. Pakrasi. 2013. *Synechococcus* sp. UTEX 2973, a new cyanobacterial chassis for synthetic biology and metabolic engineering applications. Poster presented at the *16th International Conference on Photosynthesis Research*, St. Louis, MO.

3 Microalgae Strain Isolation, Screening, and Identification for Biofuels and High-Value Products

Peter Neofotis, Andy Huang, William Chang, Floral Joseph, and Juergen E.W. Polle

CONTENTS

3.1 INTRODUCTION

Microalgae, as yet a mostly untapped biological resource, could be cultivated on nonarable land and utilize industrial flue gas as a carbon source for biofuel production (Gouveia and Oliveira 2009; U.S. Department of Energy 2010; Benemann 2013; Borowitzka and Moheimani 2013; Passell et al. 2013; Wijffels et al. 2013). With more than 100,000 species described (Guiry et al. 2014), eukaryotic microalgae are classified into several major lineages, which are more widely separated genetically than humans are from fungi (Falkowski and Knoll 2007; Chan et al. 2011; Cavalier-Smith 2013; Keeling 2013). Furthermore, with the classification of even many known

63

species tenuous, it is unknown how many microalgae species may exist (Krienitz and Bock 2012; Leliaert et al. 2014). Although this chapter focuses on eukaryotic micro-algae, the prokaryotic microalgae belonging to the cyanobacteria, formerly called blue–green algae, also display great diversity (Palinska and Surosz 2014).

Only about 15 of the currently known microalgae species are cultivated at any signif-icant scale, and these are used almost exclusively for human food supplements ("nutra-ceuticals"), cosmetics, and aquaculture feeds (Raja et al. 2008; Borowitzka 2013). Examples of green microalgae, which are already cultivated commercially belong to the freshwater genera *Chlorella* (Doucha and Livansky 2006) and *Haematococcus* (for astaxanthin production) (Guerin et al. 2003; Laurens et al. 2012). In addition, the green algal species *Dunaliella salina* is grown in large hypersaline lagoons and ponds for provitamin A production (Borowitzka 1991; Ben-Amotz et al. 2009). Moreover, the cyanobacterium *Arthrospira* sp., commonly known as *Spirulina*, is also cultivated—essentially exclusively—in raceway-type ponds, which are mixed by paddle wheels (Belay 2013). Worldwide, autotrophic (on sunlight and CO_2) microalgae production is currently no more than 20,000 tons globally, with about the same amount produced heterotrophically (Benemann 2013). The as yet untapped diversity of microalgae has attracted considerable interest in the large-scale production of biofuels (Elliott et al. 2012; Ratha and Prasanna 2012; Ratha et al. 2012), animal feed (Benemann 2013), and higher-value products (Borowitzka 2013; Draaisma et al. 2013; Wijffels et al. 2013).

Bioprospecting—the identification of superior, "platform" strains, which exhibit high growth rates and culture stability and create the desired product—is required to realize the potential of algal biotechnology (Davis et al. 2011; Mutanda et al. 2011; Barclay and Apt 2013). These strains could then be developed further through genetic improvements to achieve cost and production goals (Georgianna and Mayfield 2012; Larkum et al. 2012; Gimpel et al. 2013; Manandhar-Shrestha and Hildebrand 2013; Leu and Boussiba 2014). Microalgal bioprospecting for novel strains for production of biofuels was first carried out systematically during the 1980s U.S. Department of Energy's "Aquatic Species Program" (Sheehan et al. 1998). It was, and remains, a very labor-intensive undertaking, even with the newer high-throughput automated screening technologies now being applied in this field.

This chapter describes our process of isolation, screening, identification, charac-terization, and maintenance of microalgal strains. Although the focus was on obtain-ing strains suited for producing biofuels, many discovered strains also have potential applications in the future development of food, feed, and high-value products (e.g., carotenoids and long-chain polyunsaturated omega-3 fatty acids).

3.2 STRAIN ISOLATION

As outlined in Figure 3.1, microalgal strains can either be obtained from existing cul-ture collections or be isolated from environmental samples. Our approach employed environmental sampling from different habitats followed by strain isolation. Strains from culture collections with known characteristics were used only in our screens as benchmarks to assess the performance of the newly isolated strains. In the following, we describe the different steps involved in our approach to isolate and screen novel microalgal strains.

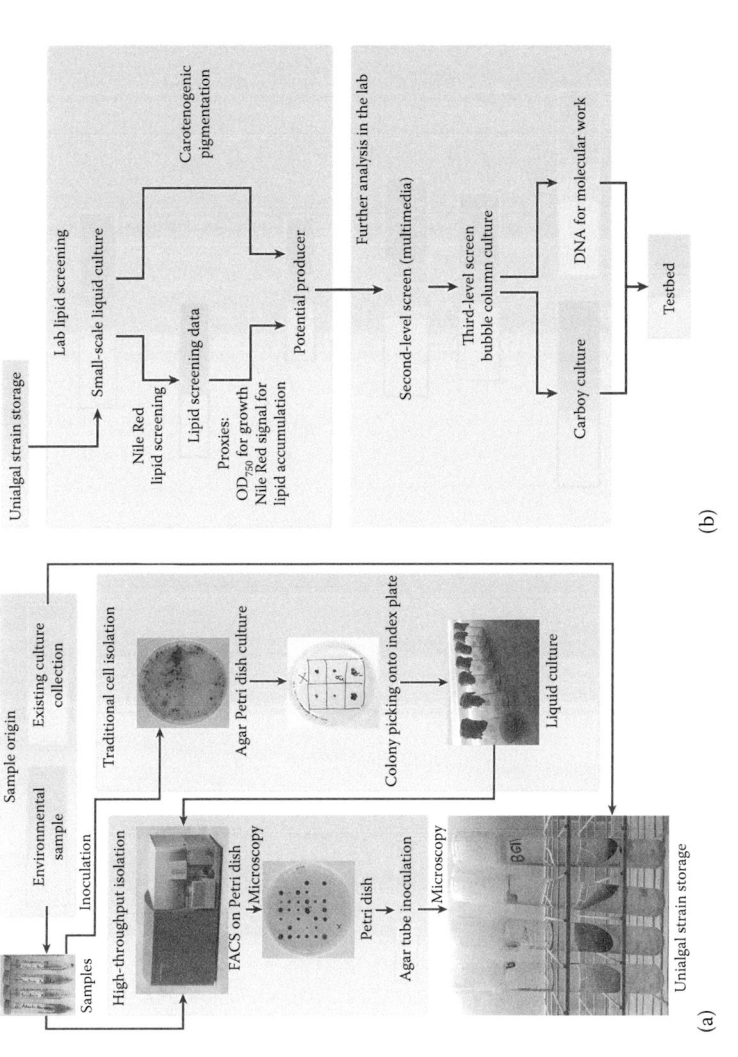

FIGURE 3.1 An overview of the bioprospecting approach from sampling to screening and testing of the strains is shown. (a) Samples were collected from various habitats, aliquots plated, and colonies picked, transferred into a liquid medium, and then cell sorted onto agar-media culture plates. (b) Strains were screened by growth in liquid batch cultures. Strains showing good growth characteristics and/or high Nile Red fluorescence were deemed candidates and grown in bubble columns for biomass productivity measurements.

3.2.1 SAMPLING

More than 1000 environmental samples were collected over a period of approximately 5 years (2007–2011) from a variety of habitats across the United States, including both dry (e.g., soil, lichen) and wet (e.g., fresh, brackish, and saltwater) samples. To isolate both summer and winter strains, sampling was performed in all seasons. When possible, multiple samples were taken from the same habitat at different times to also cover annual succession of algae in a location. Examples of sampling habitats are shown in Figure 3.2. The sampling efforts focused on California, New Mexico, and Texas, which are considered prime potential future sites for large-scale algal operations (Maxwell et al. 1985; Sheehan et al. 1998; Pate et al. 2011). Environments sampled included shallow and temporary aqueous habitats, often with extreme temperature variations and exposure to full sunlight. Freshwater habitats included birdbaths, roadside ditches, fountains, shallow ponds, creeks, rivers, and lakes. Saline water habitats included estuaries, inland salt lakes, and general coastal waters in southern Texas, New York, and Connecticut.

Whole-water (rather than concentrated) samples were collected and stored in clean containers until plating, as they have been found to provide viable cells when concentrated samples fail (Anderson 2005). As natural samples often contain zooplankton that feed upon algae, water samples were passed through 70 μm filters. Often, even some predators—such as amoeba or small ciliates—remained. The filters also, unfortunately, removed algae that form large coenobia.

When possible, water samples were first inspected via microscopy to determine what species might be isolated from the environmental samples. Figure 3.3 shows an example of the diverse microalgae seen in a freshwater sample. Though it was very time consuming and could only be performed on a fraction of the samples, microscopic analysis allowed us to understand what species were lost during the isolation process. It was noted that colonial green algae of the genera *Volvox* and *Pleodorina* were observed in many samples by microscopy but were not isolated with our approach. In contrast, many filamentous green algae present in the samples could be isolated.

3.2.2 MEDIA AND PLATING

We observed that viable algal cells were still present in many original water samples that had been kept for several years in the laboratory. Nevertheless, since many kinds of algae, unless maintained and properly handled, may die from anywhere between a few hours to a few days after collection, we plated aliquots of samples as soon as possible on solidified growth media (1.5% Agar) (Figure 3.4). On longer sampling trips, prepared Petri dishes allowed plating of aliquots of the environmental samples within minutes or hours. Once colonies developed, the plates were then shipped back to the laboratory for further processing.

With current technology, complete recovery of all microalgal species present within a given environmental sample is not possible as many freshwater and marine algal strains do not produce colonies on agar plates, despite this being a

FIGURE 3.2 Photos showing representative habitats of the wide range of environments from which algae strains could be isolated. *Left panel from top to bottom*—Freshwater algal bloom in a roadside ditch in Texas; microalgal bloom in the Brooklyn College freshwater pond in Brooklyn, New York; Pacific Ocean at San Diego, California; diverse lichens on a rock in Arizona. *Right panel from top to bottom*—Freshwater birdbath in San Diego; diatom bloom in the Salton Sea, California; green algae of the genus *Dunaliella* blooming in a brine pond adjacent to the Great Salt Lake in Utah; biofilm of microalgae at a roadside location in Louisiana.

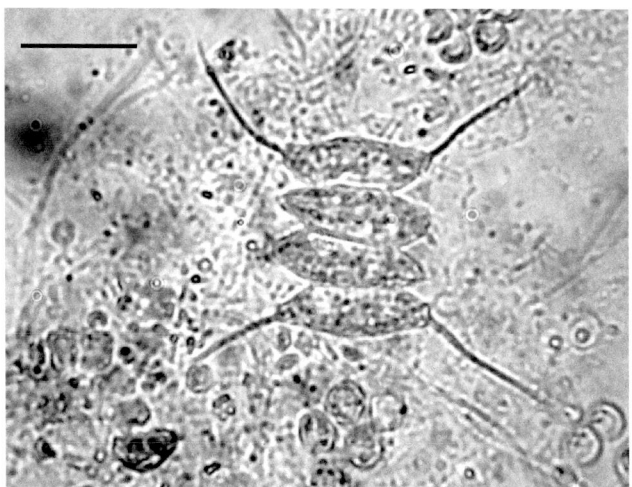

FIGURE 3.3 Light microscopic image of an environmental freshwater sample originating from the Brooklyn College pond in Brooklyn, New York. Based on morphological characters, the sample contained green algae such as *Desmodesmus*, *Ankistrodesmus*, *Kirchneriella*, *Chlamydomonas*, and *Chlorella* type of cells as well as a diatom species. In addition, there were some unknown algal species visible. The black bar represents 10 μm.

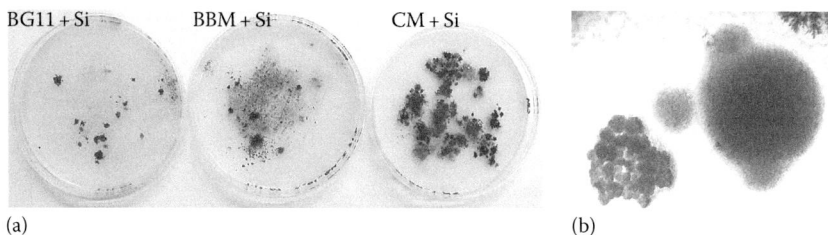

(a)   (b)

FIGURE 3.4   (See color insert.) Algal colonies grown on agar-solidified media. (a) Aliquots of the same environmental freshwater sample were placed on agar-solidified media in regular standard 90 mm Petri dishes. Three different media as indicated on the top left of each plate were used, supplemented with silicate to allow for growth of diatoms. (b) An example of algal colonies grown on an agar-solidified medium shown with a magnification of 60×.

well-established method (Anderson 2005). Yet, plating on agar-solidified media still allows for the isolation of large numbers of clonal strains (deriving from a single original cell) in situations where newer high-throughput techniques are not available. Another advantage of this technique is that during plating, algal grazers and most infectious agents are removed.

Inoculated plates were sealed with paraffin film to slow drying and were then incubated at about 50 μE m^{-2} s^{-1} continuous light provided by daylight fluorescent lamps. As some algae require day/night cycles and will not grow, or grow poorly, in continuous light (Andersen and Kawachi 2005), this introduced another selection. For some diatoms under continuous lighting on the plates, a ring growth pattern was observed, due to alternating growth and resting stages. Although continuous illumination reduced the diversity of algae isolated, many microalgae, including some of those currently mass-cultivated, such as species of the green algae *Chlorella* and *Dunaliella*, as well as the cyanobacterium *Arthrospira*, do not require day/night cycles. Continuous illumination also achieved rapid algal growth and saved time in the strain isolation process. Ultimately, our isolation and screening procedure resulted in new strains of microalgae that could be cultivated successfully in the greenhouse of Brooklyn College and in outdoor ponds (NAABB 2014), thus validating this isolation procedure.

In addition to plating aliquots of liquid samples, initially, a transfer though a liquid culture for enrichment of fast-growing cells, followed by plating, was also tested. But this resulted in the isolation of many essentially identical strains and thus was discontinued with only the traditional plating method used as the primary isolation tool.

Subaerial and aerial algae were isolated via enrichment cultures as recommended (Andersen and Kawachi 2005). In brief, soil and lichen samples were first placed into a liquid growth medium. Generally after 2 weeks, some algal growth was observed, as indicated by coloration of the liquid medium. An aliquot was then plated and colonies of algae were then observed on the plates after about another 2 weeks. Samples derived from lichen often contained cyanobacteria.

The culture medium was a critical factor in the isolation of microalgae from natural sources. Because this screen was designed primarily for eukaryotic algae, which grow photoautotrophically, freshwater samples were plated on several suitable known minimal media: C medium, Bold's basal medium (BBM), and BG11 medium (Andersen 2005) (see Figure 3.4). These media contain nitrogen in the form of nitrate. As it contains ammonium as the fixed nitrogen source for growth, an HS medium (Harris 2009) was also used initially, as with the TAP medium (Harris 2009), to allow for the isolation of strains that would be selected for preferential photoheterotrophic growth on acetate. However, the latter was abandoned early on due to overgrowth of bacteria and fungi, which often displaced the microalgal colonies. Marine strains were placed in F/2 media (Guillard and Ryther 1962; Guillard 1975) and artificial seawater media (Keller et al. 1987). Vitamins were not used for any of the freshwater media and only used initially for the F/2 media, as we were screening for strains suitable for large-scale cultivation, and vitamins can be prohibitively expensive at such scales. To allow for growth of diatoms, silica was added to all the different media to a final concentration of 0.4 mM. To isolate strains from inland highly saline water samples, a minimum medium known to allow the growth of green algae of the genus *Dunaliella*, as well as of diatoms and cyanobacteria, was used (Pick et al. 1986), as we found that many such algae (e.g., from hypersaline soils) did not grow in F/2 (Sheehan et al. 1998).

Although some cyanobacteria can form lipid droplets (Peramuna and Summers 2014), in general, cyanobacteria are not known to accumulate significant amounts of lipids as triacylglycerides (Hu et al. 2008). Due to their expected low triacylglyceride content, in our initial screen, cyanobacteria were downselected in this biofuel-focused project based on the blue–green color of colonies on the plates during the visual inspection. Still, a number of cyanobacteria were recovered later due to not all colonies of cyanobacteria having the more typical vibrant blue–green color.

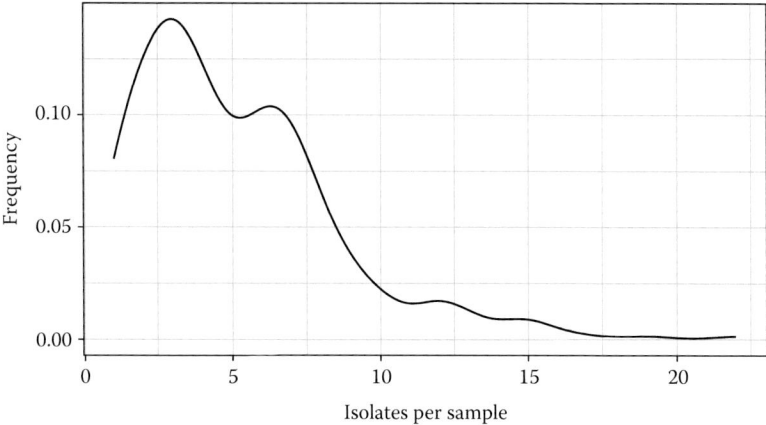

FIGURE 3.5   The frequency of the number of microalgae strains isolated per sample from a total of 350 environmental samples.

Figure 3.4 shows examples of colonies on plates. This comparison of growth on plates demonstrates that using different media increased overall diversity of strains recovered from one water sample. Colonies for transfer were chosen based on differences in colony color and overall appearance, to increase the number of unique types of algal strains thus isolated. Analysis of the number of isolated strains from over 300 samples showed that most environmental samples yielded an average of 5 unique strains, typically ranging from 1 to 10 (Figure 3.5).

3.2.3   Flow Cytometry and Cell Sorting

In recent years, flow cytometry—with its single-cell sorting capability—has become a method of choice for high-throughput isolation of algal strains (Sieracki et al. 2005; Pereira et al. 2011; Hyka et al. 2013). In fluorescence-aided cell sorting (FACS), flow cytometry is coupled with detection of fluorescence of cells (Shapiro 2004). As algal cells have pigments such as chlorophylls and carotenoids, fluorescence of these pigments can be easily detected and thus be used as sorting criteria for algal cell isolation. FACS can be used at different stages in algal strain isolation. For example, sorting of cells can be performed directly from water samples (Figure 3.1) either based on pigment fluorescence or based on staining with marker dyes, such as boron-diphyrromethane (BODIPY) or Nile Red for oil bodies (Cooper et al. 2010; Montero et al. 2011; Cirulis et al. 2012; Elliott et al. 2012).

In comparing the number of strains obtained using the plating technique to the number that successfully passed the cell sorting process, we found that about 1/3 of all strains isolated by plating did not survive the flow cytometry process. Therefore, we decided to continue with a strain isolation approach where first traditional isolation methods were utilized, as described earlier, followed by high-throughput FACS cell sorting.

In our combined approach, following the development of colonies on sampling plates (Figures 3.1 and 3.4), individual colonies were picked and transferred to index plates (Figure 3.1). On these index plates, the colonies were allowed to grow further, accumulating biomass for 2–4 weeks, before being transferred into liquid culture. In the liquid media transfer stage, the strains were grown in culture tubes for about 2–4 weeks before cell sorting. Growing cells in liquid media allowed determination of whether or not the algal strain grew well in such media, an essential characteristic for large-scale cultivation in outdoor raceway ponds. The 2–4-week growth period was required as cells from shorter incubation periods did not grow well after cell sorting, indicating that the growth phase that the cells are in impacts cell survival. Prior to sorting, 2 mL aliquots of the liquid cultures were filtered through 70 µm or 100 µm filters to remove cell clumps or large colonies from the liquid samples.

The filtered algal suspension was then cell sorted using a FACSAria System II cell sorter from BD Biosciences. The nozzle size was generally set to 100 µm or to 120 µm. If larger cells or colonies ("coenobia") of cells were present, a 120 µm nozzle was used. Pigments in algal cells were excited using a 488 nm blue Argon laser, and, based on the chlorophyll fluorescence signal (695 nm) and the carotenoid fluorescence (530 nm), the algae were sorted onto agar-solidified growth media. In addition, we concomitantly monitored the forward and side scatter. Figure 3.6 shows an example of clusters of events in plots of chlorophyll fluorescence at 695/20 nm against the carotenoid fluorescence at 530/20 nm obtained from the FACS. Clusters of events indicate populations of cells often consisting of cells from

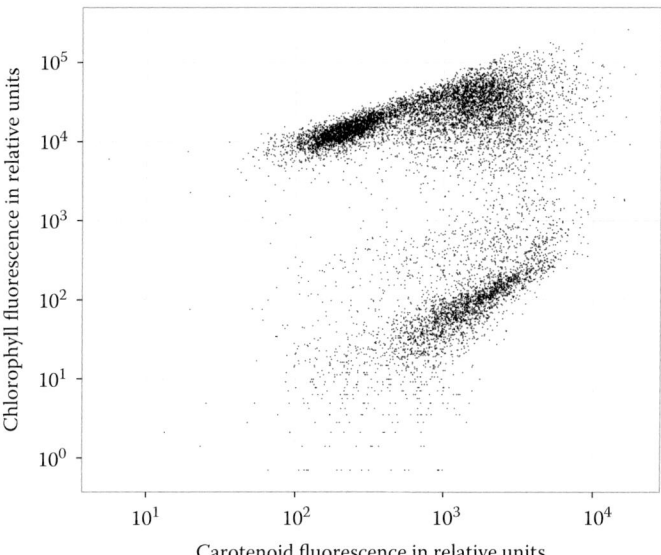

FIGURE 3.6 An exemplar scatterplot of a mixed sample, with the dots representing single events as visualized by the FACSAria System II cell sorter. The different populations were selected and single events sorted onto agar-solidified media in Petri dishes. The parameter chlorophyll fluorescence at 695 nm was used to select populations of algal cells via cell sorting. Bacterial cells may display apparent chlorophyll fluorescence, but usually at values below 10^1 relative fluorescence.

a single original cell (or clone, e.g., strain). Events can be single cells, clusters of cells from one original cell, clusters of cells from multiple clones, and/or algal cells or clusters with attached bacteria.

In the FACS experiments, populations were initially sorted into liquid media in 96-well plates and later onto agar-solidified media, with generally higher survival rates in the liquid media than on the agar plates. We successfully recovered sorted cells of the fragile cell-wall-less green alga *D. salina* grown in a medium containing 1.0 M NaCl (about threefold the salinity of seawater). Figure 3.7 shows a 96-well plate

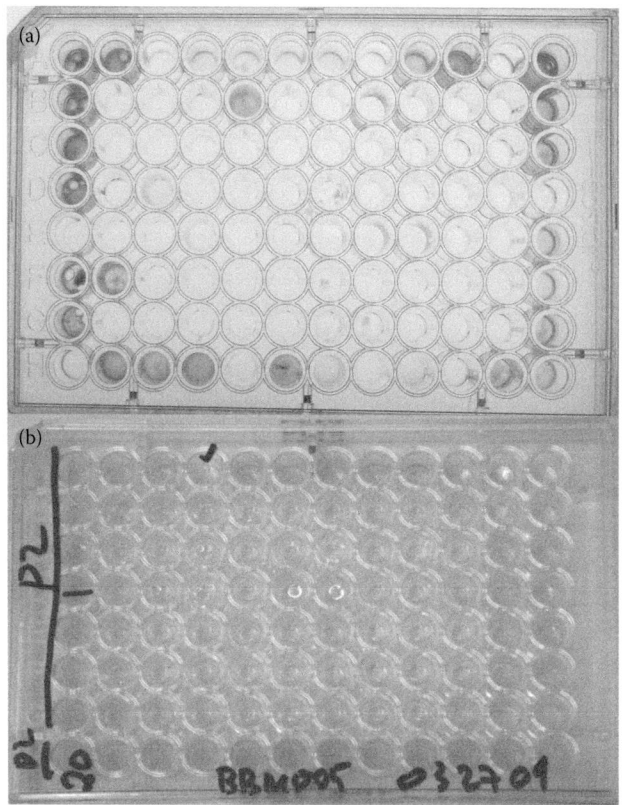

FIGURE 3.7   FACS cell sorting of events from liquid cultures into 96-well plates. (a) Sorted here was the green alga *Dunaliella salina* strain CCAP19/18. Cells of *D. salina* are fragile, as they do not have cell walls. Furthermore, the sorted culture was grown in a 1.0 M NaCl medium. Nevertheless, it was possible to sort cells and recover viable cultures in a liquid medium. In contrast, it was not possible to sort cells of *D. salina* onto regular plates with a solidified medium. (b) Sorted here were populations of a new, unknown, and carotenogenic isolate. All wells contained cultures, indicating that the sorting process had a high efficiency and that the cells survived sorting. On this plate, it is evident that the conditions in the wells were not uniform. Cultures in the outer wells had already turned orange indicating that they were under stress, whereas the cultures in the inner wells were still green. Also, a number of the inner wells show heavy condensation on the cover plate.

with cultures of *D. salina* growing in a number of wells. Only a minority of sorted cells grew up to form visible cultures, indicating that cell survival was low. To achieve even this low survival rate, the cell sorting process had to be modified, including the use of a large 120 µm nozzle, low flow rate, and increasing the salinity of the sheath fluid to 1.0 M NaCl. Higher salinities led to clogging of the fluid lines and prevented sorting. Nevertheless, this example of *D. salina* demonstrates that once the FACS sorting conditions are found that allow for survival, even very fragile cells can be FACS cell sorted. That is, however, not always a simple matter, and FACS may lose its advantage for fast isolation where isolation conditions are not standard.

Figure 3.7 shows that 96 well plates also had some major limitations. Uneven gas exchange and evaporation rates, as well as condensation, led to reduced growth of cultures in the wells at the plate center (Figure 3.7b). In the future, the use of membrane covers rather than regular plastic lids might alleviate such issues. Cells also settled in the wells and growth could not be readily followed. More importantly, only by sorting onto plates could colonies be easily determined as unialgal.

As shown in Figure 3.8, sorting cells onto agar-solidified media allowed for easy examination by microscopy of colonies to identify unialgal colonies. Overall, about 5%–10% of all colonies did not appear to have originated from a single algal cell or coenobium. Figure 3.8 shows examples of colonies obtained that appeared axenic. Lack of visible bacterial contamination on agar plates, however, is no assurance of axenic cultures. Figure 3.8 also shows algal colonies that were not unialgal and/or

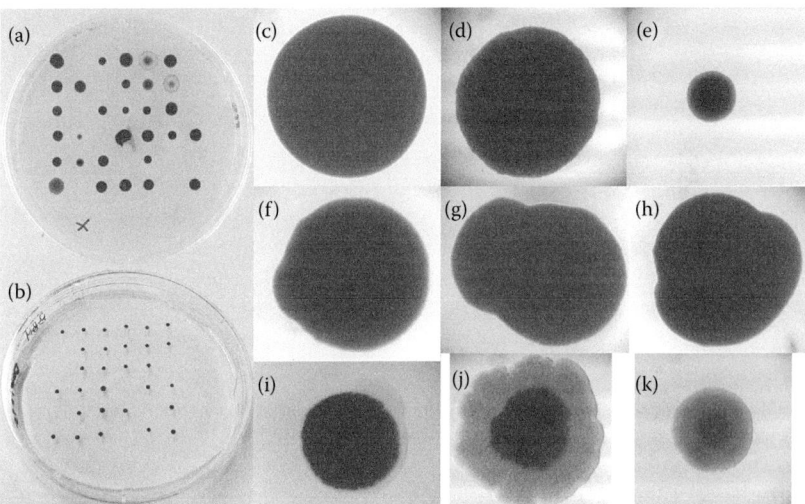

FIGURE 3.8 (See color insert.) Examples of microalgal colonies growing on agar-solidified media in Petri dishes after FACS cell sorting. (a) Plate from a non-unialgal liquid culture. Although one strain seemed to dominate, several different colony colors and types indicate the presence of at least four different species. (b) One unialgal strain only, with most colonies being axenic. (c–k) 60× magnification images of algal colonies sorted from a single event that are (c–e) unialgal and axenic, (f–h) non-unialgal, and (i–k) containing bacteria as indicated by the white, beige, or orange rings around the colonies.

had rings of white or slightly pinkish color around them, indicating heavy bacterial contamination. Different unialgal colonies were picked from each plate and transferred for storage into tubes with an agar-solidified medium overlaid with a liquid growth medium. Strains were not made axenic on a routine basis because the process is too time- and labor-consuming.

In summary, our combined approach of traditional cell culture followed by FACS cell sorting allowed us to isolate approximately 1000 strains per year, a several-fold higher yield than if just the traditional isolation approach involving plating and streaking had been used.

3.3 STRAIN MAINTENANCE

Following isolation as described earlier, strains were assembled into a catalogued culture collection. The strains were stored in glass test tubes, which contained growth media as a solidified 1.0%–1.5% agar slant, overlaid with a liquid growth medium. However, sometimes the growth medium was changed based on further growth results, such as the *multimedia* screen described in the following texts.

The storage tubes were capped and sealed with plastic paraffin film. They were then maintained with a 12-hour day–night cycle, under 50 µmol photons m^{-2} s^{-1} light, though in some cases less light was more optimal. Strains were stored at room temperature for at least 12 months. After 1 year most strains had to be transferred to new tubes with a fresh medium, although a number of strains could be stored up to 2 years without transfer. Nevertheless, for longer-term storage without such maintenance, cryopreservation should be considered (Benson 2008). In that regard, 30 of the best performing strains isolated were transferred to the UTEX culture collection for cryopreservation (NAABB 2014).

3.4 CLASSIFICATION OF MICROALGAL STRAINS

Working with thousands of isolates, any comprehensive molecular classification effort including DNA isolation, PCR, sequencing, and sequence analysis would have been time consuming and costly. Consequently, newly isolated algal strains were classified only after they passed at least the first screening step.

The most promising strains obtained from the second-level screen were subjected to molecular classification. Sequences of nuclear rDNA internal transcribed spacer 2 (ITS2) regions of algal strains have been used frequently for classification and barcoding because the ITS2 has been shown to be an excellent phylogenic marker (Koetschan et al. 2010). It has been used for species identification in environmental samples (Landis and Gargas 2007; Engelmann et al. 2009) and as a target molecule for barcoding diatoms (Moniz and Kaczmarska 2009). With regard to the green algae, the ITS2 has been used for phylogenies of orders and families within the green algal class Chlorophyceae (Hegewald et al. 2010; Buchheim et al. 2012) as well as in automated reconstruction of the green algae tree of life (Buchheim et al. 2011). Recently, a large downloadable database set of sequences became available that allows for alignment with, and comparison to, a large number of ITS2 sequences for green algae (http://its2.bioapps.biozentrum.uni-wuerzburg.de/).

For strain classification and barcoding, the DNA of the candidate microalgae was extracted using the MOBIO DNA Isolation Kit (United States). The PCR products were sent out for purification and sequencing to Eurofins (MWG Operon now Eurofins Genomics, Alabama). Amplification of the ITS2 regions can be achieved by PCR using universal ITS2 primers (forward primer ITS2-F GCATCGATGAAGAACGCAGC and reverse primer ITS2-R TCCTCCGCTTATTGATATGC) (White et al. 1990). We also conducted microscopic analysis of cell morphology. Phylogenic analysis of the best performing strains from our bioprospecting research is underway (Neofotis et al. 2015).

3.5 SCREENING FOR BIOMASS PRODUCTIVITY AND LIPID CONTENT

Following isolation, strains were subjected to a multilevel laboratory screening protocol. The screen was designed to identify strains with the highest lipid productivities. Lipid productivity is defined as the product of biomass productivity and the lipid content of the biomass. The two main current proxies used for lipid content in algae, other than the actual chemical analysis, are fluorescent dyes (Cooksey et al. 1987; Cooper et al. 2010) and, at least for green algae, their capacity to overaccumulate carotenoids (Goodwin 1980). Figure 3.1 shows the general pipeline of the screen, which is described in more detail in the following sections.

3.5.1 First-Level Screening

First, a high-throughput screening process using the lipophilic fluorescent dye Nile Red (9-diethylamino-5H-benzo[α]phenoxazine-5-one) was used to qualitatively estimate the relative cellular content of lipids (Cooksey et al. 1987). Another lipid dye, boron-diphyrromethane (BODIPY), could not be used for our assay because its background fluorescence was too high. In a study looking at the optimization of staining conditions for microalgae with three lipophilic dyes (Nile Red, BODIPY, and DiO) to reduce precipitation and fluorescence variability, Nile Red was found to have a stable fluorescence intensity that was unaffected by the broadest range of conditions and could be correlated to cellular lipid content (Cirulis et al. 2012).

Because of the previously noted limitations of 96-well plates and even 12-well or 6-well plates, ultimately, 120 batch cultures of unialgal strains were grown in their respective liquid media types in Erlenmeyer flasks in batch culture on a shaker (Innova Platform) at room temperature under continuous illumination of about 50 μmol photons m^{-2} s^{-1} provided by daylight fluorescent lighting (Figure 3.9). The cultures were then screened for their Nile Red fluorescence and optical density (OD_{750}) at weekly intervals. Depending on the strain, these batch cultures grew for about 3–5 weeks.

For the weekly measurements of growth and relative lipid content, aliquots of 200 μL of algal culture were taken from the Erlenmeyer flasks and placed into wells of black, clear-bottom, surface nonbinding 96-well polystyrene plates (Corning). A set of 12 strains was tested at a time. For each strain, six wells were used. If

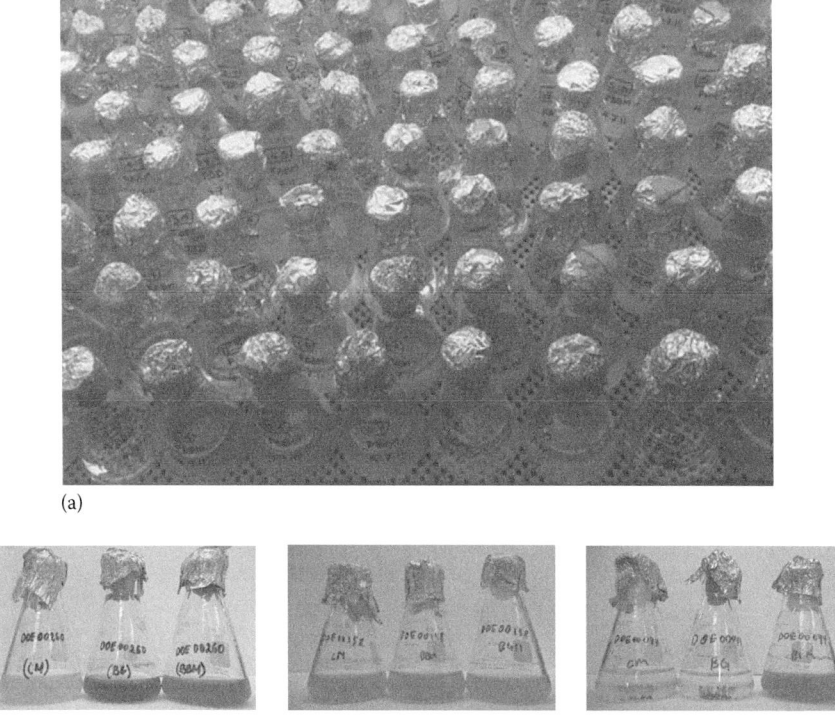

(a)

(b)   (c)   (d)

FIGURE 3.9 Screening of strains of microalgae for growth and lipid accumulation was performed in batch cultures using Erlenmeyer flasks on a shaker. (a) Shown is an example for the first-level screen of growth of microalgal cultures in batch cultures on a shaker. (b-d) Examples of second-level multimedia screen outcomes in (from L-R) C, BG11, BBM media. (b) The strain shown was isolated from a C medium plate, but grew best in a liquid BG11 medium. (c) The strain was isolated from a BG11 medium plate, but grew well on all three media tested. (d) The strain was isolated from a BG11 medium plate, but the best growth occurred in a liquid BBM medium. Note the clumpy growth of this strain in the C medium and BG11 medium.

necessary, the 200 μL aliquots of the samples were diluted to an OD_{750} of 0.1 to about 0.3 to ensure consistency in measurements. Into three of the six wells for each strain, Nile Red was added to a 3.0 μM final concentration. The other three wells, without Nile Red, were the negative controls. Six wells of the plate were used as negative controls with only medium in them. As standard controls for comparison, six wells were also loaded with aliquots of a culture of the unicellular green alga *Chlamydomonas reinhardtii* strain CW15 in a TAP medium, a low lipid producer with known characteristics in this protocol.

After allowing 15 minutes for the dye to penetrate in the intracellular bodies and bind to the cellular lipids, the assay plate was screened with the Biotek Synergy 2 microplate reader with an excitation at 485/20 nm and emissions at 590/35 nm for Nile Red fluorescence and absorbance at 750 nm to assess optical density. The OD_{750} was used as a proxy for biomass accumulation with its values representing the growth of the culture. The level of fluorescence of Nile Red at 590 nm indicated the level of lipids in a culture (Cooksey et al. 1987). Nile Red is only suitable for quantification of lipids with saturated fatty acids, and thus strains accumulating larger amounts of lipids with polyunsaturated fatty acids may not have been detected in our screen. Then, the Nile Red fluorescence values were normalized to the biomass as measured by the OD_{750}. After 4–5 weeks, progressive optical density values and normalized Nile Red readings per unit optical density were plotted against time. As the bar for the low lipid–producing strain (Figure 3.10), fluorescence due to *C. reinhardtii* was compared to that of the newly isolated strains. The cutoff for selecting novel strains for further study was either a Nile Red fluorescence above 60,000 relative units (compared to 5,000 for *C. reinhardtii*) or an exceptional growth based on the OD_{750}. Overall, about 20% of all isolated and screened strains passed the first-level screen. One of the first strains identified with excellent growth and lipid accumulation properties was a strain of *Scenedesmus* (*Acutodesmus*) *obliquus* (EN0004) (Neofotis et al. 2015) (Figures 3.10 and 3.11).

In our experience, although many strains grow well in an HS medium which contains ammonium as the only nitrogen source, none of the hundreds of strains examined in the HS medium accumulated lipids. Indeed, subsequent tests with several

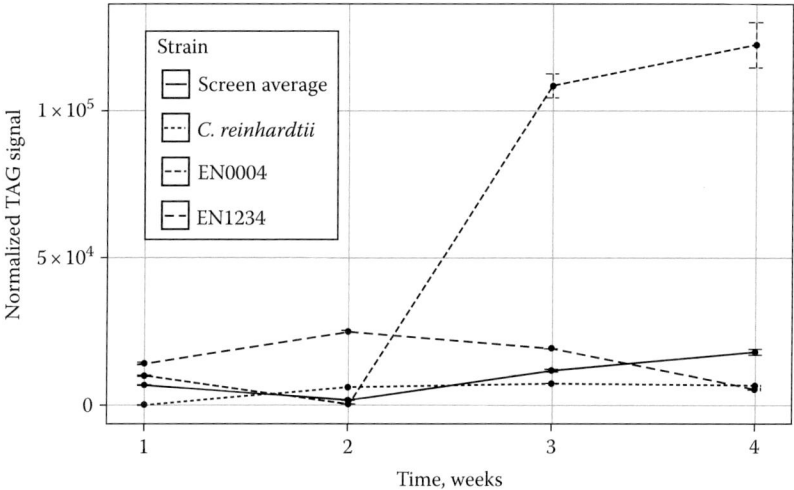

FIGURE 3.10 Normalized Nile Red fluorescence representing the lipid contents of cells. The Nile Red signal is shown for three strains during the screening period. Error bars represent the standard deviation among triplicates. Although strain EN1234 is carotenogenic and overaccumulates triacylglycerols in cytosolic oil bodies (see Figure 3.11e, f), it did not display high relative Nile Red values in our first-level screen. Low Nile Red fluorescence levels for carotenogenic strains were often seen in the first-level screen. This may be due, in part, to thick cell walls prohibiting efficient penetration of Nile Red into the cells. In addition, the carotenoids within the oil bodies might quench Nile Red fluorescence. Although such carotenogenic strains did not pass the Nile Red screen, their orange and reddish culture colors indicated triacylglycerol accumulation and so were advanced to the second level of screening.

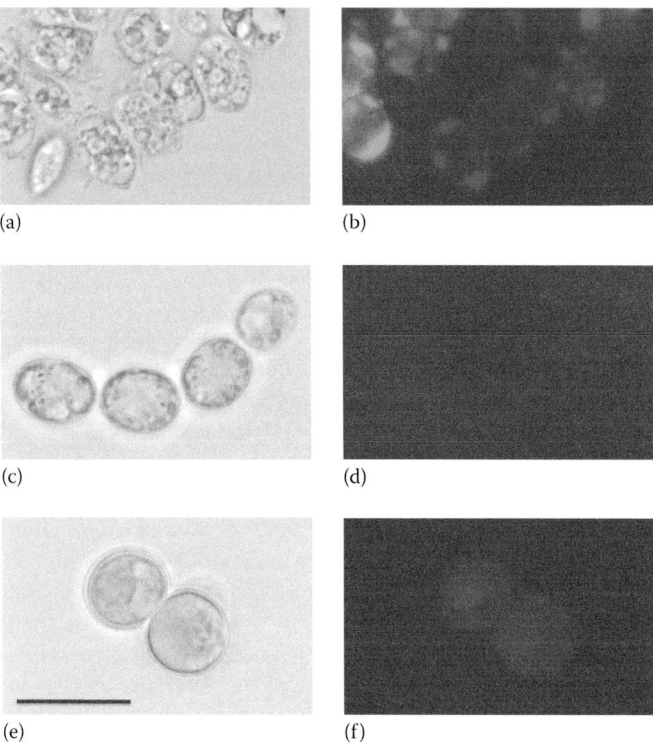

FIGURE 3.11   (See color insert.) Cells from two of the strains deemed to be potential high producers, showing bright fields (a, c, e) and Nile Red fluorescence images from the same pairwise field of view at 530 nm (b, d, f). (a, b) *Scenedesmus obliquus* strain EN0004 with oil bodies visible with fluorescence microscopy. (c, d) Strain EN1234 (a putative *Coelastrum*-related Desmid) from a young culture with few significant oil bodies visible. (e, f) Stressed cells of strain EN1234 taken from an older culture. Note the orange coloration under the bright field indicative of carotenoid accumulation (e) and apparent Nile Red fluorescence (f). The bar indicates 10 μm.

strains, each grown in multiple media, showed that algae grown in an HS medium did not accumulate lipids. When grown on a nitrate-containing C medium, a BG11 medium, and BBM media, these same strains accumulated lipids. We also noticed that many newly isolated *Haematococcus* strains grew well in liquid HS media, but never accumulated the characteristic red astaxanthin carotenoid. Based on these observations, the use of HS media was discontinued.

3.5.2   Screening for Carotenoid Overaccumulation

As a complementary method of screening (Figure 3.1), all cultures of strains were also examined for changes in color. In particular, yellowish, orange, and red-dish cultures, indicating overaccumulation of carotenoids (Figures 3.11 through 3.13), were further inspected through bright field and Nile Red fluorescence

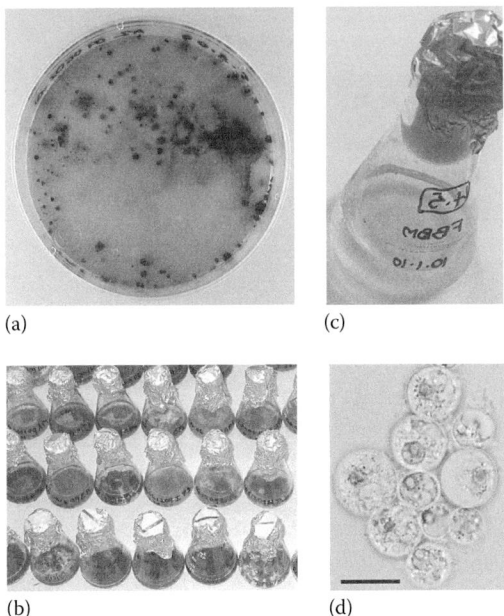

(a)   (c)

(b)   (d)

FIGURE 3.12   **(See color insert.)** Identification of strains that overaccumulate carotenoids. (a) Sometimes, colonies of microalgae grown on plates would display a yellowish, orange, or reddish color indicating that the cells overaccumulated carotenoids. (b) More often, strains of liquid cultures from the first-level screen would display a yellowish, orange, or reddish color upon aging. (c) A culture displaying an apricot color indicating accumulation of carotenoids. (d) Microscopy of cells from culture are shown in (c). The cells have an apricot color originating from very large oil bodies that sequester the carotenoids (data not shown). The bar represents 10 μm.

microscopy (Olympus BX51). Algae strains observed to accumulate carotenoids were presumed to also have the capability to produce significant levels of triacylglycerols, due to the known sequestration of carotenoids in oil bodies. Cells of one carotenogenic strain (EN1234), a *Coelastrum*-related species of the family Scenedesmaceae within the class Chlorophyceae of the green algae (Neofotis et al. 2015), are shown in Figure 3.11 as an example. Staining of cells with Nile Red or BODIPY (Cooper et al. 2010) confirmed the location of carotenoids within triacylglycerol-containing cytosolic oil bodies.

Sometimes, based on the colony color on the initial plates from environmental samples, strains could already be identified as carotenogenic (Figure 3.12a). However, in batch cultures, the majority of strains identified as carotenoid overaccumulating only exhibited such characteristics in the first-level screen when they reached the stationary phase (Figure 3.11e and Figure 3.12b).

Currently, *D. salina* and *Haematococcus pluvialis* are the only species cultivated on commercial scale for carotenoid production. However, we isolated various strains of other green algae genera able to accumulate carotenoids (Neofotis et al. 2015). The pigment profiles of two of these strains, EN1234 and *H. pluvialis* strain Haema001, are

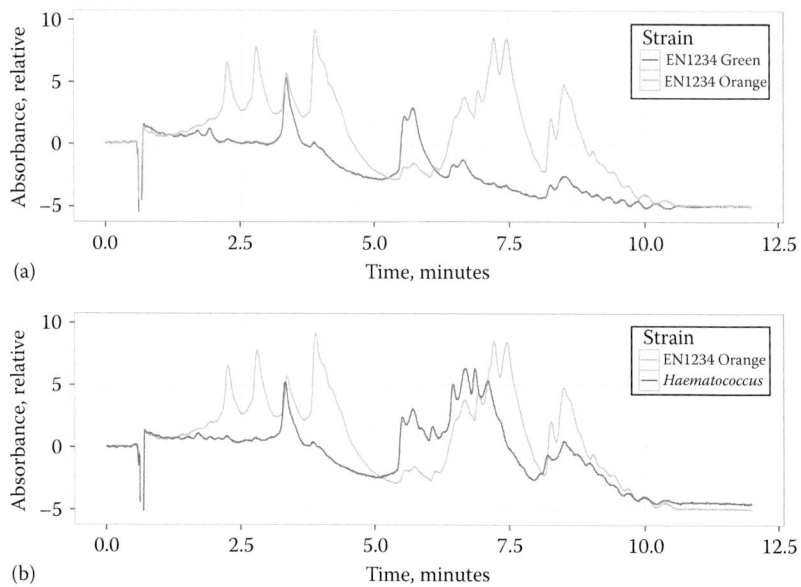

(a)

(b)

FIGURE 3.13 Displayed are UHPLC traces that show the absorbance of pigments from cell extracts plotted as the 450 nm/550 nm difference. Different pigments such as chlorophylls and carotenoids elute at known times and the relative level of absorbance represents the amount of these pigments present in the strains. (a) Profiles of extracts from strain EN1234 from young cultures (green) and older cultures (orange). (b) Profiles of extracts from carotenoid overaccumulating cells of strain EN1234 (orange) and *Haematococcus* sp. strain Haema001.

compared in Figure 3.13. To obtain the profiles, dichloromethane/methanol 2:1 (v/v) was used to extract the pigments. The extracts were then separated using a C18 reverse-phase column during ultrahigh-performance liquid chromatography (UHPLC). The extracts were then run through a photodiode array detector (Flexar FX-15 UHPLC system from Perkin Elmer). Different carotenoids within the spectrum were identified based on absorbance spectrum and based on concomitant molecular mass determination by application of mass spectrometry using atmospheric pressure chemical ionization (APCI) in connection with a time-of-flight mass spectrometer (Perkin Elmer, Massachusetts). Similar to the green alga *H. pluvialis*, strain EN1234 contained astaxanthin, although it was only a minor component of its total carotenoids. Further analysis of the carotenoid profiles of strain EN1234 and some other carotenoid overaccumulating strains showed that they accumulate precursors of astaxanthin in different ratios as well as possibly novel carotenoids that may be of commercial interest for feed formulations or nutraceuticals (Neofotis et al. 2015).

3.5.3 SECOND-LEVEL STRAIN SCREENING

Strains that had passed the first-level screen were then subjected to a second-level screen to determine their growth in media other than the initial isolation medium. For freshwater strains, a C medium, BG11 medium, and BBM were all used as these media have very different nutrient concentrations (i.e., nitrate, phosphate, and iron). Color and density of algal cultures were inspected visually (Figure 3.9) to estimate the growth of the cultures. In many cases, strains would grow in a broad range of media (example shown in Figure 3.9c). But often, one medium was preferred by a strain (examples in Figure 3.9b, d). Based on the results obtained from this *multimedia* screen, the storage medium for strains was adjusted to the liquid medium that the strains grew in best. The best growth medium as determined in the second-level *multimedia* screen was then used in the third-level screening.

3.5.4 THIRD-LEVEL SCREENING FOR BIOMASS AND LIPID PRODUCTIVITIES OF STRAINS OF MICROALGAE

Potential future platform strains that had undergone the *multimedia* screen were then tested for their biomass and lipid productivities in bubble columns (500 mL, 3 cm inner diameter) at 28°C under continuous illumination with fluorescent daylight lamps of about 200 μmol photons m^{-2} s^{-1} (Figure 3.14). So that the cultures would not have CO_2 limitations or major pH changes, CO_2 was provided to cultures at a concentration of 1% in air. This allowed, in a shorter period of time, the production of enough biomass for gravimetric and lipid content analysis. This third-level screen evaluated the growth potential of the strains under conditions of sufficient CO_2, constant pH, and no O_2 accumulation, in contrast to the CO_2 limiting and uncontrolled pH, and potentially O_2 inhibitory conditions present in the second-level screening. Strains that grow well under both conditions are of potential interest in process scale-up.

When precultures were in their light-limited growth phase, they were used to start new batch cultures (Figure 3.14) from which growth curves based on ash-free dry weight (AFDW) were obtained (Figure 3.15). Figure 3.14 shows the succession of cultures of different newly isolated strains in each one of the bubble columns. As visible from the photos of the cultures, some, but far from all strains, grew well under these laboratory conditions. Overall, more than 100 freshwater and brackish/marine strains were tested in bubble columns. More than 50 of these newly isolated strains grew well. When the cultures reached their stationary phase, the total lipid content of the biomass was determined by

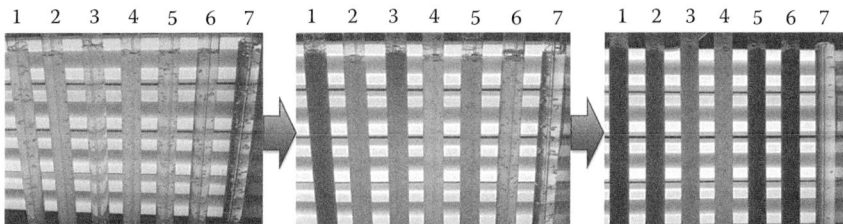

FIGURE 3.14 Bubble columns with each containing a unique, newly isolated strain. *Scenedesmus obliquus* strain EN0004 was grown in the leftmost column (#1). The sequence of photos visualizes the different growth rates of the various strains. Photos were taken 24, 48, and 120 hours after culture inoculation.

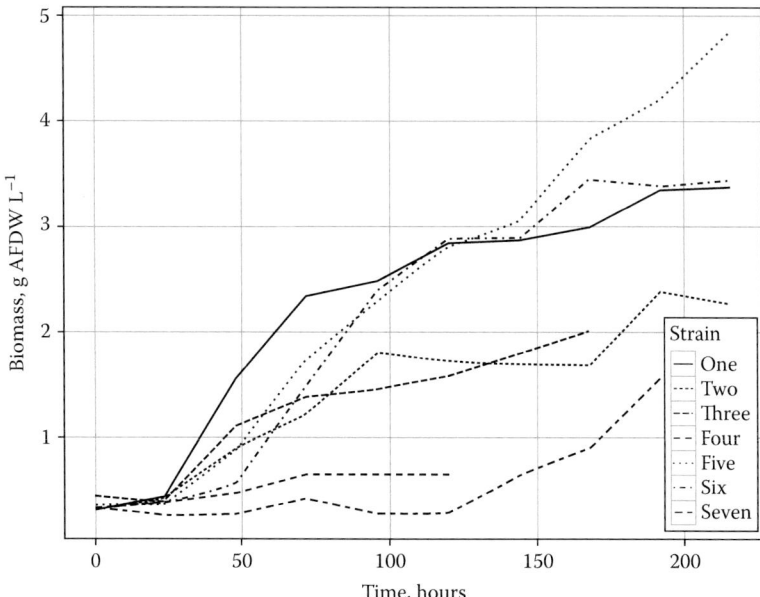

FIGURE 3.15   Growth curves are plotted based on biomass for the strains shown in Figure 3.14. Some strains such as #5 did not show any growth inhibition during our experiment and could reach biomass concentrations of more than 5 g L⁻¹.

a gravimetric method (Bigogno et al. 2002). Under the aforementioned growth conditions in bubble columns in a batch culture mode, the average biomass productivity of the best performing strains during the light-limited phase was in the range of 0.7–1.0 g AFDW L^{-1} day^{-1}. We compared our newly isolated strains to the Eustigmatophyte *Nannochloropsis salina* strain CCMP1776, a strain used in outdoor cultivation trials in raceway ponds (Unkefer et al. 2011; NAABB 2014) and also in outdoor photobioreactors (Quinn et al. 2012). In our laboratory bubble column tests, *N. salina* reached only about 0.5–0.7 g AFDW L^{-1} day^{-1}. Most of the strains in the third-level screen reached the stationary phase after about 5–6 days. At that time, the new strain's total lipid content was about 15%–25%, with lipid productivities of about 0.2 g total lipid L^{-1} day^{-1}. Although growing slower, the lipid content of *N. salina* could reach over 30% within the 5 days in batch culture, thus ultimately reaching a similar lipid productivity as the strains we isolated. However, these newly isolated strains did exhibit shorter initial lag phases and higher daily biomass productivities.

Recently, for biofuels applications, hydrothermal liquefaction (HTL) technologies, which convert the entire biomass into crude oils rather than depending only on extractable lipids, have come to the forefront in this field (Barreiro et al. 2013; Stephens et al. 2013; Zhu et al. 2013). Consequently, higher biomass productivity is now considered a viable alternative to lipid productivity or content in algal liquid fuel production.

In our third-level screen, the strains with the highest biomass productivities belonged to the green algal families Scenedesmaceae and the Chlorellaceae.

Visual examples of the rapid growth of some of the Scenedesmaceae are shown in Figure 3.14. Culture #1 contained the *S. obliquus* strain EN0004, which was one of the first and best biomass and lipid producers. Culture #5 contained a *Desmodesmus* species and culture #6 contained a second *Scenedesmus* species. In addition to strain EN0004, later during our screening, several other strains of the species *S. obliquus*—isolated from a variety of different habitats—were identified in our *multilevel* screen as having high biomass and lipid productivities. *S. obliquus* EN0004 and the *Desmodesmus* sp., shown in Figure 3.14 as #5, performed success-fully in small-scale raceway ponds at the NAABB (National Alliance for Advanced Biofuels and Bioproducts) testbed facility of the Texas A&M University at Pecos, Texas (NAABB 2014).

3.6   DISCUSSION

More than 5000 strains were isolated during this project, and dozens of newly iso-lated strains were identified as potential new platform strains, for either biofuels applications or carotenoid production. However, only relatively few strains went through the entire three-stage screening process, and we focus here only on two strains of the larger group of freshwater Scenedesmaceae belonging to the green algae, *S. obliquus* EN0004 and strain EN1234, the latter being one example of a *Coelastrum*-related, but as of yet unknown species.

EN0004 was classified as *S. obliquus* based on its molecular marker rDNA–ITS2 sequence—confirmed via its morphology (see Figure 3.11a). This strain was chosen for further characterization because it exhibited one of the highest growth rates and lipid productivities in the *multilevel* screen. Others have also consid-ered strains of *S. obliquus* for use in potential biofuels applications (Mandal and Mallick 2009; Breuer et al. 2012; Vigeolas et al. 2012), and strains of this species were previously found suitable for production in outdoor mass cultures (Becker 1984; Grobbelaar et al. 1990). The isolation of *S. obliquus* by our bioprospecting method, as described earlier, confirmed that the approach taken can select for species, such as this, which are potentially well suited for outdoor cultivation for biofuels.

The isolation of the second strain, EN1234, demonstrates that novel strains could be identified as potential production strains for biofuels and/or carotenoid applica-tions. Although this strain had low Nile Red fluorescence readings in the first-level screen (Figure 3.10), cultures of this strain turned orange, a characteristic of a con-comitant lipid and carotenoid producing green alga. Light microscopy confirmed that cells had high amounts of carotenoids in cytosolic oil bodies during the lipid accumulation phase (Figure 3.11). Since carotenoids are known fluorescence quench-ers, there was the possibility that the presence of carotenoids in Nile Red–labeled oil bodies could have reduced the Nile Red signal during the screening process. Overall, about 5%–10% of all strains screened did not exhibit high Nile Red fluo-rescence levels. But many of these strains could be identified as high lipid producers based on color changes due to their carotenoid coaccumulation with TAGs.

Based on molecular sequences and microscopic investigation, strain EN1234 is a green alga belonging to the family Scenedesmaceae (Figure 3.11). In general, cells of strain EN1234 were coccoid with cells from older batch cultures containing large orange oil bodies. However, cells of EN1234 were clearly different from those of *H. pluvialis*, and further molecular phylogenic analysis will help clarify the taxonomic position of this novel strain.

Regarding biomass and total lipid productivities, strain EN1234 performed as well as *S. obliquus* strain EN0004 in bubble columns, but unlike strains of *S. obliquus*, strain EN1234 has not yet been tested in outdoor culture. In addition to the poorly understood green algal strain EN1234, a number of other carotenogenic strains with similar morphology and physiology were identified. Some of these high-producing strains were also successfully tested outdoors in small-scale high-rate ponds (NAABB 2014).

The aforementioned bioprospecting approach, with a combination of traditional cell cultivation and FACS cell sorting, allowed us to isolate about 1000 strains per year. FACS cell sorting greatly enhanced the speed of strain isolation to a level that many more strains were isolated than could be screened. Because of manpower and time constraints, only about 2/3 of all isolated strains could ultimately be put through the first-level screen. Furthermore, only about 20% of the strains that passed the first-level and second-level screen could then be screened at the third level. Additionally, of the candidate platform strains coming out of our laboratory screening process, only a handful of strains could be taken to the final stage of outdoor ponds or large photobioreactors. Outdoor testing should be carried out as soon as possible after initial isolation. This is due to potential deterioration of strain characteristics including adaptations to low light during growth in the laboratory (Polle et al. 2004). In our case, the turnaround time from isolation to larger-scale testing in ponds or bioreactors was about 12 months. However, in a test to determine the fastest turnaround time, it took less than 3 months from strain isolation to completion of indoor screening.

Another limitation faced was the classification of isolated strains. As mentioned earlier, obtaining taxonomic data for thousands of strains was not possible. Therefore, we focused on classification of only the best performing strains. Based on microscopic analysis and using molecular markers of potential future platform strains, it was found that in our collection some algal species are represented by multiple strains—isolated from different environmental samples (Neofotis et al. 2015). For example, several strains of the green alga *S. obliquus* emerged as top performers. Having multiple strains of the same species could be valuable, as it increases the diversity in germplasm of species, which might be exploited in future strain development.

3.7 CONCLUSIONS

There is little reason to believe that the natural organisms most suitable for further algae biofuel development work have already been isolated (Mutanda et al. 2011; Wilkie et al. 2011). The majority of algae species have yet to be identified or characterized. Therefore, bioprospecting for novel natural platform strains suitable for algal mass cultivation and the elucidation of their potential for generation of products is an essential first step for advancing microalgal biotechnology. Of course, multiple strategies can be applied to obtain novel promising strains including univariate or multivariate approaches (Barclay and Apt 2013). We employed an approach that consisted of isolation of a large number of new strains that were then screened for use in biofuels applications. To that end, our bioprospecting project resulted in 30 potential platform strains, some of which had been tested successfully in outdoor ponds (NAABB 2014). By no means should the bioprospecting of microalgae be viewed as a completed process. Any new microalgal-based technology would be expected to go through a phase of identification of the "best" platform strains for that application, which might include new bioprospecting projects.

Following the identification of proper platform strains, the next step should be strain development to improve existing qualities of the strain or to introduce new metabolic characteristics. As with higher plants, a number of techniques such as classical breeding, mutagenesis, directional selection, and genetic engineering could be applied. Applying all these techniques to a strain, which has already evolved in the natural environment to be well suited to algal mass culturing, offers the most promise. Nevertheless, strain bioprospecting is the first step in making microalgal technologies commercially successful.

ACKNOWLEDGMENTS

This research was supported by the Air Force Office of Scientific Research (Grants #FA9550-08-1-0170 and #FA9550-08-1-0403) and by the U.S. Department of Energy (Grants #DE-EE0003129 and #DE-EE0003046-28302B). The authors greatly appreciate the support by their collaborators within the National Alliance for Advanced Biofuels and Bioproducts, specifically Dr. J. Olivares and Dr. R. Sayre. The authors are grateful to Dr. D. Tran for his support in the early work of strain isolation. The authors also thank Dr. R. Ovalle for the use of his plate reader and Dr. J. Nishiura for his aid in fluorescent microscopy as well as for his helpful consultation. The authors thank the members of Boy Scout Troop 1949 in Katy, Texas, and the high school students working with Dr. C. McEntee at Brooklyn College in New York, for the sampling and isolation of strains. Thanks also go to the editors, Dr. S. Slocombe and Dr. J. Benemann, for their comments and suggestions for the improvement of the chapter.

REFERENCES

Andersen, R. A. 2005. *Algal Culturing Techniques.* Amsterdam, the Netherlands: Elsevier.

Andersen, R. A. and M. Kawachi. 2005. Chapter 6: Traditional microalgae isolation and techniques. In *Algae Culturing Techniques* ed. R. A. Anderson. Amsterdam, the Netherlands: Elsevier.

Barclay, W. and K. Apt. 2013. Strategies for bioprospecting microalgae for potential commercial applications. In *Handbook of Microalgal Culture: Applied Phycology and Biotechnology*, eds. R. Richmond and Q. Hu. Oxford, UK: Wiley-Blackwell.

Barreiro, D. L., W. Prins, F. Ronsse, and W. Brilman. 2013. Hydrothermal liquefaction (HTL) of microalgae for biofuel production: State of the art review and future prospects. *Biomass Bioenergy* 53:113–127.

Becker, E. W. 1984. Biotechnology and exploitation of the green alga *Scenedesmus obliquus* in India. *Biomass* 4:1–19.

Belay, A. 2013. Biology and industrial production of *Arthrospira* (*Spirulina*). In *Handbook of Microalgal Culture: Applied Phycology and Biotechnology*, eds. R. Richmond and Q. Hu. Oxford, UK: Wiley-Blackwell.

Ben-Amotz, A., J. E. W. Polle, and D. V. Subba Rao. 2009. *The Alga Dunaliella: Biodiversity, Physiology, Genomics and Biotechnology.* Enfield, NH: Science Publishers.

Benemann, J. 2013. Microalgae for biofuels and animal feeds. *Energies* 6:5869–5886.

Benson, E. E. 2008. Cryopreservation of phytodiversity: A critical appraisal of theory practice. *CRC Crit. Rev. Plant. Sci.* 27:141–219.

Bigogno, C., I. Khozin-Goldberg, S. Boussiba, A. Vonshak, and Z. Cohen. 2002. Lipid and fatty acid composition of the green oleaginous alga *Parietochloris incisa* the richest plant source of arachidonic acid. *Phytochemistry* 60:497–503.

Borowitzka, L. J. 1991. Development of western biotechnology's algal β-carotene plant. *Bioresour. Technol.* 38:251–252.

Borowitzka, M. A. 2013. High-value products from microalgae-their development and commercialisation. *J. Appl. Phycol.* 25:743–756.

Borowitzka, M. A. and N. R. Moheimani. 2013. Sustainable biofuels from algae. *Mitig. Adapt. Strateg. Glob. Change* 18:13–25.

Breuer, G., P. P. Lamers, D. E. Martens, R. B. Draaisma, and R. H. Wijffels. 2012. The impact of nitrogen starvation on the dynamics of triacylglycerol accumulation in nine microalgae strains. *Bioresour. Technol.* 124:217–226.

Buchheim, M. A., A. Keller, C. Koetschan, F. Forster, B. Merget, and M. Wolf. 2011. Internal transcribed spacer 2 (nu ITS2 rRNA) sequence-structure phylogenetics: Towards an automated reconstruction of the green algal tree of life. *PLoS One* 6:e16931.

Buchheim, M. A., D. M. Sutherland, T. Schleicher, F. Forster, and M. Wolf. 2012. Phylogeny of Oedogoniales, Chaetophorales and Chaetopeltidales (Chlorophyceae): Inferring from sequence-structure analysis of ITS2. *Ann. Bot.* 109:109–116.

Cavalier-Smith, T. 2013. Symbiogenesis: Mechanisms, evolutionary consequences, and systematic implications. *Annu. Rev. Ecol. Evol. Syst.* 44:145–172.

Chan, C. X., J. Gross, H. S. Yoon, and D. Bhattacharya. 2011. Plastid origin and evolution: New models provide insights into old problems. *Plant Physiol.* 155:1552–1560.

Cirulis, J. T., B. C. Strasser, J. A. Scott, and G. M. Ross. 2012. Optimization of staining conditions for microalgae with three lipophilic dyes to reduce precipitation and fluorescence variability. *Cytometry A* 81:618–626.

Cooksey, K. E., J. B. Guckert, S. A. Williams, and P. R. Callis. 1987. Fluorometric-determination of the neutral lipid-content of microalgal cells using Nile Red. *J. Microbiol. Methods* 6:333–345.

Cooper, M. S., W. R. Hardin, T. W. Petersen, and R. A. Cattolico. 2010. Visualizing "green oil" in live algal cells. *J. Biosci. Bioeng.* 109:198–201.

Davis, R., A. Aden, and P. T. Pienkos. 2011. Techno-economic analysis of autotrophic microalgae for fuel production. *Appl. Energ.* 88:3524–3531.

Doucha, J. and K. Livansky. 2006. Productivity, CO_2/O_2 exchange and hydraulics in outdoor open high density microalgal (*Chlorella* sp.) photobioreactors operated in a Middle and Southern European climate. *J. Appl. Phycol.* 18:811–826.

Draaisma, R. B., R. H. Wijffels, P. M. Slegers, L. B. Brentner, A. Roy, and M. J. Barbosa. 2013. Food commodities from microalgae. *Curr. Opin. Biotechnol.* 24:169–177.

Elliott, L. G., C. Feehan, L. M. L. Laurens, P. T. Pienkos, A. Darzins, and M. C. Posewitz. 2012. Establishment of a bioenergy-focused microalgal culture collection. *Algal Res.* 1:102–113.

Engelmann, J. C., S. Rahmann, M. Wolf et al. 2009. Modelling cross-hybridization on phylogenetic DNA microarrays increases the detection power of closely related species. *Mol. Ecol. Resour.* 9:83–93.

Falkowski, P. and A. Knoll. 2007. *Evolution of Primary Producers in the Sea.* Burlington, MA: Elsevier.

Georgianna, D. R. and S. P. Mayfield. 2012. Exploiting diversity and synthetic biology for the production of algal biofuels. *Nature* 488:329–335.

Gimpel, J. A., E. A. Specht, D. R. Georgianna, and S. P. Mayfield. 2013. Advances in micro-algae engineering and synthetic biology applications for biofuel production. *Curr. Opin. Chem. Biol.* 17:489–495.

Goodwin, T. W. 1980. *The Comparative Biochemistry of the Carotenoids*. London, UK: Chapman & Hall.

Gouveia, L. and A. C. Oliveira. 2009. Microalgae as a raw material for biofuels production. *J. Ind. Microbiol. Biotechnol.* 36:269–274.

Grobbelaar, J. U., C. J. Soeder, and E. Stengel. 1990. Modeling algal productivity in large outdoor cultures and waste treatment systems. *Biomass* 21:297–314.

Guerin, M., M. E. Huntley, and M. Olaizola. 2003. *Haematococcus* astaxanthin: Applications for human health and nutrition. *Trend. Biotechnol.* 21:210–216.

Guillard, R. R. L. 1975. Culture of phytoplankton for feeding marine invertebrates. In *Culture of Marine Invertebrate Animals*, eds. W. L. Smith and M. H. Chanley, pp. 26–60. New York: Plenum Press.

Guillard, R. R. L. and J. H. Ryther. 1962. Studies of marine planktonic diatoms. I. *Cyclotella nana* Hastedt and *Detonula confervacea* Cleve. *Can. J. Microbiol.* 8:229–239.

Guiry, M. D., G. M. Guiry, L. Morrison et al. 2014. AlgaeBase: An on-line resource for algae. *Cryptogam. Algol.* 35:105–115.

Harris, E. H. 2009. *The Chlamydomonas Sourcebook: Introduction to Chlamydomonas and Its Laboratory Use*, 2nd edn. Amsterdam, the Netherlands: Elsevier.

Hegewald, E., M. Wolf, A. Keller, T. Friedl, and L. Krienitz. 2010. ITS2 sequence-structure phylogeny in the Scenedesmaceae with special reference to *Coelastrum* (Chlorophyta, Chlorophyceae), including the new genera *Comasiella* and *Pectinodesmus*. *Phycologia* 49:325–335.

Hu, Q., M. Sommerfeld, E. Jarvis et al. 2008. Microalgal triacylglycerols as feedstocks for biofuel production: Perspectives and advances. *Plant J.* 54:621–639.

Hyka, P., S. Lickova, P. Pribyl, K. Melzoch, and K. Kovar. 2013. Flow cytometry for the development of biotechnological processes with microalgae. *Biotechnol. Adv.* 31:2–16.

Keeling, P. J. 2013. The number, speed, and impact of plastid endosymbioses in eukaryotic evolution. *Annu. Rev. Plant Biol.* 64:583–607.

Keller, M. D., R. C. Selvin, W. Claus, and R. R. L. Guillard. 1987. Media for the culture of oceanic ultraphytoplankton. *J. Phycol.* 23:633–638.

Koetschan, C., F. Forster, A. Keller et al. 2010. The ITS2 database III-sequences and structures for phylogeny. *Nucleic Acids Res.* 38:D275–D279.

Krienitz, L. and C. Bock. 2012. Present state of the systematics of planktonic coccoid green algae of inland waters. *Hydrobiologia* 698:295–324.

Landis, F. C. and A. Gargas. 2007. Using ITS2 secondary structure to create species-specific oligonucleotide probes for fungi. *Mycologia* 99:681–692.

Larkum, A. W. D., I. L. Ross, O. Kruse, and B. Hankamer. 2012. Selection, breeding and engineering of microalgae for bioenergy and biofuel production. *Trends Biotechnol.* 30:198–204.

Laurens, L. M. L., M. Quinn, S. Van Wychen, D. W. Templeton, and E. J. Wolfrum. 2012. Accurate and reliable quantification of total microalgal fuel potential as fatty acid methyl esters by *in situ* transesterification. *Anal. Bioanal. Chem.* 403:167–178.

Leliaert, F., H. Verbruggen, P. Vanormelingen et al. 2014. DNA-based species delimitation in algae. *Eur. J. Phycol.* 49:179–196.

Leu, S. and S. Boussiba. 2014. Advances in the production of high-value products by micro-algae. *Ind. Biotechnol.* 10:169–183.

Manandhar-Shrestha, K. and M. Hildebrand. 2013. Development of flow cytometric procedures for the efficient isolation of improved lipid accumulation mutants in a *Chlorella* sp. microalga. *J. Appl. Phycol.* 25:1643–1651.

Mandal, S. and N. Mallick. 2009. Microalga *Scenedesmus obliquus* as a potential source for biodiesel production. *Appl. Microbiol. Biotechnol.* 84:281–291.

Maxwell, E. L., A. G. Folger, and S. E. Hogg. 1985. Resource evaluation and site selection for microalgae production systems. In *SERI Aquatic Species Program 1984 Annual Report*, ed. Solar Energy Research Institute, SERI/TR-215-2484. Golden, CO: Solar Energy Research Institute. http://www.nrel.gov/docs/legosti/old/2484.pdf (accessed March 5, 2015).

Moniz, M. B. J. and I. Kaczmarska. 2009. Barcoding diatoms: Is there a good marker? *Mol. Ecol. Res.* 9:65–74.

Montero, M. F., M. Aristizabal, and G. G. Reina. 2011. Isolation of high-lipid content strains of the marine microalga *Tetraselmis suecica* for biodiesel production by flow cytometry and single-cell sorting. *J. Appl. Phycol.* 23:1053–1057.

Mutanda, T., D. Ramesh, S. Karthikeyan, S. Kumari, A. Anandraj, and F. Bux. 2011. Bioprospecting for hyper-lipid producing microalgal strains for sustainable biofuel production. *Bioresour. Technol.* 102:57–70.

NAABB. 2014. National Alliance for Advanced Biofuels and Bioproducts (NAABB) synopsis final report, Washington, DC. http://www.energy.gov/eere/bioenergy/downloads/national-alliance-advanced-biofuels-and-bioproducts-synopsis-naabb-final (accessed March 5, 2015).

Neofotis, P., A. Huang, K. Sury et al. 2015. Characterization and classification of highly productive microalgae strains discovered for biofuel and bioproduct generation. *Algal Res.*, in press.

Palinska, K. A. and W. Surosz. 2014. Taxonomy of cyanobacteria: A contribution to consensus approach. *Hydrobiologia* 740:1–11.

Passell, H., H. Dhaliwal, M. Reno et al. 2013. Algae biodiesel life cycle assessment using current commercial data. *J. Environ. Manage.* 129:103–111.

Pate, R., G. Klise, and B. Wu. 2011. Resource demand implications for U.S. algae biofuels production scale-up. *Appl. Energy* 88:3377–3388.

Peramuna, A. and M. L. Summers. 2014. Composition and occurrence of lipid droplets in the cyanobacterium *Nostoc punctiforme*. *Arch. Microbiol.* 196:881–890.

Pereira, H., L. Barreira, A. Mozes et al. 2011. Microplate-based high-throughput screening procedure for the isolation of lipid-rich marine microalgae. *Biotechnol. Biofuels* 4:61.

Pick, U., L. Karni, and M. Avron. 1986. Determination of ion content and ion fluxes in the halotolerant alga *Dunaliella salina*. *Plant Physiol.* 81:92–96.

Polle, J. E., A. Belay, J. C. Weissman, M. Huesemann, P. Pedroni, and J. Benemann. 2004. Microalgae culture maintenance for greenhouse gas abatement and other applications. In *Proceedings of the 10th International Congress for Culture Collections*, ed. B. Watanabe. Tsukuba, Japan: Japan Symposia Algae.

Quinn, J. C., T. Yates, N. Douglas et al. 2012. *Nannochloropsis* production metrics in a scalable outdoor photobioreactor for commercial applications. *Bioresour. Technol.* 117:164–171.

Raja, R., S. Hemaiswarya, N. A. Kumar, S. Sridhar, and R. Rengasamy. 2008. A perspective on the biotechnological potential of microalgae. *Crit. Rev. Microbiol.* 34:77–88.

Ratha, S. K. and R. Prasanna. 2012. Bioprospecting microalgae as potential sources of "Green Energy"—Challenges and perspectives. *Appl. Biochem. Microbiol.* 48:109–125.

Ratha, S. K., R. Prasanna, V. Gupta, D. W. Dhar, and A. K. Saxena. 2012. Bioprospecting and indexing the microalgal diversity of different ecological habitats of India. *World J. Microbiol. Biotechnol.* 28:1657–1667.

Shapiro, H. M. 2004. The evolution of cytometers. *Cytometry A* 58:13–20.

Sheehan, J., T. Dunahay, J. Benemann, and P. Roessler. 1998. A look back at the US Department of Energy's aquatic species program—Biodiesel from algae. In *Report NREL/TP-580-24190*, ed. Golden, CO: National Renewable Energy Laboratory. http://www.nrel.gov/biomass/pdfs/24190.pdf (accessed March 5, 2015).

Sieracki, M. E., N. Poulton, and N. Crosbie. 2005. Automated isolation techniques for microalgae. In *Algal Culturing Techniques*, ed. R. A. Anderson. Amsterdam, the Netherlands: Elsevier.

Stephens, E., I. L. Ross, and B. Hankamer. 2013. Expanding the microalgal industry—Continuing controversy or compelling case? *Curr. Opin. Chem. Biol.* 17:444–452.

U.S. Department of Energy. 2010. *U.S. DOE Roadmap: National Algal Biofuels Technology Roadmap*, eds. D. Fishman, R. Majumdar, J. Morello, R. Pate, and Y. Yang. Washington, DC: Office of Energy Efficiency and Renewable Energy Biomass Program. http://www1. eere.energy.gov/bioenergy/pdfs/algal_biofuels_roadmap.pdf (accessed March 5, 2015).

Unkefer, C. J., S. R. Starkenburg, S. N. Twary, M. S. Park, P. S. Chain, and P. J. Unkefer. 2011. The National Alliance for Advanced Biofuels and Bioproducts (NAABB). Abstracts of Papers of the American Chemical Society, No. 242. http://www.obpreview2011. govtools.us/presenters/public/InsecureDownload.aspx?filename=NAABB-DOE-PeerReview-Apr2011-Final-sm-3.pdf (accessed March 5, 2015).

Vigeolas, H., F. Duby, E. Kaymak et al. 2012. Isolation and partial characterization of mutants with elevated lipid content in *Chlorella sorokiniana* and *Scenedesmus obliquus*. *J. Biotechnol.* 162:3–12.

White, T. J., T. Bruns, S. J. Lee, and J. Taylor. 1990. Amplification and direct sequencing of fungal ribosomal RNA genes for phylogenetics. In *PCR Protocols*, eds. M. A. Innis, D. H. Gelfand, J. J. Sninsky, and T. J. White, pp. 315–324. San Diego, CA: Academic Press.

Wijffels, R. H., O. Kruse, and K. J. Hellingwerf. 2013. Potential of industrial biotechnology with cyanobacteria and eukaryotic microalgae. *Curr. Opin. Biotechnol.* 24:405–413.

Wilkie, A. C., S. J. Edmundson, and J. G. Duncan. 2011. Indigenous algae for local bioresource production: Phycoprospecting. *Energy Sustain. Dev.* 15:365–371.

Zhu, Y., K. O. Albrecht, D. C. Elliott, R. T. Hallen, and S. B. Jones. 2013. Development of hydrothermal liquefaction and upgrading technologies for lipid-extracted algae conversion to liquid fuels. *Algal Res.* 2:455–464.

4 Screening and Improvement of Marine Microalgae for Oil Production

Stephen P. Slocombe, QianYi Zhang, Michael Ross, Michele S. Stanley, and John G. Day

CONTENTS

4.1 INTRODUCTION

The microalgal biotech sector began in the 1940s, where the first modern attempts to grow microalgae focused on finding alternative sources of chemicals for use in munition manufacturing and fuels during World War II, by examining the production of lipids by various microalgae (Harder and Witsch 1942; Burlew 1953). In the postwar years, microalgae were seen as one solution to a shortfall in food production before the advent of the "Green Revolution." Later, during the oil crisis of the 1970s, when the price of petroleum was high, microalgae were revisited for their potential use in biofuels based on their ability to accumulate oil, which is usually in the form of tria-cylglycerols (Borowitzka 2013a). Microalgae have high theoretical productivities,

partly due to their unicellular nature, which leads to an efficient resource allocation into storage products as opposed to structure, their rapid growth rate, and their
carbon-concentrating mechanisms, which increase the efficiency of CO_2 utilization
(Giordano et al. 2005).

The global economy is to a great extent dependent on the use of oil, gas, and coal,
and disinvestment in fossil fuels, and their replacement with sustainable alternatives
appears to be required to avoid worsening climate change. In this context, economic
conditions need to become more favorable, or the costs of production must be lowered, for microalgal biofuels to deliver solutions (Stephens et al. 2010). A variety of
strategies have been proposed including coproduction of biofuels with high-value
products (Williams and Laurens 2010). The technologies developed for the production of intermediate-value products at larger scale such as niche market "healthy"
vegetable oils and industrial oils could act as steppingstones toward this goal.

Microalgae are currently grown for high-value biomass destined for health food
or aquaculture feed, as well as specific high-value products. The latter are mostly
lipid-based nutraceuticals or cosmeceuticals such as carotenoids and omega-3 polyunsaturated fatty acids. Some of these such as the beta-carotene precursor are currently produced profitably at large scale in artificial lagoons (Borowitzka 2013b).

4.2 EXPLOITING THE DIVERSITY OF MARINE STRAINS

Relatively few microalgae are grown commercially at large scale. One of those most
frequently produced commercially is the cyanobacterium *Spirulina*, which has low
levels of oil. Strains making up the bulk of the harvested algal biomass market are
mostly extremophiles (*Spirulina*, *Dunaliella*) or freshwater (*Chlorella*) (Benemann
2013). Microalgae growing in extreme environments (such as temperature or pH
extremes or high salinity) are less likely to be outcompeted by faster growing strains
or subject to predators and pathogens (Day et al. 2012). Furthermore, salt-tolerant
and marine strains have an advantage in reducing competition with agriculture for
land and freshwater resources (Day et al. 2012). Such strains are more likely to tolerate increases in ionic strength due to evaporative losses, which is a common problem
in open ponds (Sheehan et al. 1998).

There has long been an interest in addressing the lack of robust strains for growth
in outdoor ponds by searching for new oil-producing strains from the environment,
often from saline water bodies. This was one of the main goals of the U.S. DOE
Aquatic Species Program (ASP), which ran from 1980 to 1996 (Sheehan et al. 1998).
Although there are relatively few microalgal species/strains in commercial use
for growing in saline conditions or seawater, there is no apparent diversity short-
age in marine or saline environments as well as freshwater (Parker et al. 2008).
Up to 72,500 algal species exist in total, according to the most conservative projections but estimated figures for diatoms alone range up to 200,000 (Guiry 2012).
To date, about 44,000 species have been documented and 73% of these are described
in AlgaeBase (Guiry 2012).

Another approach is to screen existing cultures held in collections worldwide.
Together, the major service culture collections hold more than 20,000 publicly available isolates and are listed in Table 4.1. The largest marine collection is held at the

TABLE 4.1
List of the Major Microalgal Culture Collections

Collection	Location	No. of Available Strains	Information
UTEX	Austin, Texas	>3000	http://www.utex.org/default.aspx
NIES	Tsukuba, Japan	2746	http://mcc.nies.go.jp/
NCMA (CCMP)	Bigelow, Maine	2727	https://ncma.bigelow.org/products/algae/all-algae/show-180
CCAP	SAMS, Oban, United Kingdom	2707	http://www.ccap.ac.uk/key_to_strain_data.htm
SAG	Göttingen, Germany	~2400	http://www.uni-goettingen.de/en/184982.html
CCAC	Cologne, Germany	~1750	http://www.ccac.uni-koeln.de/geninform.htm
ATCC	Manassas, Virginia	1147	http://www.lgcstandards-atcc.org/?geo_country=gb
CSIRO (ANACC)	Hobart, Tasmania, Australia	>1000	http://www.csiro.au/en/Research/Collections/ANACC/About-our-collection
NIVA	Oslo, Norway	995	https://niva-cca.no/
PCC	Paris, France	>750	http://cyanobacteria.web.pasteur.fr/
CCALA	Třeboň, Czech Republic	>700	http://www.butbn.cas.cz/
CPCC	Waterloo, Ontario, Canada	~500	https://uwaterloo.ca/canadian-phycological-culture-centre/
MBA (PLY)	Plymouth, United Kingdom	~400	http://www.mba.ac.uk/culture-collection/
CCCM (NEPCC, FWAC)	Vancouver, British Columbia, Canada	282	http://www3.botany.ubc.ca/cccm/

National Center for Marine Algae and Microbiota (NCMA) (formerly the Culture Centre for Marine Plankton [CCMP]) at Bigelow. Other large repositories such as the CCAP collection (Culture Collection of Algae and Protozoa) at SAMS (the Scottish Association for Marine Science) have a high proportion of saline/marine taxa (Gachon et al. 2013). Useful data resources include AlgaeBase (http://www.algaebase.org/) that lists the world's known algal species and strains, and StrainInfo (http://www.straininfo.net/) that links publicly available strains to the main collections. Additional web resources provided by the collections are listed in Table 4.1.

Strain-to-strain variations in yields are often substantial; thus, it is sensible to test multiple strains from the given taxa (Day et al. 2012). However, even without isolating additional new strains from the environment, the screening of existing collections is a substantial undertaking; therefore, development of methods and strategies is an ongoing requirement to isolate promising strains for further analysis.

Since the publication of the ASP closeout report (Sheehan et al. 1998), a few large-scale screens for oil production have been published, although some have been carried out by commercial companies. There are reports of a number of smaller-scale analyses (<50 microalgae) in the literature, for example, Huerlimann et al. (2010), Rodolfi et al. (2009), and MacDougall et al. (2011). Larger published screens include those that were undertaken by Doan et al. (2011), who screened 96 marine strains isolated from the coastal waters of Singapore. A screen of 2076 strains of the Sammlung von Algenkulturen (SAG) collection in Göttingen focused on fatty acid composition but did not provide data on oil content or biomass production (Lang et al. 2011). In the case of the ASP screen itself, over 3000 strains were isolated and later narrowed down to about 300, with 51 studied in closer detail. As of 2009, about 23 of the latter were still extant and had been re-established at the NREL (Knoshaug et al. 2009). A large screen of strains collected from a range of environments in the United States is described in Chapter 3. In addition, about one-third of the marine strains in the CCAP collection have recently been screened, and examples from this are given in this chapter (Slocombe et al. 2015).

4.3 PRACTICAL APPROACHES TO SCREENING MICROALGAE

Improving oil productivity has been the main focus of microalgal screening, but other parameters may be examined in pursuit of improvements in production and harvesting (e.g., flue gas utilization, autoflocculation, coproduction with value products such as carotenoids, carbohydrate for biogas, and protein for feed). Whichever screening method is chosen, it must be of sufficiently high processivity to deal with high numbers of strains (Table 4.2). It is also important to recognize the need for robust measurements of productivity, as well as for the content of oil or other products (Sheehan et al. 1998; Griffiths and Harrison 2009; Borowitzka 2013c).

Some thought needs to be given to the choice of growth conditions in the screen. A standard set of laboratory conditions can be applied to all strains, but the final use envisaged for the strains needs to be considered. For the marine strains, media based on filtered or artificial seawater is a logical choice (e.g., f/2) (Slocombe et al. 2015). Sources of inorganic nitrogen and phosphorus need to be considered, usually nitrate and phosphate, and some attention needs to be paid to the N/P ratio, which will influence nutrient limitation (Leonardos and Geider 2004). For instance, f/2 has an N/P ratio of 24:1, which is relatively high (c.f. the Redfield ratio, 19:1) and could be phosphate limiting. Provision of nitrate can lead to elevation of medium pH, which in turn can lead to CO_2 limitation but is usually counteracted by a supply of CO_2 at 1%–5% in air, which also serves to increase productivity (Scherholz and Curtis 2013) (see Chapter 10). Potential sources of organic N and C from urea or waste remediation might also be considered. Provision of urea tends to lead to a balanced medium pH, whereas ammonium can result in a pH decrease (Scherholz and Curtis 2013). Whether or not to supply CO_2 could depend on the envisaged use of the strains, which could range from photobioreactors, where this is normally supplied, to large lagoons without any form of artificial gas exchange (Benemann 2013). Supply of air alone might favor strains with less susceptibility to carbon limitation and with active carbon-concentrating mechanisms (CCMs) (Giordano et al. 2005; Borowitzka 2013c) (see Chapter 1). Batch mode culture is the simplest approach to screening, but many end-use cultivation systems will be based on semicontinuous growth or two-phase growth (see Chapters 11 through 13).

TABLE 4.2

Comparison of High Processivity Methods for Total Lipid Measurement in Microalgae

Method	Harvest and Freeze-Dry Step	Material Required	Handling Time/ Sample	Run Time	Accuracy	Compositional Data
CN elemental analysis[a]	Yes	<1 mg DW	2 min	8 min	High	C, N, total lipid or hydrocarbon, protein
GC-FID/direct esterification[b]	Yes	2–5 mg DW	10 min	15–30 min	High	TFA content and composition
FTIR spectroscopy[c]	Yes	3–5 mg DW	10 min: need fine powder	15 min	High	Lipid, protein, carbohydrate
Fluorescent dye plate assay (e.g., Nile Red)[d]	Not needed	0.2 mL culture	10 min/96 wells	5 min/96 wells	Semiquantitative	Oil content
Flow cytometry[e]	Not needed	1 mL culture	10 min	5–15 min	Semiquantitative	Oil content (e.g., Nile Red)
Cell flotation (centrifugation)[f]	Not needed	1 mL culture	5 min	5 min	Low and condition-/ species-specific	None

[a] Slocombe et al. 2015.
[b] Slocombe et al. 2013b.
[c] Mayers et al. 2013.
[d] Chen et al. 2009.
[e] Doan et al. 2011.
[f] Sheehan et al. 1988.

4.3.1 METHOD FOR SCREENING MARINE STRAINS

The following procedure was developed from screening of the marine strains in the CCAP collection at SAMS (Slocombe et al. 2015). The organization of the screen is shown in Figure 4.1. The culture medium used in this screen was based on f/2 using a propriety source of artificial seawater and with relatively low levels of nitrate (0.88 mM) and phosphate (0.036 mM) for growth in batch culture. The aim of this screen was to select for strains for large-scale production with a limited nutrient supply, and air was supplied rather than CO_2. This was assumed to favor strains that were productive for oil at lower nutrient concentrations and that would cope better with rises in culture pH that leads to carbon limitation (Borowitzka 2013c).

Starter cultures of 100 mL were incubated in 250 mL conical flasks to late-log phase and subbed into 400 mL cultures at 5% vol. into 3 × 500 mL aerated Erlenmeyer

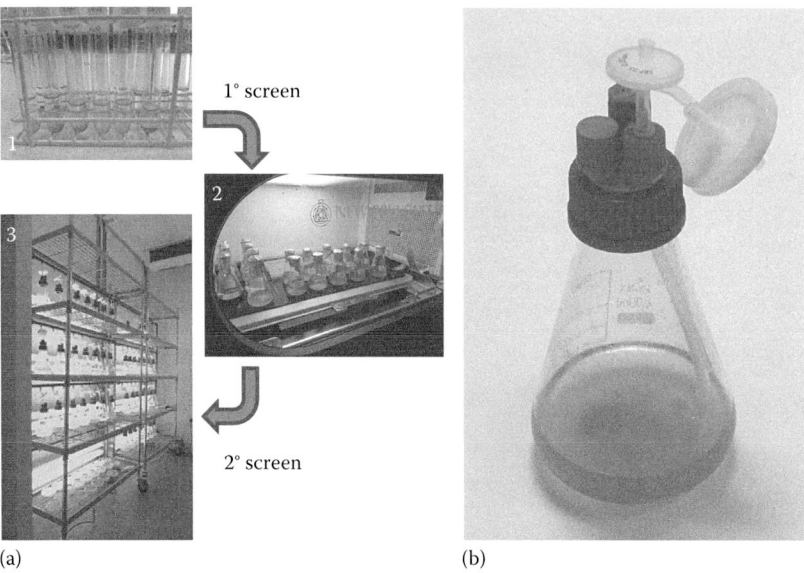

(a)   (b)

FIGURE 4.1   A schematic diagram (a) showing the screening process in terms of the three culturing steps: (1) a fresh inoculum from a maintained culture, (2) a primary screen for growth in f/2, and (3) a secondary screen for growth and composition in triplicate cultures in ported 500 mL Erlenmeyer flasks (b).

flasks (Pyrex, Corning, United States) capped with a ported GL45 connection system (Duran, Mainz, Germany) enabling sterile filtration of input air (Figure 4.1). Aeration and culture mixing were achieved by introducing air at 60 mL/min through a 4 mm bore silicone tube to the base of the flask. f/2 media was made up with artificial seawater at 33.5 g/L (Instant Ocean, Aquarium Systems, France) (Slocombe et al. 2015). Cultures were monitored by dual measurement of in vivo chlorophyll fluorescence and cell turbidity (e.g., $A_{735 \text{ or } 750}$). The light regime was 150 ± 10 µmol photons/m$^2 \cdot$ s for 16 h: 8 h L/D, at 20°C with warm and cool fluorescence strip lighting. Setting up starter cultures served as a preliminary screen for growth where two attempts to grow the strain were made. Inoculation dates for the starters and triplicate cultures, along with biomass harvest dates for the latter, were recorded. This enabled a simple measure of biomass productivity to be estimated over the course of batch culture from inoculation to stationary phase harvest. Entry into stationary phase was taken as the point for biomass harvest and was defined as no change ($\pm$10%) in a 1–2-day period of the biomass proxy readings. The aim was to avoid leaving cultures for too long in stationary phase due to the risk of bacterial growth or culture crash.

Cell material was recovered by centrifugation, taking into account cell size and fragility: with *Nannochloropsis* (diameter 3 µm), up to 4000 g is required, whereas for large cells such as the pennate diatom *Cylindrotheca* (length 100 µm), $\leq$1000 g was used. Certain high-oil content strains may not completely sediment by centrifugation due their lowered density. This was not a common occurrence in the screen as described earlier, probably due to harvest on entry into stationary phase and hence avoidance of a prolonged incubation stationary phase. It can be overcome by diluting the saline media with water to reduce media density, provided that the cells are sufficiently robust. An alternative to centrifugation is filtration (Moheimani et al. 2013). Depending on the filter matrix, mechanical disruption of the filter or incorporation of the filter into the assays may be required, both of which can impact processing time and assay scale. Overall, there are pros and cons associated with both centrifugation and filtration methods. Harvested material sedimented by centrifugation was freeze-dried and stored in hermetically sealed vials under nitrogen at $-80°C$ to avoid oxidation of unsaturated fatty acids.

4.3.2 MEASURING MICROALGAL COMPOSITION

In the screen procedure described earlier (Figure 4.1), a 300 mL harvest at stationary phase provided material ranging from 0.02 to 1.50 (mean 0.31) g DW/L, which was more than adequate for separate assays for total glycerolipid content (GC-FID, or Flame Ionization Detector, and GC-MS of FAMES, or Fatty Acid Methyl Esters), protein, total carbohydrate, and combustive mass spectrometry (MS) elemental analysis (for C, N). Although biomass can be expressed as g dry weight (DW), organic content can vary widely so measurement of C provides a more robust measure. Alternatively, measurement of ash-free dry weight (AFDW) is often used (Moheimani et al. 2013). In protein assay methods, there is often a balance to be struck between maximizing extraction and preservation of the peptide substrate for the assay. To address this, a method based on a hot-TCA/Lowry assay was developed further to optimize extraction of lyophilized material from the more recalcitrant algal strains (*Chlorella* and eustigmatophytes) (Slocombe et al. 2013a). Carbohydrates can be conveniently assayed using the Dubois method from lyophilized material (Slocombe et al. 2015). Multi-assays have been developed for microalgal biomass that have the advantage of reducing labor and allowing simultaneous measure of total fatty acids, protein, carbohydrate, and pigments (chlorophyll and carotenoids) from a single extract through spectrophotometric assays based on 1–2 mL of culture or 1 mg DW (Chen and Vaidyanathan 2012, 2013). Lipid, protein, and carbohydrate can be assayed by Fourier transform infrared (FTIR) spectroscopy as demonstrated for *Nannochloropsis* (Mayers et al. 2013) or by near-infrared (NIR) spectroscopy (Laurens and Wolfrum 2013).

4.3.3 HIGH-THROUGHPUT MEASUREMENT OF OIL CONTENT

Microalgae often accumulate lipid to high levels, usually as nonpolar glycerolipids in the form of triacylglycerol (TAG), and this constituent can account for up to 80% of total lipids (Hu et al. 2008). Neutral lipid can also be in the form of diacylglycerol (DAG) or monoacylglycerol (MAG) or in some species, other nonpolar lipids or hydrocarbons. Direct derivatization of glycerolipids followed by GC is frequently used and also the most effective way to extract fatty acid–derived lipids, of which there are several methods (reviewed in Laurens et al. 2012; Slocombe et al. 2013b). GC-FID analysis is the most quantitatively accurate measure of total glycerolipid levels. Comparative work has shown that unlike direct derivatization, solvent extraction methods often fail to extract lipids completely from microalgae such as *Chlorella* with recalcitrant cell walls (Laurens et al. 2012; Slocombe et al. 2013b).

Considering the range of fatty acids found in marine microalgae, tritricosanoic acid (23:0 TAG) stored in chloroform at 5 mg/mL is recommended as a standard.

In high-throughput screening, rapid measurements are needed and a comparison of potential methods is shown in Table 4.2. Although direct transesterification combined with GC is a standard rapid approach, sample preparation can still be relatively labor intensive (Table 4.2). Combustive elemental analysis by MS for carbon (C) and nitrogen (N) is a robust measure, which provides an indication of biomass value but can also give a measure of lipid content. This procedure requires minimal DW and handling time compared with the other methods and can be scaled to lower culture volumes than GC analysis or FTIR spectroscopy (Table 4.2). By comparing the elemental analysis and GC data from an algal screen, it was found that >90% of strains with high total fatty acid (TFA) (>30% DW) could be identified through >48% C and <3% N (Slocombe et al. 2015). The C/N ratio is often calculated in the microalgal field, but without the absolute data, this information is insufficient to assess oil content. High-C content and low-N content is indicative of abundant glycerolipids or hydrocarbons (75%–85% C) as opposed to proteins/amines (53% C, 16% N) or storage carbohydrates or cell wall (44% C).

Dye-based methods such as Nile Red and boron-dipyrromethene (BODIPY) that penetrate algal cells and interact with oil bodies such as Nile Red can be carried out rapidly by fluorescent plate assay for a semiquantitative measure of oil content (Chen et al. 2009). However, when comparing different strains as in screening, there are likely to be differences in dye uptake, fluorescence quenching, and cell size that will influence the data generated. These dyes have also been used with flow cytometry to carry out screens (Doan et al. 2011) (Table 4.2).

Probably the simplest method to identify high oil-content strains is to exploit their tendency to float rather than sediment with centrifugation and this has been noted for *Nannochloropsis* (Sheehan et al. 1998). This approach is likely to be influenced by the precise growth conditions (e.g., provision of CO_2 versus air, media composition, and density) and strain-specific factors such as cell structure. In the screen of the CCAP collection, as described in Section 4.3.1, few strains displayed this trait. Of the two strains with the highest oil content, each having 55% DW TFA, one of these, a *Nannochloropsis* sp., showed this characteristic whereas the second, a marine *Chlorella* strain, did not. This trait becomes more prevalent in both strains with continued incubation at stationary phase, with further oil accumulation (data not shown). Extended incubation at stationary phase in order for this effect to manifest itself in screens might not always be good practice, however, since oil can also be degraded over time and cultures can crash or become subject to bacterial contamination. This method could evidently be applied in screening for high-oil variants (or mutants) within the genus *Nannochloropsis*.

4.3.4 MINIATURIZATION IN SCREENING

Given the methods available for high-throughput analysis for oil content that use small amounts of culture or DW material, there is a strong case for miniaturizing the culturing process to improve processivity (Table 4.2). Rotary carousels holding test tubes have been used (Sheehan et al. 1998). Handling a large number of small cultures of different species simultaneously is challenging in terms of maintaining standards of culture maintenance and also in the monitoring of growth. Microalgae can sometimes be grown in conventional assay plates (either 96-well plates, or those with a lesser number of wells), but measures must be taken to ensure adequate mixing and gas exchange (Van Wagenen et al. 2014). There have been recent steps toward small culture volume multi-PBR modules; using a polymethyl methacrylate–molded system (Taleb et al. 2015) and 3D printing technology might allow further advances. In future, such systems might shift the preliminary focus of screens from product content measurement toward a more holistic approach including measurements of biomass productivity and max growth rates in response to changed parameters such as temperature, pH, [CO_2], light intensity etc. Further miniaturization of analysis down to the level of lab-on-a-chip is also beginning to be exploited in algal research and could be effective in screening with high processivity (Kim et al. 2014).

4.4 MARINE MICROALGAE FOR BIOFUELS

Developing large-scale production is a challenging necessity for synthesis of industrial feedstocks or biofuels from microalgal biomass. From a cost perspective, it is essential to maximize oil content and productivity in large-scale cultures as well as making improvements in ease of cultivation, harvesting, and oil extraction (Griffiths and Harrison 2009; Stephens et al. 2010; Borowitzka et al. 2012; Borowitzka 2013b). Hence, there is a good case for identifying the most productive oil-producing marine strains. In a screen of marine strains isolated from the coastal waters of Singapore, the majority of the top lipid-yielding strains were found to be *Nannochloropsis* (Doan et al. 2011). This was also found to be the case in the screen of marine strains from the CCAP collection (Slocombe et al. 2015). Within the *Nannochloropsis* genus, strains of *N. oceanica* species were found to have the highest lipid content and productivities (Vieler et al. 2012; Slocombe et al. 2015). The two highest lipid-producer strains found in the CCAP screen were *N. oceanica* CCAP 849/10 and a marine strain of *Chlorella vulgaris* CCAP 211/21A, with a total fatty acid (TFA) content of over 50% DW (Figure 4.2). Both strains warrant further testing in saline open-air ponds given the former was isolated from a fish hatchery and the latter belongs to a species that is already commercially exploited, albeit freshwater (Sandnes et al. 2006; Benemann 2013; Slocombe et al. 2015). Although the vast majority of the strains examined in the CCAP collection were originally collected from the natural environment, three isolated from mariculture facilities appeared in the top 10 most productive strains (Slocombe et al. 2015). This supports the notion that the commercial environment is likely to select robust and productive strains (Borowitzka 2013c). In terms of biomass productivity and yields as assessed by DW carbon content, the best marine strains from the CCAP collection were of broad phylogenetic origin. These included *N. oceanica* strain, a *Tetraselmis* sp. CCAP 66/60, and *Dunaliella polymorpha*, CCAP 19/14, the latter being the most productive strain for carbohydrate (Slocombe et al. 2015).

Timing of oil accumulation is also important when considering model strains. Among the higher oil-producing marine strains in the CCAP screen, the green algal genera (*Chlorella, Dunaliella, Chloroidium,* and *Tetraselmis*) and eustigmatophycean

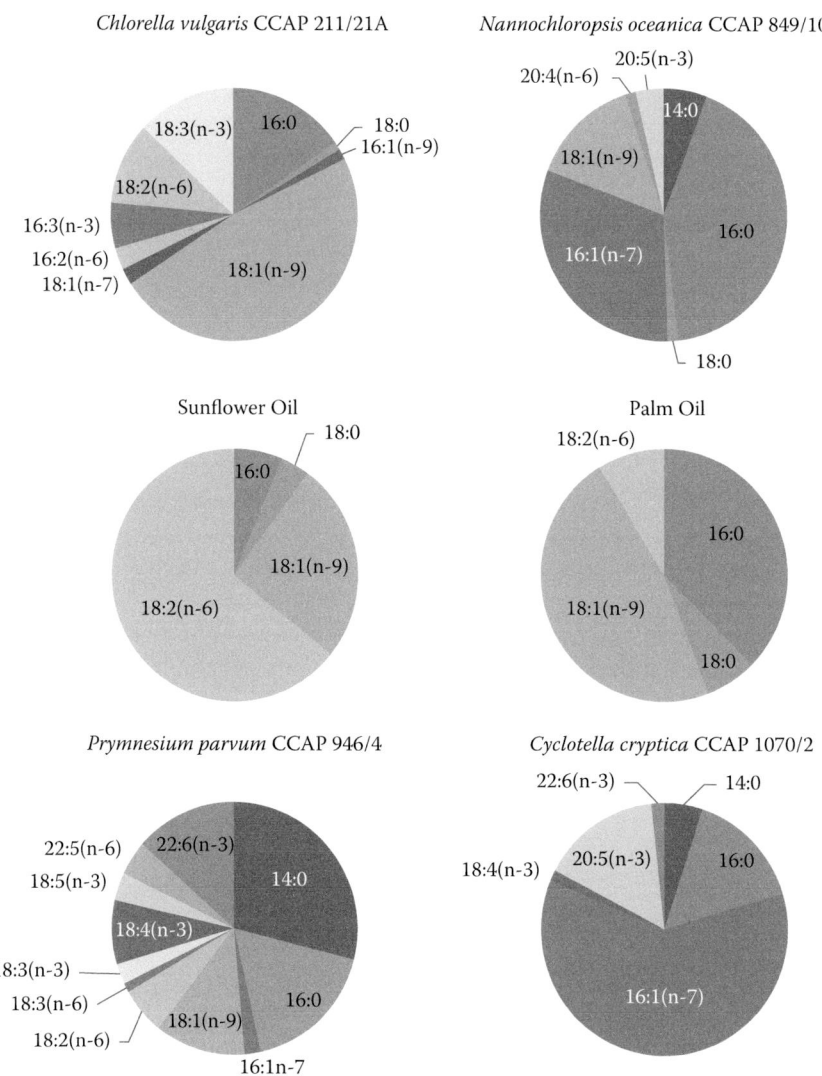

FIGURE 4.2 FA compositions are depicted for productive marine microalgal strains isolated from the screen of the CCAP collection in comparison with those of exemplary seed oils used for biofuel and comestibles. The microalgal data shown were measured at the stationary phase in a batch culture. The uppermost marine *Chlorella* and *Nannochloropsis* strains were the most productive for oil, also having high oil content (~55% TFA). *Prymnesium parvum* CCAP 946/4 was an example of a strain showing moderate oil (TFA) production combined with a relatively high C_{14} FA content but also with high DHA (22:6n–3). The *Cyclotella cryptica* strain was the most productive for the valuable FA EPA (20:5n–3). FA < 1% total are excluded from the pie charts, which were drawn from tabulated data from Ramos et al. (2009) for the terrestrial plant oils and from supplementary tabulated data from Slocombe et al. (2015) for microalgae.

Nannochloropsis strains tended to accumulate more oil in the stationary phase, whereas the haptophytes and diatoms did not show this degree of temporal variation (Slocombe et al. 2015). This can impact the choice of cultivation mode; the latter group might be more suited for continuous harvest.

The composition of algal oil also has implications for end use, be this for biofuels or industrial oils. In the case of biodiesel produced by the usual method of FA transesterification, FA composition can be important depending on the application. Polyunsaturated FAs (PUFAs) are considered to be less desirable in lipids destined for biodiesel due to poor oxidative stability (Knothe 2009; Ramos et al. 2009). Conversely, cold-flow issues associated with high saturate levels of medium-chain length and above ($C_{\geq 16}$) can also be problematic (e.g., palm oil) (Knothe 2009; Ramos et al. 2009). Properties at low temperatures limit the potential use of standard bio-diesels in aviation to <5% in blends (Wardle 2003). By comparison, biofuels rich in short-chain C_{10-14} methyl esters (ME) (e.g., from coconut oil) have been successfully tested in jet-fuel blends (Llamas et al. 2012). Although PUFA levels are, on average, high in marine algal oils compared with most current biodiesels, levels were shown to vary widely (Slocombe et al. 2015). The FA composition of *Nannochloropsis* appears to be reasonably favorable for biodiesel in having relatively low PUFA (at stationary phase) and high levels of 16:1n–7 combined with high 16:0, and some 14:0 (Figure 4.2). This compares with palm oil where 16:0 and 18:1n–9 are the two major FA (Knothe 2009; Ramos et al. 2009) (Figure 4.2). Several diatom and haptophyte species showed the capacity for producing shorter chain FA, having very high C_{14} saturate levels (<40% TFA), but this was mostly associated with high C_{20-22} PUFA levels and lower oil content and productivity (Slocombe et al. 2015). A haptophyte, *Prymnesium parvum* CCAP 946/4, had 28% 14:0 TFA with a moderate lipid content of 16% DW (Figure 4.2) (Slocombe et al. 2015). Overall, there is a great deal of diversity in fatty acid composition that may be exploited in further screens to find desirable fatty acid compositions to match desired biofuel properties. Ultimately, genetic engineering of FA composition of high oil-producing model strains may be required to achieve this objective.

Generating tailored biofuels by downstream processing of biomass constituents could be a valid alternative approach. Long-chain hydrocarbons can be amenable to the hydrocracking processes used in the petroleum industry. These are secreted from the colonial microalga *Botryococcus*, but these slow-growing freshwater species are difficult to cultivate (Stephens et al. 2010). Among the marine microalgae, certain haptophytes can synthesize C_{37-39} alkenes, alkenones, and alkenoates (Rieley et al. 1998; Eltgroth et al. 2005; Rontani et al. 2006). These long-chain molecules are characterized by unusually spaced double bonds that are particularly suited to catalytic cracking, for example, into jet-fuel range molecules by butenolysis (O'Neil et al. 2012, 2015). The biosynthetic pathways of these molecules are not fully understood, but a model based on malonyl/methyl-malonyl-SCoA C_2-chain elongation has been proposed (Rontani et al. 2006). The alkenones, like TAG, appear to accumulate in lipid bodies and act as a mobilizable store (Eltgroth et al. 2005). The biomass productivities of some of the relevant haptophyte strains (e.g., *Isochrysis* and *Chrysotila*) are good, and they are robust in cultivation (O'Neil et al. 2012; Slocombe et al. 2015).

4.5 HIGH-VALUE FATTY ACIDS FROM MARINE MICROALGAE

Marine microalgae have received much interest as a source of omega-3 very long-chained FA for use in dietary supplements or for sequestering in the food chain as feed for aquaculture or fisheries (Harwood and Guschina 2009; Borowitzka 2013b; Lenihan-Geels et al. 2013). The principal fatty acids of commercial value are the omega-3 FA's eicosapentanoic acid (EPA or 20:5n–3) and docosahexaenoic acid (DHA or 22:6n–3). Although not strictly "essential fatty acids" (such as linoleic acid (LA) and α-linolenic acid (ALA)), they have many reported dietary health benefits (Horrocks and Yeo 1999; Lenihan-Geels et al. 2013; Molfino et al. 2014). Their precursors such as stearidonic acid (SDA or 18:4n–3) can also be beneficial in human metabolism, where conversion of ALA to SDA is a bottleneck in EPA/DHA production (Lenihan-Geels et al. 2013). Apart from the insufficiency of EPA and DHA and overconsumption of fat, there is some indication in the literature that low overall omega-3 to omega-6 dietary ratio is detrimental to health (Han et al. 2012; Lenihan-Geels et al. 2013). This ratio is thought to influence relative flux into omega-3 or omega-6 FA precursors for pro- and anti-inflammatory agonists, respectively, where the latter might promote cardiovascular disease (CVD) (Lenihan-Geels et al. 2013; Gautam et al. 2014). Consequently, the levels of nonpremium value omega-3 FAs such as α-linolenic acid (ALA) that feature in marine green algae but are very low in some vegetable oils also need to be considered (Figure 4.2). Since *Chlorella* is already cultivated for health products, the productive halotolerant *C. vulgaris* strain (CCAP 211/21A) identified in the CCAP screen might have potential in niche market "pro-health" vegetable oils (Figure 4.2) (Slocombe et al. 2015).

Levels of FA unsaturation in marine microalgae are relatively high compared with the oilseeds of terrestrial plants that provide the bulk of biodiesels and comestible oil (Ramos et al. 2009; Slocombe et al. 2015). Despite this observation, when comparing species and strains, an inverse relationship is evident between PUFA content and TFA content in biomass. This has been attributed to flux competition for pathway intermediates in FA biosynthesis between the desaturase pathways and acyltransferases for TAG production (Roessler 1990; Slocombe et al. 2015). The tendency to divert saturated and monounsaturated FA into TAG could be an obstacle in maximizing both the desired omega-3 FA content in oil and ultimately productivity (Roessler 1990). On the other hand, the amount of these PUFAs that become incorporated into TAG varies widely and this diversity could be exploited in screening (Tonon et al. 2002). For instance, heterotrophic marine thraustrochytrids can accumulate particularly high levels of DHA, and in *Schizochytrium*, a distinct pathway for DHA synthesis by polyketide synthases (PKS) has been acquired from bacteria (Guschina and Harwood 2006). Out of a screen of 175 photoautotrophic strains from the CCAP collection, a *Cyclotella cryptica* strain was identified that matched the productivity levels for EPA of known benchmark strains that were already exploited in aquaculture/mariculture (Figure 4.2) (Slocombe et al. 2015). Therefore, upscaling screening could be enough to identify novel and possibly superior strains for high-value omega-3 FA, although some development at the genetic level may be required.

4.6 FURTHER IMPROVEMENT OF MODEL STRAINS

Once model strains have been selected by screening, optimization of the growth conditions to increase productivity might be considered along with further strain improvement through artificial selection, breeding, conventional mutagenesis, or genetic engineering. Forward genetic screening of mutagenized or random gene-tagged libraries could then be carried out using the same screening methods as described earlier.

4.6.1 MUTAGENESIS IN MICROALGAE

Mutagenesis procedures can generate sources of genetic variation within a chosen model strain that might then be screened for increased levels of oil content or secondary products such as carotenoids. Furthermore, modification of PUFA composition has been achieved in *Nannochloropsis* using mutagenesis, then screening by GC (Schneider et al. 1995). Mutagenesis in microalgae has been carried out using UV or gamma irradiation or chemical mutagens (Schneider et al. 1995; Ishikawa et al. 2004; Vigeolas et al. 2012; Cazzaniga et al. 2014; Nakanishi and Deuchi 2014). Important considerations include the ploidy level of the alga. Algae only observed in the haploid phase such as *Nannochloropsis* or *Chlorella* will reveal a phenotypic effect immediately, whereas species diploid in the vegetative phase (e.g., diatoms) may require recombination via sexual reproduction. Correctly assigning a phenotype to a given mutation can be problematic if many other mutations are present without a readily applicable sexual cycle that might be exploited to segregate these out (Cazzaniga et al. 2014). However, high mutational frequencies are often utilized to minimize the screening effort to find mutants (Schneider et al. 1995).

It is unclear whether or not sexual reproduction occurs in *Nannochloropsis* (Pan et al. 2011) although it might occur cryptically in *Chlorella* sp., as shown in *Ostreococcus* (Blanc et al. 2010). It is a necessary part of the life cycle in most diatoms (Chepurnov et al. 2004) and this is also the case in certain chlorophytes, such as *Chlamydomonas* and *Volvox* (Frenkel et al. 2014). Haplodiploidy is associated with heteromorphic life cycles seen in certain coccolithophore species and has been speculated to be ubiquitous in the haptophytes (Houdan et al. 2004).

Another important practical consideration is whether or not the alga can produce discrete colonies on agar plates. This is the case for many green algae (e.g., *Chlorella*, *Scenedesmus*, *Chlamydomonas*), eustigmatophytes (*Nannochloropsis*), and some diatoms (e.g., *Phaeodactylum tricornutum*), whereas others do not grow on agar plates (e.g., haptophyte *Isochrysis galbana*)—or do so but are motile and cannot form colonies (e.g., benthic diatoms such as *Cylindrotheca*). The option of employing pour plates can overcome this and discrete colonies are readily obtained, but subsequent manipulation of individual colonies without cross-contamination is extremely challenging.

Screening of mutagenized algae is readily performed on agar plates. The simplest screen to apply is positive selection, such as resistance to a treatment that is normally inhibitory to growth such as high salinity, temperature, or an herbicide, or the enhanced production of a colored compound such as carotenoid or chlorophyll (Ishikawa et al. 2004; Nakanishi and Deuchi 2014). Negative selection, that is, enhanced sensitivity to a condition or a reagent generally requires replica plating so susceptible strains can be rescued (also needed where screening destroys the colony, e.g., iodine staining). Such approaches might include the failure to breakdown oil reserves, a desirable trait that could maximize oil content (Li et al. 2012). An alternative to identifying high-oil strains is to locate starch-deficient mutants, among the chlorophyte algae, using techniques such as iodine staining (Work et al. 2010; Vigeolas et al. 2012).

4.6.2 Genetic Transformation and Engineering

Genetic variation for screening can also be generated by gene transformation. For instance, random libraries of knocked out genes can be produced following gene-tagging strategies and screened in similar ways as mutagenized libraries. Reverse genetic approaches might also be taken to knockout gene candidates, chosen according to their gene function, which might improve oil content or productivities. An alternative is "tilling" for mutations in the gene of interest in mutated libraries of microalgae using a PCR-based approach, which avoids generation of GMOs (McCallum et al. 2000). Genome editing, or genome rewriting and synthetic biology strategies might also be employed to redirect metabolic pathways or add novel regulatory networks and even to improve GMO containment (Urnov et al. 2010; Burgess 2013; Hwang et al. 2013) (described in more detail in Chapter 6).

Many microalgae appear to be amenable to transformation by electroporation, or other relatively fast methods such as glass-bead abrasion (e.g., *Chlamydomonas*) (Neupert et al. 2012). Transformation of marine strains by electroporation was pioneered in *Nannochloropsis* and has been shown to work well with a number of species/strains. Site-specific recombination can also be exploited for creating targeted gene knockouts in this genus (Kilian et al. 2011; Radakovits et al. 2012; Vieler et al. 2012). Within *Nannochloropsis*, transformation frequency and susceptibility to antibiotics required for transformant selection varies between species and might be a function of cell wall properties (Vieler et al. 2012). In diatoms, cell structure and the presence of silica might impede transformation, which might necessitate ballistic approaches, for example, *Thalassiosira* (Poulsen et al. 2006). Finally, in some green algae agrobacterium-mediated transformation has been reported (Sørensen et al. 2014).

4.6.3 Strategies for Increasing Oil Production

A good understanding of oil production in microalgae is required to take full advantage of the molecular tools that are becoming available to improve strains, and this requires continued research. Oil productivity is invariably dependent on a trade-off between growth and allocation of resources elsewhere, such as into oil (Beardall and Raven 2013). The key is to achieve this balance and channel more nongrowth resources into oil. Although some microalgae accumulate oil during optimal growth, many do so with the onset of stresses such as nutrient depletion (e.g., N-depletion) that are typically associated with stationary phase in batch culture (Hu et al. 2008).

In addition to *de novo* synthesis of fatty acids from fixed CO_2, there appears to be resource remobilization toward fatty acid synthesis (e.g., from carbohydrate, protein catabolism) and also recycling of membrane fatty acids into storage lipid (Burrows et al. 2012; Hockin et al. 2012; Valenzuela et al. 2012; Yoon et al. 2012; Dong et al. 2013). In *Nannochloropsis*, recycling of EPA from plastid galactolipids into TAG appears later in response to N-deprivation than the appearance of TAG containing solely C_{16-18} FA, although much of this EPA appears to be catabolized (Li et al. 2014; Martin et al. 2014). In fact with prolonged N-deprivation, there is a downregulation of photosynthesis, deconstruction of the plastid thylakoid during N-deprivation, and autophagy of the plastid (Dong et al. 2013). Photosynthesis is highly dependent on nitrogen for protein and chlorophyll; therefore, when N supply is diminished, reducing the photosynthetic apparatus can provide nitrogen for recycling (Valenzuela et al. 2012; Dong et al. 2013). This strategy of recycling and storage lipid accumulation allows remobilization of the photosynthetic apparatus when nutrient resources return to the environment. Some facets of the survival strategy will not necessarily be relevant to increased oil generation in an industrial setting, however. For instance, oil remobilization or cell wall thickening could be suppressed genetically with the aim of putting a greater proportion of reserves into oil (Wang et al. 2005; Tsai et al. 2014). A better understanding of the cell wall composition of model marine strains such as *Nannochloropsis* could also enable improvements in product extraction and in minimizing resource allocation away from oil (Scholz et al. 2014).

Gains in oil content and production levels can be made in theory by gene alterations that (1) increase FA flux into oil or (2) decrease FA mobilization from TAG or FA breakdown by beta-oxidation. In the first case, increasing expression of key regulatory steps in fatty acid synthesis (e.g., ACC) has met with limited success and may be a consequence of the already high flux into membrane fatty acids associated with optimal growth or a need to increase other FAS steps concomitantly (Roessler et al. 1994). In fact, it has been suggested in the marine diatom *P. tricornutum* that under nutrient limitation carbon might be "pushed" into fatty acid synthesis, and hence into oil, by a surfeit of metabolites derived from photosynthesis, such as acetyl CoA and NADPH, rather than "pulled" by increases in fatty acid synthesis capacity (Valenzuela et al. 2012). Another target for gene alteration is increasing levels of the various acyltransferases that are known to produce the end product TAG (Roessler et al. 1994). Increases in these enzymes (e.g., DGAT, DGGT, and PDAT in *Chlamydomonas reinhardtii*, and several DGAT genes implicated in *Nannochloropsis*) during TAG accumulation induced by stresses such as N-deprivation have been reported (Boyle et al. 2012; Blaby et al. 2013; Li et al. 2014). The latter enzyme, PDAT, is also thought to play a role in membrane lipid turnover and remodeling, which is required to maintain membrane structure and mobility. Interestingly, it appears to play a greater role in TAG accumulation during logarithmic growth in *C. reinhardtii* (Yoon et al. 2012). An alternative to attempting to increase flux in to oil directly is to shutdown other routes for partitioning of photosynthetic carbon such as starch in green algae (Li et al. 2010; Work et al. 2010; Jaeger et al. 2014). Many marine strains such as *Nannochloropsis* produce beta-glucans in place of starch as carbohydrate reserves, and pathway gene candidates have been identified in this species (Volkman et al. 1993; Li et al. 2014).

As described earlier, instead of increasing oil biosynthesis, an alternative is to decrease mobilization of storage lipid, for instance, by targeting lipases involved in

TAG turnover such as CrLIP1, which is downregulated during TAG accumulation in *C. reinhardtii* (Li et al. 2012). In plants, mutation in the beta-oxidation pathway and peroxisomal fatty acid transport genes gives rise to a block in breakdown of fatty acids from membrane lipid turnover and TAG lipolysis, causing accumulation of oil in vegetative tissues, particularly during senescence (Slocombe et al. 2009). Disruption of CGI-58 type genes involved in glycerolipid and energy maintenance in eukaryotic cells is associated with ectopic production of lipid droplets in mammals and plants. This multifunctional protein, which possesses hydrolase and acyltransferase activities, interacts with lipid droplets and peroxisomal ABC fatty acid transporters associated with beta-oxidation (Park et al. 2013). In the marine diatom *Thalassiosira pseudonana*, knockdown of this gene has led to increases in logarithmic and stationary phase TAG levels, without apparently compromising growth rates (Trentacoste et al. 2013). Regulatory factors such as transcription factor CHT7 appear to control cellular quiescence in relation to nutrient status in *C. reinhardtii*. Knockdown of this gene results in a slowdown in TAG reserve mobilization in response nutrient resupply and could thus lead to increased yields (Tsai et al. 2014).

4.7 CONCLUSIONS

Screening and strain selection for enhanced oil content and productivity will continue to play an important role in improving productivity and specific compositional qualities for biofuels and value products. There is a great deal of phylogenetic variation among unicellular algae in the natural environment, in collections, and also in man-made environments or commercial settings where selection of suitable microalgae might have occurred. High-throughput methodology is already available, but could benefit from further developments in miniaturization at the cultivation level. This in turn could lead to a more "holistic" approach that will examine parameters of strain performance and oil production for modeling growth at large scale (see Chapter 5). Selection of model strains is likely to be followed by further improvement based on mutagenesis and further rounds of screening or by genetic engineering (described in more detail in Chapter 6). A greater understanding of the regulatory mechanisms of oil production is already enabling the improvement of strains based on the specific targeting of genes for modification. Taken together, improved oil productivity strains are likely to require multiple gene modifications, with proven long-term genetic and functional stability (Stacey and Day 2014) to bring about an increase in cost effectiveness and viability in industrial applications.

ACKNOWLEDGMENTS

This work was carried out as part of the BioMara project (www.biomara.ac.uk), receiving support from the European Regional Development Fund through the INTERREG IVA Programme, Highlands and Islands Enterprise, Crown Estate, Northern Ireland Executive, Scottish Government, and Irish Government, also National Capability funding for the Culture Collection for Algae and Protozoa (CCAP) from the UK Natural Environment Research Council (NERC).

REFERENCES

Beardall, J. and J. Raven. 2013. Limits to phototrophic growth in dense culture: CO_2 supply and light. In *Algae for Biofuels and Energy* SE-5, eds. M. A. Borowitzka and N. R. Moheimani, Vol. 5: Developments in Applied Phycology, pp. 91–97. Dordrecht, the Netherlands: Springer.

Benemann, J. 2013. Microalgae for biofuels and animal feeds. *Energies* 6:5869–5886.

Blaby, I. K., A. G. Glaesener, T. Mettler et al. 2013. Systems-level analysis of nitrogen starvation-induced modifications of carbon metabolism in a *Chlamydomonas reinhardtii* Starchless Mutant. *Plant Cell* 25:4305–4323.

Blanc, G., G. Duncan, I. Agarkova et al. 2010. The *Chlorella variabilis* NC64A genome reveals adaptation to photosymbiosis, coevolution with viruses, and cryptic sex. *Plant Cell* 22:2943–2955.

Borowitzka, M. A. 2013a. Energy from microalgae: A short history. In *Algae for Biofuels and Energy* SE-1, eds. M. A. Borowitzka and N. R. Moheimani, Vol. 5: Developments in Applied Phycology, pp. 1–15. Dordrecht, the Netherlands: Springer.

Borowitzka, M. A. 2013b. High-value products from microalgae—Their development and commercialisation. *J. Appl. Phycol.* 25:743–756.

Borowitzka, M. A. 2013c. Species and strain selection. In *Algae for Biofuels and Energy* SE-4, eds. M. A. Borowitzka and N. R. Moheimani, Vol. 5: Developments in Applied Phycology, pp. 77–89. Dordrecht, the Netherlands: Springer.

Borowitzka, M. A., B. J. Boruff, N. R. Moheimani, N. Pauli, Y. Cao, and H. Smith. 2012. Identification of the optimum sites for industrial-scale microalgae biofuel production in WA using a GIS model. http://www.murdoch.edu.au/_document/News/CRST-AlgaeBiofuelsGIS-FinalReportt.pdf (accessed December 22, 2015).

Boyle, N. R., M. D. Page, B. Liu et al. 2012. Three acyltransferases and nitrogen-responsive regulator are implicated in nitrogen starvation-induced triacylglycerol accumulation in *Chlamydomonas. J. Biol. Chem.* 287:15811–15825.

Burgess, D. J. 2013. Technology: A CRISPR genome-editing tool. *Nat. Rev. Genet.* 14:80.

Burlew, J. 1953. *Algal Culture from Laboratory to Pilot Plant.* Washington, DC: Carnegie Institution of Washington.

Burrows, E. H., N. B. Bennette, D. Carrieri et al. 2012. Dynamics of lipid biosynthesis and redistribution in the marine diatom *Phaeodactylum tricornutum* under nitrate deprivation. *Bioenerg. Res.* 5:876–885.

Cazzaniga, S., L. Dall'Osto, J. Szaub et al. 2014. Domestication of the green alga *Chlorella sorokiniana*: Reduction of antenna size improves light-use efficiency in a photobioreactor. *Biotechnol. Biofuels* 7:1–13.

Chen, W., C. Zhang, L. Song, M. Sommerfeld, and Q. Hu. 2009. A high throughput Nile Red method for quantitative measurement of neutral lipids in microalgae. *J. Microbiol. Met.* 77:41–47.

Chen, Y. and S. Vaidyanathan. 2012. A simple, reproducible and sensitive spectrophotometric method to estimate microalgal lipids. *Anal. Chim. Acta* 724:67–72.

Chen, Y. and S. Vaidyanathan. 2013. Simultaneous assay of pigments, carbohydrates, proteins and lipids in microalgae. *Anal. Chim. Acta* 776:31–40.

Chepurnov, V. A., D. G. Mann, K. Sabbe, and W. Vyverman. 2004. Experimental studies on sexual reproduction in diatoms. *Int. Rev. Cytol.* 237:91–154.

Day, J. G., S. P. Slocombe, and M. S. Stanley. 2012. Overcoming biological constraints to enable the exploitation of microalgae for biofuels. *Bioresour. Technol.* 109:245–51.

Doan, T. T. Y., B. Sivaloganathan, and J. P. Obbard. 2011. Screening of marine microalgae for biodiesel feedstock. *Biomass Bioenerg.* 35:2534–2544.

Dong, H.-P., E. Williams, D. Wang et al. 2013. Responses of *Nannochloropsis oceanica* IMET1 to long-term nitrogen starvation and recovery. *Plant Physiol.* 162:1110–1126.

Eltgroth, M. L., R. L. Watwood, and G. V. Wolfe. 2005. Production and cellular localization of neutral long-chain lipids in the Haptophyte algae *Isochrysis galbana* and *Emiliania huxleyi*. *J. Phycol.* 41:1000–1009.

Frenkel, J., W. Vyverman, and G. Pohnert. 2014. Pheromone signaling during sexual reproduction in algae. *Plant J.* 79:632–644.

Gachon, C. M. M., S. Heesch, F. C. Küpper et al. 2013. The CCAP knowledgebase: Linking protistan and cyanobacterial biological resources with taxonomic and molecular data. *Syst. Biodivers.* 11:407–413.

Gautam, M., A. Izawa, Y. Shiba et al. 2014. Importance of fatty acid compositions in patients with peripheral arterial disease. *PLoS One* 9:e107003.

Giordano, M., J. Beardall, and J. A. Raven. 2005. CO_2 Concentrating mechanisms in algae: Mechanisms, environmental modulation, and evolution. *Annu. Rev. Plant Biol.* 56:99–131.

Griffiths, M. J. and S. T. L. Harrison. 2009. Lipid productivity as a key characteristic for choosing algal species for biodiesel production. *J. Appl. Phycol.* 21:493–507.

Guiry, M. D. 2012. How many species of algae are there? *J. Phycol.* 48:1057–1063.

Guschina, I. and J. L. Harwood. 2006. Lipids and lipid metabolism in eukaryotic algae. *Prog. Lipid Res.* 45:160–186.

Han, S. N., A. H. Lichtenstein, L. M. Ausman, and S. N. Meydani. 2012. Novel soybean oils differing in fatty acid composition alter immune functions of moderately hypercholesterolemic older adults 1–3. *J. Nutr.* 142:2182–2187.

Harder, R. and H. V. Witsch. 1942. Über massenkultur von diatomeen. *Ber. Deut. Bot. Ges.* 60:146–152.

Harwood, J. L. and I. A. Guschina. 2009. The versatility of algae and their lipid metabolism. *Biochimie* 91:679–684.

Hockin, N. L., T. Mock, F. Mulholland, S. Kopriva, and G. Malin. 2012. The response of diatom central carbon metabolism to nitrogen starvation is different from that of green algae and higher plants. *Plant Physiol.* 158:299–312.

Horrocks, L. A. and Y. K. Yeo. 1999. Health benefits of docosahexaenoic acid (DHA). *Pharmacol. Res.* 40:211–225.

Houdan, A., C. Billard, D. Marie et al. 2004. Holococcolithophore-heterococcolithophore (Haptophyta) life cycles: Flow cytometric analysis of relative ploidy levels. *Syst. Biodivers.* 1:453–465.

Hu, Q., M. Sommerfeld, E. Jarvis et al. 2008. Microalgal triacylglycerols as feedstocks for biofuel production: Perspectives and advances. *Plant J.* 54:621–639.

Huerlimann, R., R. de Nys, and K. Heimann. 2010. Growth, lipid content, productivity, and fatty acid composition of tropical microalgae for scale-up production. *Biotechnol. Bioeng.* 107:245–257.

Hwang, W. Y., Y. Fu, D. Reyon et al. 2013. Efficient genome editing in zebrafish using a CRISPR-Cas system. *Nat. Biotechnol.* 31:227–229.

Ishikawa, E., H. Sansawa, and H. Abe. 2004. Isolation and characterization of a *Chlorella* mutant producing high amounts of chlorophyll and carotenoids. *J. Appl. Phycol.* 16:385–393.

Jaeger, L. D., R. E. M. Verbeek, R. B. Draaisma, D. E. Martens, J. Springer, and G. Eggink. 2014. Superior triacylglycerol (TAG) accumulation in starchless mutants of *Scenedesmus obliquus*: (I) mutant generation and characterization. *Biotechnol. Biofuels* 7:69.

Kilian, O., C. S. E. Benemann, K. K. Niyogi, and B. Vick. 2011. From the cover: High-efficiency homologous recombination in the oil-producing alga *Nannochloropsis* sp. *Proc. Natl. Acad. Sci. USA* 108:21265–21269.

Kim, H. S., T. L. Weiss, H. R. Thapa, T. P. Devarenne, and A. Han. 2014. A microfluidic photobioreactor array demonstrating high-throughput screening for microalgal oil production. *Lab Chip* 14:1415–1425. http://www.ncbi.nlm.nih.gov/pubmed/24496295.

Knoshaug, E. P., R. Sestric, E. Jarvis, Y.-C. Chou, P. T. Pienkos, and A. Darzins. 2009. Current status of the department of energy's aquatic species program lipid-focused algae collection. Poster presented at the *31st Symposium on Biotechnology for Fuels and Chemicals*, San Francisco, CA. http://www.nrel.gov/docs/fy09osti/45788.pdf (accessed December 21, 2015).

Knothe, G. 2009. Improving biodiesel fuel properties by modifying fatty ester composition. *Energy Environ. Sci.* 2:759–766.

Lang, I., L. Hodac, T. Friedl, and I. Feussner. 2011. Fatty acid profiles and their distribution patterns in microalgae: A comprehensive analysis of more than 2000 strains from the SAG culture collection. *BMC Plant Biol.* 11:124.

Laurens, L. M. L., M. Quinn, S. Van Wychen, D. W. Templeton, and E. J. Wolfrum. 2012. Accurate and reliable quantification of total microalgal fuel potential as fatty acid methyl esters by *in situ* transesterification. *Anal. Bioanal. Chem.* 403:167–178.

Laurens, L. M. L. and E. J. Wolfrum. 2013. High-throughput quantitative biochemical characterization of algal biomass by NIR spectroscopy; multiple linear regression and multivariate linear regression analysis. *J. Agric. Food Chem.* 61:12307–12314.

Lenihan-Geels, G., K. S. Bishop, and L. R. Ferguson. 2013. Alternative sources of omega-3 fats: Can we find a sustainable substitute for fish? *Nutrients* 5:1301–1315.

Leonardos, N. and R. J. Geider. 2004. Responses of elemental and biochemical composition of *Chaetoceros muelleri* to growth under varying light and nitrate:phosphate supply ratios and their influence on critical N:P. *Limnol. Oceanogr.* 49:2105–2114.

Li, J., D. Han, D. Wang et al. 2014. Choreography of transcriptomes and lipidomes of *Nannochloropsis* reveals the mechanisms of oil synthesis in microalgae. *Plant Cell* 26:1645–1665.

Li, X., C. Benning, and M. H. Kuo. 2012. Rapid triacylglycerol turnover in *Chlamydomonas reinhardtii* requires a lipase with broad substrate specificity. *Eukaryotic Cell* 11:1451–1462.

Li, Y., D. Han, G. Hu et al. 2010. *Chlamydomonas* Starchless Mutant defective in ADP-glucose pyrophosphorylase hyper-accumulates triacylglycerol. *Metab. Eng.* 12:387–391.

Llamas, A., M. García-Martínez, A.-M. Al-Lal, L. Canoira, and M. Lapuerta. 2012. Biokerosene from coconut and palm kernel oils: Production and properties of their blends with fossil kerosene. *Fuel* 102:483–490.

MacDougall, K. M., J. McNichol, P. J. McGinn, S. J. B. O'Leary, and J. E. Melanson. 2011. Triacylglycerol profiling of microalgae strains for biofuel feedstock by liquid chromatography-high-resolution mass spectrometry. *Anal. Bioanal. Chem.* 401:2609–2616.

Martin, G. J. O., D. R. Hill, I. L. D. Olmstead et al. 2014. Lipid profile remodelling in response to nitrogen deprivation in the microalgae *Chlorella* sp. (Trebouxiophyceae) and *Nannochloropsis* sp. (Eustigmatophyceae). *PLoS One* 9:e103389.

Mayers, J. J., K. J. Flynn, and R. J. Shields. 2013. Rapid determination of bulk microalgal biochemical composition by Fourier-Transform Infrared Spectroscopy. *Bioresour. Technol.* 148:215–220.

McCallum, C. M., L. Comai, E. A. Greene, and S. Henikoff. 2000. Targeted screening for induced mutations. *Nat. Biotechnol.* 18:455–457.

Moheimani, N., M. Borowitzka, A. Isdepsky, and S. Sing. 2013. Standard methods for measuring growth of algae and their composition. In *Algae for Biofuels and Energy* SE-16, eds. M. A. Borowitzka and N. R. Moheimani, Vol. 5: Developments in Applied Phycology, pp. 265–284. Dordrecht, the Netherlands: Springer.

Molfino, A., G. Gioia, F. R. Fanelli, and M. Muscaritoli. 2014. The role for dietary omega-3 fatty acids supplementation in older adults. *Nutrients* 6:4058–4072.

Nakanishi, K. and K. Deuchi. 2014. Culture of a high-chlorophyll-producing and halotolerant *Chlorella vulgaris*. *J. Biosci. Bioeng.* 117:617–619.

Neupert, J., N. Shao, Y. Lu, and R. Bock. 2012. Genetic transformation of the model green alga *Chlamydomonas reinhardtii*. *Methods Mol. Biol.* 847:35–47.

O'Neil, G. W., C. A. Carmichael, T. J. Goepfert et al. 2012. Beyond fatty acid methyl esters: Expanding the renewable carbon profile with alkenones from *Isochrysis* sp. *Energy Fuels* 26:2434–2441.

O'Neil, G. W., A. R. Culler, J. R. Williams et al. 2015. Production of jet fuel range hydrocarbons as a coproduct of algal biodiesel by butenolysis of long-chain alkenones. *Energy Fuels* 29:992–930.

Pan, K., J. Qin, S. Li et al. 2011. Nuclear monoploidy and asexual propagation of *Nannochloropsis oceanica* (Eustigmatophyceae) as revealed by its genome sequence. *J. Phycol.* 47:1425–1432.

Park, S., S. K. Gidda, C. N. James et al. 2013. The α/β hydrolase CGI-58 and peroxisomal transport protein PXA1 coregulate lipid homeostasis and signaling in arabidopsis. *Plant Cell* 25:1726–1739.

Parker, M. S., T. Mock, and E. V. Armbrust. 2008. Genomic insights into marine microalgae. *Annu. Rev. Genet.* 42:619–645.

Poulsen, N., P. M. Chesley, and N. Kröger. 2006. Molecular genetic manipulation of the diatom *Thalassiosira pseudonana* (Bacillariophyceae). *J. Phycol.* 42:1059–1065.

Radakovits, R., R. E. Jinkerson, S. I. Fuerstenberg, H. Tae, R. E. Settlage, J. L. Boore, and M. C. Posewitz. 2012. Draft genome sequence and genetic transformation of the oleaginous alga *Nannochloropsis gaditana*. *Nat. Commun.* 3:686.

Ramos, M. J., C. M. Fernández, A. Casas, L. Rodríguez, and A. Pérez. 2009. Influence of fatty acid composition of raw materials on biodiesel properties. *Bioresour. Technol.* 100:261–268.

Rieley, G., M. A. Teece, A. M. R. Torren et al. 1998. Long-chain alkenes of the Haptophytes *Isochrysis galbana* and *Emiliania huxleyi*. *Lipids* 33:617–625.

Rodolfi, L., G. Chini Zittelli, N. Bassi et al. 2009. Microalgae for oil: Strain selection, induction of lipid synthesis and outdoor mass cultivation in a low-cost photobioreactor. *Biotechnol. Bioeng.* 102:100–112.

Roessler, P. G. 1990. Environment control of glycerolipid metabolism in microalgae: Commercial implications and future research directions. *J. Phycol.* 26:393–399.

Roessler, P. G., L. M. Brown, T. G. Dunahay et al. 1994. Genetic engineering approaches for enhanced production of biodiesel fuel from microalgae. In *Enzymatic Conversion of Biomass for Fuels Production*, eds. M. E. Himmel, J. O. Baker, and R. P. Overend, Vol. 566, pp. 255–270. New York: American Chemical Society.

Rontani, J.-F., F. G. Prahl, and J. K. Volkman. 2006. Re-examination of the double bond positions in alkenones and derivatives: Biosynthetic implications. *J. Phycol.* 42:800–813.

Sandnes, J. M., T. Ringstad, D. Wenner, P. H. Heyerdahl, T. Källqvist, and H. R. Gislerød. 2006. Real-time monitoring and automatic density control of large-scale microalgal cultures using near Infrared (NIR) optical density sensors. *J. Biotechnol.* 122:209–215.

Scherholz, M. L. and W. R. Curtis. 2013. Achieving pH control in microalgal cultures through fed-batch addition of stoichiometrically balanced growth media. *BMC Biotechnol.* 13:39.

Schneider, J. C., A. Livne, A. Sukenik, and P. G. Roessler. 1995. A mutant of *Nannochloropsis* deficient in eicosapentaenoic acid production. *Phytochemistry* 40:807–814.

Scholz, M. J., T. L. Weiss, R. E. Jinkerson et al. 2014. Ultrastructure and composition of the *Nannochloropsis gaditana* cell wall. *Eukaryot. Cell.* 13:1450–1464.

Sheehan, J., T. Dunahay, J. Benemann, and P. Roessler. 1998. A look back at the U.S. Department of Energy's Aquatic Species Program: Biodiesel from algae. Close-Out Report. Golden, CO. http://www.nrel.gov/biomass/pdfs/24190.pdf (accessed December 21, 2015).

Slocombe, S. P., J. Cornah, H. Pinfield-Wells et al. 2009. Oil accumulation in leaves directed by modification of fatty acid breakdown and lipid synthesis pathways. *Plant Biotechnol. J.* 7:694–703.

Slocombe, S. P., M. Ross, N. Thomas, S. McNeill, and M. S. Stanley. 2013a. A rapid and general method for measurement of protein in micro-algal biomass. *Bioresour. Technol.* 129:51–57.

Slocombe, S. P., Q. Zhang, K. D. Black, J. G. Day, and M. S. Stanley. 2013b. Comparison of screening methods for high-throughput determination of oil yields in micro-algal biofuel strains. *J. Appl. Phycol.* 25:961–972.

Slocombe, S. P., Q. Zhang, M. Ross et al. 2015. Unlocking nature's treasure-chest: Screening for oleaginous algae. *Sci. Rep.* 5:09844.

Sørensen, I., Z. Fei, A. Andreas, W. G. T. Willats, D. S. Domozych, and J. K. C. Rose. 2014. Stable transformation and reverse genetic analysis of *Penium margaritaceum*: A platform for studies of charophyte green algae, the immediate ancestors of land plants. *Plant J.* 77:339–351.

Stacey, G. N. and Day, J. G. 2014. Putting cells to sleep for future science. *Nat. Biotechnol.* 32:320–322.

Stephens, E., I. L. Ross, J. H. Mussgnug et al. 2010. Future prospects of microalgal biofuel production systems. *Trends Plant Sci.* 15:554–564.

Taleb, A., J. Pruvost, J. Legrand et al. 2015. Development and validation of a screening procedure of microalgae for biodiesel production: Application to the genus of marine microalgae *Nannochloropsis*. *Bioresour. Technol.* 177:224–232.

Tonon, T., D. Harvey, T. R. Larson, and I. A. Graham. 2002. Long chain polyunsaturated fatty acid production and partitioning to triacylglycerols in four microalgae. *Phytochemistry* 61:15–24.

Trentacoste, E. M., R. P. Shrestha, S. R. Smith et al. 2013. Metabolic engineering of lipid catabolism increases microalgal lipid accumulation without compromising growth. *Proc. Natl. Acad. Sci. USA* 110:19748–19753.

Tsai, C.-H., J. Warakanont, T. Takeuchi, B. B. Sears, E. R. Moellering, and C. Benning. 2014. The protein compromised hydrolysis of triacylglycerols 7 (CHT7) acts as a repressor of cellular quiescence in *Chlamydomonas*. *Proc. Natl. Acad. Sci. USA* 111:15833–15838.

Urnov, F. D., E. J. Rebar, M. C. Holmes, H. S. Zhang, and P. D. Gregory. 2010. Genome editing with engineered zinc finger nucleases. *Nat. Rev. Genet.* 11:636–646.

Valenzuela, J., A. Mazurie, R. P. Carlson et al. 2012. Potential role of multiple carbon fixation pathways during lipid accumulation in *Phaeodactylum tricornutum*. *Biotechnol. Biofuels* 5:40.

Van Wagenen, J., S. L. Holdt, D. De Francisci, B. Valverde-Pérez, B. G. Plósz, and I. Angelidaki. 2014. Microplate-based method for high-throughput screening of microalgae growth potential. *Bioresour. Technol.* 169:566–572.

Vieler, A., G. Wu, C.-H. Tsai et al. 2012. Genome, functional gene annotation, and nuclear transformation of the heterokont oleaginous alga *Nannochloropsis oceanica* CCMP1779. *PLoS Genet.* 8:e1003064.

Vigeolas, H., F. Duby, E. Kaymak et al. 2012. Isolation and partial characterization of mutants with elevated lipid content in *Chlorella sorokiniana* and *Scenedesmus obliquus*. *J. Biotechnol.* 162:3–12.

Volkman, J. K., M. R. Brown, G. A. Dunstan, and S. W. Jeffrey. 1993. The biochemical composition of marine microalgae from the class eustigmatophyceae. *J. Phycol.* 29:69–78.

Wang, S. B., F. Chen, M. Sommerfeld, and Q. Hu. 2005. Isolation and proteomic analysis of cell wall-deficient *Haematococcus pluvialis* Mutants. *Proteomics* 5:4839–4851.

Wardle, D. A. 2003. Global sale of green air travel supported using biodiesel. *Renew. Sustain. Energy Rev.* 7:1–64.

Williams, P. J. le B. and L. M. L. Laurens. 2010. Microalgae as biodiesel and biomass feedstocks: Review and analysis of the biochemistry, energetics and economics. *Energy Environ. Sci.* 3:554–590.

Work, V. H., R. Radakovits, R. E. Jinkerson et al. 2010. Increased lipid accumulation in the *Chlamydomonas reinhardtii* sta7–10 Starchless Isoamylase mutant and increased carbohydrate synthesis in complemented strains. *Eukaryot. Cell* 9:1251–1261.

Yoon, K., D. Han, Y. Li, M. Sommerfeld, and Q. Hu. 2012. Phospholipid:diacylglycerol acyltransferase is a multifunctional enzyme involved in membrane lipid turnover and degradation while synthesizing triacylglycerol in the unicellular green microalga *Chlamydomonas reinhardtii. Plant Cell* 24:3708–3724.

5 Estimating the Maximum Achievable Productivity in Outdoor Ponds
Microalgae Biomass Growth Modeling and Climate-Simulated Culturing

*Michael Huesemann, Mark Wigmosta,
Braden Crowe, Peter Waller, Aaron Chavis,
Samuel Hobbs, Scott Edmundson, Boris Chubukov,
Vincent J. Tocco, and André Coleman*

CONTENTS

5.1   INTRODUCTION

In response to mounting concerns about global climate change and the need to produce carbon-neutral transportation fuels, there has been renewed and increasing interest in utilizing microalgae for the generation of drop-in biodiesel and jet biofuel (U.S. DOE 2010). In order to develop an economically viable microalgae biofuels production process, it is imperative to identify strains that exhibit high annual biomass productivities (average >30 g/m^2·day) in outdoor culture systems (U.S. DOE 2012). Significant campaigns have been initiated by industry, academia, and other research organizations to find promising new microalgae strains by either prospecting or genetic engineering, which might be suitable for large-scale, economic, biofuel production.

Just because a strain grows well in laboratory cultures under particular incubation conditions (e.g., room temperature and relatively low light intensities), there is no guarantee that it will perform satisfactorily in outdoor pond cultures that are subjected to daily and seasonal water temperature and light fluctuations. Furthermore, unless experiments are specifically designed to generate data for input to a predictive growth model, it is impossible to extrapolate the findings from the laboratory to identify the operating conditions (i.e., culture depth, dilution rate) or the geographic pond location where biomass productivities (monthly or annual) for a given strain are optimal.

To address the challenge of determining the maximum achievable monthly, seasonal, and annual biomass productivities for selected promising strains in outdoor ponds at any geographic location of interest, we developed the following four-step integrated strategy (Figure 5.1). First, the strain is characterized in terms of its specific growth rate response to temperature and light, including darkness, as well as its light attenuation characteristics. Second, these species-specific parameters serve as inputs to Pacific Northwest National Laboratory's (PNNL) biomass growth model (Huesemann et al. 2013) that, in conjunction with the sunlight and pond water temperature predictions provided by PNNL's Biomass Assessment Tool (BAT) (Wigmosta et al. 2011; Venteris et al. 2013a,b; Coleman et al. 2014), is used to predict the respective biomass productivities for tens of thousands of hypothetical pond locations in the United States. Third, monthly and annual biomass productivity maps are created to identify the geographic location(s) of optimum biomass productivity and to generate light and water temperature "scripts" (time series) for

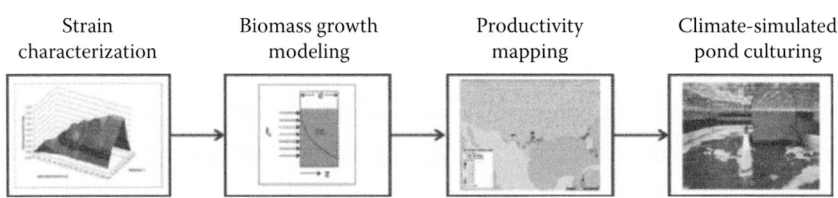

| Strain characterization | Biomass growth modeling | Productivity mapping | Climate-simulated pond culturing |

FIGURE 5.1   Integrated strategy for screening strains for high biomass productivity potential.

hypothetical ponds at the best site(s). Finally, the maximum achievable biomass productivity for the given strain is confirmed by conducting culture experiments in PNNL's indoor LED-lighted and temperature-controlled ponds under climate-simulated conditions, using the previously determined light and temperature scripts. This integrated strategy provides an efficient and low-risk approach to screen promising strains for their potential to exhibit high biomass productivities and to validate their model-predicted performance in climate-simulation ponds, before transitioning to cultivation in outdoor raceways.

5.1.1 LIGHT AND TEMPERATURE

Since light and temperature are the key determinants of biomass productivity in outdoor pond cultures operated under nutrient-replete (N, P, CO_2, etc.) and well-mixed conditions, it is necessary to know the selected strain's specific growth rate as a function of these two key variables. There is a large body of literature on the effects of increasing light intensity on a microalgae species' maximum specific growth rate or photosynthetic oxygen evolution rate, resulting in useful knowledge about the quantum efficiency, the saturating light intensity and potential photoinhibition (Platt and Jassby 1976; Falkowski et al. 1985; Geider et al. 1985). There are also a few studies on the relationship between specific growth rate and temperature for multiple (Eppley 1972; Goldman and Carpenter 1974) or individual species (Yoder 1979; Coles and Jones 2000; Montagnes and Franklin 2001; Boyd et al. 2013). However, this information is of limited use when faced with the task of identifying suitable microalgae for outdoor pond cultivation, given that each strain has its own unique light and temperature response curves or functions. Consequently, for each selected strain, the maximum specific growth rate needs to be measured during the exponential growth phase in unshaded cultures as a function of light intensity and temperature.

Light and temperature also affect the rate of biomass loss overnight due to dark respiration. Biomass-specific dark respiration rates in microalgae are highly species dependent, varying from 0.01 to 0.6 day^{-1}, and can be up to 30% of the strain's maximum specific growth rate (Geider and Osborne 1989). For example, Ryther and Guillard (1962) determined the respiratory coefficient (g carbon respired per hour per gram chlorophyll) for five diatoms at 5°C–25°C, demonstrating a significant difference among these species. Also, Grobbelaar and Soeder (1985) measured respiratory oxygen uptake rates for two microalgae species in dark ponds and found these to be dependent upon incubation temperature, time in the dark, and on the temperature and irradiance under which the microalgae had been cultivated before. Torzillo et al. (1991) cultivated *Spirulina platensis* in outdoor tubular photobioreactors and reported that night biomass loss depended on the temperature and light irradiance at which the cultures were grown. Finally, when culturing *Chlorella pyrenoidosa*, Ogbonna and Tanaka (1996) observed that the magnitude of biomass loss during the dark period was positively correlated with the dark time temperature and negatively correlated with the temperature during the light period. Furthermore, cells in the light-sufficient exponential growth phase exhibited much larger biomass loss in the dark than those in the light-limited linear or stationary growth phase. In summary, in order to model biomass loss in the dark, it is necessary to know, for a given strain, how the rate of biomass loss during the dark period is affected by temperature and average light intensity during the light period.

5.1.2 Mathematical Models of Microalgae Growth

Hundreds of mathematical and numerical models for predicting phytoplankton net primary productivity (Behrenfeld and Falkowski 1997) or microalgae biomass growth in photobioreactors and ponds have been published (for more recent ones, see Packer et al. 2011, Quinn et al. 2011, Bechet et al. 2013, James et al. 2013, and references therein). These models generally share two common features: first, they estimate the light attenuation within the culture, and second, they predict the rate of biomass growth as a function of either incident or absorbed light (Bechet et al. 2013; Huesemann et al. 2013). With few exceptions (Cornet et al. 1992; Fernandez et al. 1997; Pottier et al. 2005; Cornet and Dussap 2009), most models rely on Beer–Lambert's law to determine the light intensity as a function of culture depth and biomass concentration. This, however, is problematic for high-density cultures where light scattering can be significant. In addition, models that attempt to incorporate biomass loss in the dark use assumed rather than measured dark respiration rates or maintenance energy requirements (Goldman 1979; Quinn et al. 2011; James et al. 2013).

Unfortunately, the existing microalgae biomass growth models have only limited utility in terms of applying them to screen novel promising microalgae strains for their potential to achieve high biomass productivities in outdoor pond cultures. There are at least two reasons: First, most models are complex, requiring the input of many difficult to determine parameters, which is time-consuming and expensive. Second, model validation has been either nonexistent or questionable (Huesemann et al. 2013). As Chalup and Laws (1990) wrote more than 20 years ago when reviewing existing microalgae growth models, "With a judicious choice of parameter values, the models almost always seem to give a good fit to available experimental data." If a model requires many input parameters, it is easy to make appropriate "judicious choices" and assumptions or select the "right" ones from the large literature until a good model fit to the experimental culture data is obtained. While a reasonable model fit shows that the predictions are pointing in the right direction, it is questionable whether such a model, in the absence of further rigorous validation, can be trusted to accurately forecast the performance of outdoor pond cultures.

5.2 MICROALGAE BIOMASS GROWTH MODEL

Our objective is to develop a biomass growth model based on independently measured strain-specific (specific biomass growth or loss rate as a function of light and temperature, scatter-corrected biomass light absorption coefficient), environmental (incident light intensity, water temperature), and key culture operational (depth, dilution rate) input parameters. This would eliminate the "judicious choice" problem and thus improve the model's predictive power.

The environmental input parameters, that is, light intensity and water temperature as a function of time, are generated for hypothetical pond locations via PNNL's Biomass Assessment Tool (BAT) (Wigmosta et al. 2011; Venteris et al. 2013a,b; Coleman et al. 2014), which was initially developed for a "generic" microalga parameterized with values taken from the literature and which included an assumed response to temperature and light. In order to enable the prediction of biomass productivities for a specific microalgae strain, the BAT code was modified by incorporating PNNL's microalgae growth model using the aforementioned species-specific input parameters.

It is our objective to demonstrate how the integrated strategy shown in Figure 5.1 can be used to identify strains for their potential to exhibit high biomass productivities in outdoor ponds.

5.2.1 Light Attenuation and Scattering

In the model, it is assumed that light and temperature are the key and instantaneous determinants of microalgae growth and productivity and that no other factors such as nutrients, CO_2, and mixing (i.e., mass transfer) are limiting. Furthermore, it is assumed that the culture pH remains constant via feedback-controlled CO_2 addition and that there is no growth inhibition by photosynthetic oxygen or other compounds.

The pond water temperature (T) is easily measured in outdoor cultures or predicted using the mass and energy balance pond temperature model within the Biomass Assessment Tool. According to Beer–Lambert's law, for a given biomass concentration (B), light attenuates as a function of light penetration distance (z) as follows:

$$I(z) = I_0 \cdot e^{-k_a \cdot B \cdot z} \qquad (5.1)$$

where
   I_0 is the incident light intensity at the culture surface
   $I(z)$ is the light intensity at depth (z)
   k_a is the biomass light absorption (extinction) coefficient

Since the biomass concentration B increases in a growing culture, the light attenuation becomes increasingly stronger with time.

Beer–Lambert's law states that the light absorbance ($A = \log_{10}(I_0/I)$) is directly correlated to the compound's concentration and light path length. This relationship is linear but applicable only for monochromatic, collimated (parallel) light in dilute light-absorbing solutions where scattering is negligible (Ingle and Crouch 1988). Although these conditions are generally not fulfilled for dense pond cultures illuminated by full-spectrum sunlight, most microalgae biomass growth models employ Beer–Lambert's law to estimate the light intensity as a function of depth. In order to account for possible light scattering in dense microalgae cultures, we utilized an experimentally determined scatter-corrected biomass light attenuation coefficient (k_{sca}), following a procedure adopted from Suh and Lee (2003), for the prediction of light intensity as a function of light penetration distance z and biomass concentration B (see Appendix 5A.3):

$$k_{sca} = k_a \cdot \frac{K_B}{K_B + B} \cdot \frac{K_z}{K_z + z} \tag{5.2}$$

where K_B and K_z are the light scattering coefficients related to biomass concentration (B) and light path length (z), respectively.

5.2.2 Biomass Production

If the culture volume is discretized into n *equal-sized* parallel volume layers orthogonal to incident light intensity, the biomass concentration in each volume layer is assumed to increase exponentially from $B(t)$ to $B(t + \Delta t)$ during time interval Δt as follows:

$$B(t + \Delta t) = B(t) \cdot e^{\mu \cdot \Delta t} \tag{5.3}$$

where μ is the specific growth rate (day^{-1}) in the respective volume layer. The specific growth rate is affected by the water temperature (T) and light intensity (I) in the respective volume layer. Assuming perfect mixing, the water temperature is the same in all volume layers and can be predicted as a function of time using the pond energy balance model, which is part of the Biomass Assessment Tool (BAT; see later). The light intensity in each volume layer can be predicted via Beer–Lambert's law using the experimentally determined scatter-corrected light attenuation coefficient (Equations 5.1 and 5.2).

Since the application of Equation 5.3 requires knowledge of the specific growth rate at the particular temperature and light intensity within each discretized volume layer, it is necessary to know how μ varies with temperature and light intensity. In dilute cultures with minimum self-shading, the specific growth rate of microalgae can be experimentally determined as a function of temperature and light as follows:

$$\mu = f(T, I) \tag{5.4}$$

where $f(T, I)$ is the two-dimensional array (or surface) of specific growth rates measured for different combinations of temperature and light intensity values. Since each microalgae strain has a unique response to light (i.e., compensation light intensity, saturating light intensity, photoinhibition) and temperature (i.e., optimum temperature, temperature tolerance range), the function $f(T, I)$ is species specific and must be experimentally determined prior to running the model. It is implicitly assumed that as individual cells traverse the n different layers within the culture, they respond instantaneously to the new light conditions, exhibiting the experimentally determined specific growth rate at the local light intensity per Equation 5.4. This assumption has not only been made previously by others (Goldman 1979; Grima et al. 1996; Yun and Park 2003; Bosma et al. 2007) but has also been positively confirmed by verifying that, when measuring conventional PI curves, the photosynthetic oxygen evolution rate changes immediately after changes in light intensity (Huesemann et al. 2009, 2013).

5.2.3 DARK RESPIRATION

Since biomass loss overnight due to dark respiration can have a significant, negative effect on biomass productivity (see discussion earlier), it is necessary to know the rate of biomass loss (μ_{dark}) in the absence of light ($I = 0$) as a function of temperature (T) and the average light intensity (I_{avg}) that the cells were exposed to in the mixed pond culture during the preceding day:

$$\mu_{dark} = f(T, I_{avg}) \tag{5.5}$$

I_{avg} is calculated by averaging the depth-integrated light attenuation profiles (Equations 5.1 and 5.2) in the pond culture for each time step (Δt) over the entire day preceding the night where biomass loss due to dark respiration occurs. The relationship (Equation 5.5) between biomass loss rates in the dark (μ_{dark}) and temperature (T) and average light intensity (I_{avg}) is independently determined in laboratory experiments as described in Appendix 5A.

5.2.4 MODELING PROCEDURE

In summary, the modeling procedure involves the following steps:

1. Select an initial biomass concentration, $B(t = 0)$.
2. Discretize the culture volume into n *equal-sized* parallel volume layers orthogonal to incident light intensity.
3. Calculate the light intensity at the midpoint of each of the n culture volume layers (Equations 5.1 and 5.2).
4. For daytime conditions, determine the specific growth rate (μ) in each of the n culture volume layers using experimentally species-specific growth rate data (Equation 5.4; Figure 5.2). Since cells in well-mixed dense cultures during the daytime are exposed to high-average light intensities at or near the surface of the pond while receiving little or no light during their short residence time in the bottom part for the pond and therefore may lose biomass due to dark respiration, the specific "growth" rate in the complete absence of light ($I = 0$ μmol/m$^2 \cdot$s) was assumed to be μ_{dark} for the experimentally determined case of high-average light intensity (see Figure 5.3).
5. For nighttime conditions, calculate the biomass loss rate in the dark (μ_{dark}) as a function of pond water temperature and the average light intensity during the preceding day using experimentally determined biomass loss rate data (Equation 5.5, Figure 5.3).

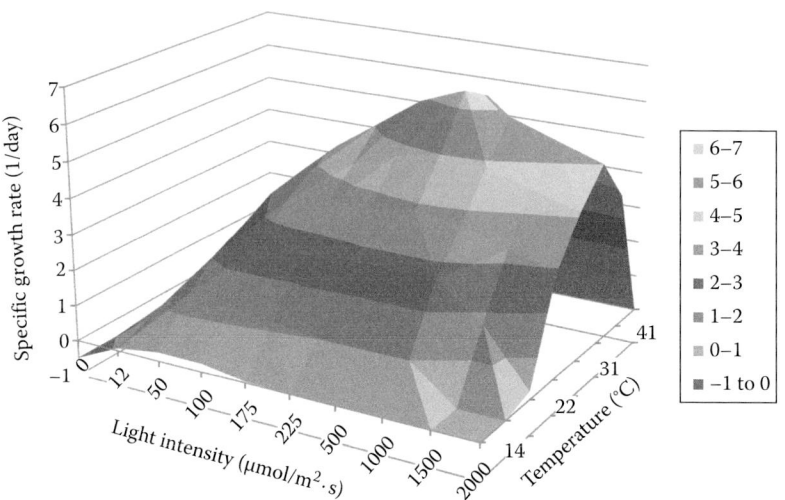

FIGURE 5.2   (See color insert.) Specific growth rate (μ) of *Chlorella sorokiniana* (DOE 1412) as a function of light intensity (0–1950 μmol/m$^2 \cdot$s) and temperature (14°C–41°C). Each value is the average of three to six measurements taken in replicate growth rate experiments.

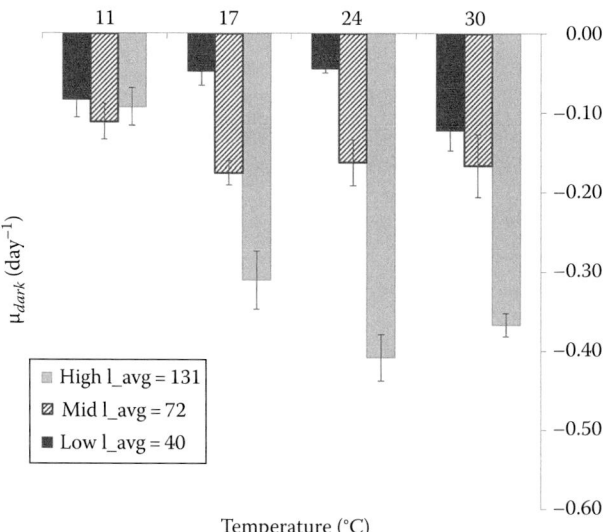

FIGURE 5.3   Specific biomass loss rate (μ_{dark}) of *Chlorella sorokiniana* (DOE 1412) in the dark as a function of temperature and the average light intensity (high, mid, low, in μmol/m$^2 \cdot$s) experienced during the preceding light period. Each value is computed from duplicate initial and final AFDW values in replicate (2–3) dark incubation experiments.

6. Calculate the increase (μ during the day) or decrease (μ_{dark} during the night) in biomass concentration in each of the n culture volume layers during time interval Δt (Equation 5.3).
7. Determine the new biomass concentration $B(t + \Delta t)$ in the entire culture by numerically averaging the biomass concentrations of all n culture volume layers ($B = \Sigma B_i/n$).
8. Repeat steps 3 through 7 for each time step Δt until having reached the desired final time.

This modeling routine was programmed in Visual Basics for Applications (VBA Microsoft Excel) and MATLAB® (for integration with the Biomass Assessment Tool), using $n = 100$ and $\Delta t = 5$ minutes.

5.2.5 BIOMASS ASSESSMENT TOOL

The Biomass Assessment Tool (Wigmosta et al. 2011; Venteris et al. 2013a,b; Coleman et al. 2014) is an integrated model, analysis, and data management research and development architecture that couples advanced spatial and numerical models to capture variable spatial scale (i.e., site-specific to national to global scales) and high-temporal-resolution (1–3 hours) environmental conditions, biomass and lipid production potential, resource requirements, and sustainability metrics for bioenergy feedstocks. The BAT makes use of the highest resolution and best available datasets to provide a comprehensive resource and production assessment for potential algal biofuel production facilities. The BAT currently includes (Coleman et al. 2014) the following components: (1) a multiscale land-suitability model; (2) open and closed mass and energy balance pond models; (3) a biomass growth model (presented herein); (4) trade-off analysis routines to evaluate biomass production potential with available land, water, and/or infrastructure resources; (5) water source and use intensity analysis under current and altered climates for freshwater, seawater, and saline groundwater; (6) nutrient and CO_2 flue gas source, availability, and demand models; (7) least-cost transport models for water, nutrients, CO_2, and refinery feedstock delivery; (8) a land valuation/acquisition model; (9) an enterprise facility scaling model; and (10) a site leveling and costing model.

5.3 APPLICATION OF THE FOUR-STEP INTEGRATED STRATEGY FOR *CHLORELLA SOROKINIANA*

5.3.1 STRAIN CHARACTERIZATION: STEP 1

Here, we summarize experimental results obtained as inputs to the biomass growth model (see Appendix 5A for a brief presentation of the methods). The maximum specific growth rates (μ_{max}) of *Chlorella sorokiniana* (DOE 1412) was measured in light-sufficient exponentially growing cultures as a function of light intensity and temperature to provide the required biomass growth model input parameters (see Equation 5.4). The highest specific growth rate of 6.5 day^{-1} (a doubling time of 2.5 hours) was observed at 36°C and 435 $\mu mol/m^2 \cdot s$ (Figure 5.2). No growth was detected at 14°C but the strain exhibited robust growth between 18°C and 41°C, the highest temperature tested. The maximum specific growth rate of *Chlorella* generally (i.e., for temperatures ≥22°C) increased with light intensity until reaching a plateau at a saturating light intensity of about 250 μmol/m²·s and declined to various degrees, depending on temperature, at higher light intensities, most likely due to photoinhibition. The reduction in specific growth rates at high-light intensities was particularly severe at lower temperatures (18°C and 22°C), most likely because the mechanisms of repairing the molecular damages of photoinhibition are not very effective at low temperatures, as has also been observed by others (Jensen and Knutsen 1993; Vonshak et al. 2001; Bosma et al. 2007).

The rate of biomass loss in the dark (μ_{dark}) as a function of temperature and the average light intensity (I_{avg}) during the light period was experimentally determined and served as another critical model input (see Equation 5.5). As shown in Figure 5.3, the magnitude of μ_{dark} increased with increasing I_{avg} for cells cultured at 17°C, 24°C, and 30°C while it was about the same and independent of I_{avg} for cells at 11°C.

The decline of light intensity with culture depth was measured for the range of biomass concentrations encountered in the Arizona outdoor pond cultures (Figure 5.4). For each light attenuation curve, the respective light scattering coefficients K_z and K_B (see Equation 5.2) were determined by curve fitting via least squares regression, using a separately determined biomass light absorption (extinction) coefficient (k_a) of 49.5 ($1/(OD_{750}\cdot m)$). Since K_z and K_B varied with biomass concentration, the required light scattering coefficient values for a given biomass concentration during the modeling routine were computed by linear interpolation.

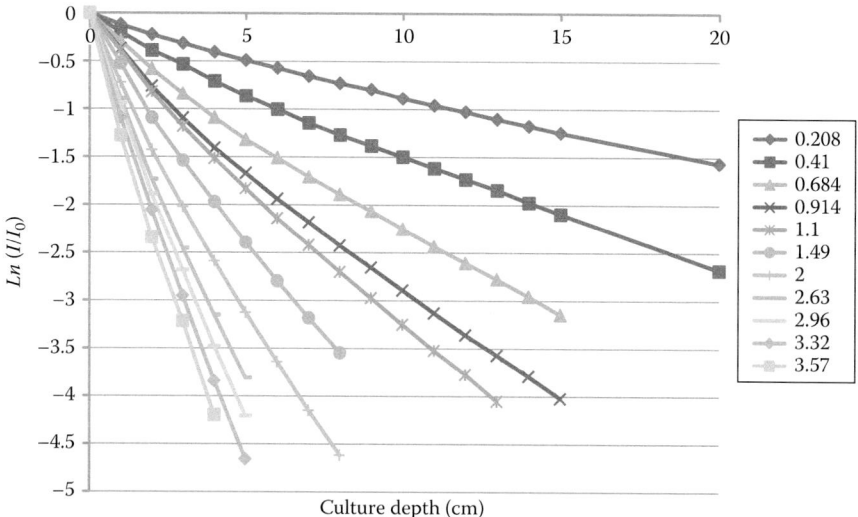

FIGURE 5.4 Natural logarithm of the ratio of light intensity (*I*) over incident light intensity (I_0) as a function of culture depth (*z*) for different biomass concentrations (OD_{750}) of *Chlorella sorokiniana* (DOE 1412).

5.3.2 Biomass Growth Model Validation: Step 2

To demonstrate that the model (see Equations 5.1 through 5.5) predicts the growth performance in actual outdoor ponds, we compared the growth performance of *C. sorokiniana* (DOE 1412) cultured in three replicated outdoor ponds under batch culture conditions at the University of Arizona (Rimrock site) to model predictions. The model used the following strain-specific input parameters: (1) the maximum specific growth rate (μ) as a function of light intensity and temperature (Figure 5.2); (2) the biomass loss rate in the dark (μ_{dark}) as a function of temperature and average light intensity (Figure 5.3); (3) the scatter-corrected biomass light absorption coefficient (k_{sca}) obtained by fitting the light attenuation data (Figure 5.4) using nonlinear regression (Equation 5.2); (4) the incident sunlight intensity and pond water temperature data recorded during the 1-month-long outdoor batch culture experiment; and (5) the pond depth ($d = 25$ cm).

As shown in Figure 5.5, the model-predicted and measured biomass concentrations compared reasonably well during the exponential and midlinear batch growth phase, that is, for the first 20 days following inoculation. The sawtooth pattern of the model-predicted concentration curve reflects the periodic increase of biomass during the day, followed by biomass loss at night due to dark respiration. Since the outdoor ponds were sampled each morning around 10 a.m., after a period of biomass loss overnight followed by very little growth in the early-morning hours due to low water temperatures, the measured biomass concentrations should be compared to the

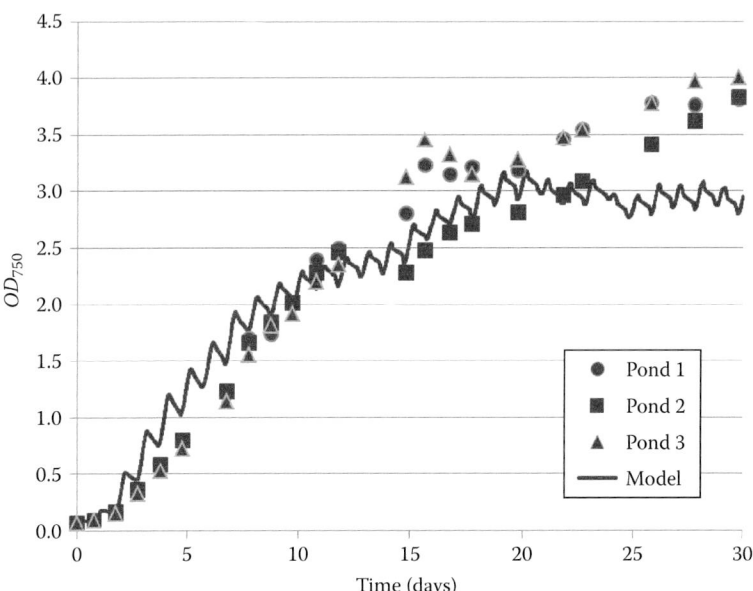

FIGURE 5.5 Predicted (line) and measured biomass concentrations (three different symbols) as a function of time in three outdoor Arizona ponds during the linear and early stationary phase of the batch culture experiment with *Chlorella sorokiniana* (DOE 1412).

predicted early-morning low biomass concentrations following the dark period (i.e., the diurnal low points of the model-predicted curve). When doing so, a markedly good agreement between measured and model-predicted concentrations is observed.

It should be noted, however, that the model underpredicted the biomass concentrations during the later phase of the batch culture experiment. A possible hypothesis is that the model overpredicted the biomass losses in the aphotic zone of dense pond cultures during the daytime because the chosen input parameters for μ_{dark} were too negative due to the aforementioned assumption that cells are exposed to high average light intensity near the surface of the pond and thus exhibit relatively large negative biomass loss rates in the darker, lower part of the pond. However, as the culture density increases with time, the average light intensity that cells are exposed to at or near the surface declines, resulting in less negative μ_{dark} values (see Figure 5.3). The use of such less negative μ_{dark} values, experimentally determined for cells grown under lower light intensities, would result in model predictions even more closely simulating the biomass growth measurements and is justified based on the aforementioned arguments.

No information is presently available in the peer-reviewed literature regarding the effects of light exposure on short-term (seconds to minutes) biomass losses due to dark respiration in the low-light and aphotic zones of the pond, and thus, this mechanism is not incorporated in the modeling algorithm. It may be that with the aforementioned adjustment in μ_{dark}, the model validation would be complete and sufficiently predictive. In other words, such short-term light variations and pond mixing effects would be anticipated to have no effect on productivity within the temperature–light regime modeled. However, this hypothesis must be tested, and thus, more research is needed on the effects of rapid cycling of microalgal cells through the light and dark zones of a well-mixed dense pond culture and how this affects the rate of biomass production in the light and loss in the aphotic zone.

In conclusion thus far, the biomass growth model presented herein can serve as a useful tool for screening promising microalgae strains for their ability to exhibit high biomass productivities in outdoor ponds operated under batch or semicontinuous culture mode, requiring knowledge only of the strain-specific parameters (μ, μ_{dark}, k_{sca}) and the sunlight and water temperature conditions. Obtaining these parameters for each strain is presently a laborious undertaking, but with greater adoption of this model, it will be possible to rationalize, miniaturize, and automate much of this work, allowing for rapid throughput of promising strains. This will greatly reduce the requirement for outdoor pond tests, which are presently the bottleneck in the development of microalgae biomass and biofuels production.

5.3.3 Biomass Productivity Mapping and Generation of Light Intensity and Water Temperature Scripts: Step 3

The other required model inputs are the site-specific temperature and light intensities throughout the year under which the algae growth needs to be modeled. Within the previously described BAT (Biomass Assessment Tool), a two-dimensional hydrodynamic and water quality model, MASS2 (Perkins and Richmond 2007; Wigmosta et al. 2011), was used to estimate 30 years of site-specific hourly pond water temperature and freshwater evaporative water loss based on air temperature, incoming solar radiation, humidity, wind speed, and calculated net longwave radiation for a 25 cm pond depth for locations throughout the United States. The U.S. Department of Agriculture's CLIGEN stochastic weather generator (Nicks and Lane 1989; Meyer et al. 2008) was used to produce a 30-year meteorological record at 2637 locations throughout the United States (89 in Arizona). These averaged daily data were disaggregated to hourly values using a physics-based approach (Waichler and Wigmosta 2003) and were used to drive the MASS2 model. Figure 5.6 shows an example of the predicted hourly incident sunlight intensities and pond water temperatures for a hypothetical pond located in Phoenix, Arizona, during the month of July.

The modeled pond water temperature (T) and light intensity (I) data were used as inputs to the biomass growth model run in continuous culture mode (dilution rate = 0.3 day^{-1}) at large-scale production sites (>100 ha) located throughout Arizona using 89 different meteorological station locations. The results, shown in Figure 5.7 (seasonal averages) and Figure 5.8 (annual average), demonstrate both the spatial variability in predicted biomass productivities and their seasonal patterns. Sites in northern Arizona are generally less productive, in part due to the generally higher elevations (ca. 1500–2600 m) and resulting lower daily minimum temperatures. The period of productivity for the selected strain (*C. sorokiniana*) is highest in the summer (June–August), followed by fall (September–November), where temperature and light availability are favorable. Productivity performance in the winter (December–February) and spring (March–May) for most sites throughout Arizona are minimal to none. For less thermotolerant strains such as *Nannochloropsis salina*, the summer pond temperatures may actually inhibit growth, and thus, higher productivities would be predicted in the fall and spring seasons (data not shown).

5.3.4 Climate-Simulated Culturing: Step 4

In order to determine whether the outdoor pond culture experiment that had been conducted for *C. sorokiniana* in Arizona could be replicated using the indoor LED-lighted and temperature-controlled ponds, we operated two indoor raceway cultures under climate-simulated conditions by subjecting them to the sunlight and water temperature fluctuations that had been recorded during the outdoor study. As shown in Figure 5.9, the biomass concentration (OD_{750} and ash-free dry weight [AFDW]) in the two indoor climate-simulation LED ponds increased at about the same rate as in the three outdoor raceways in Arizona during the first 15 days of linear batch growth. For a more quantitative comparison of biomass productivities in outdoor and indoor ponds, the slopes of each biomass growth curve were determined by linear least squares regression. The variability of the OD_{750} versus time slopes ($R^2 \geq 0.98$ in all cases, Figure 5.9a) in the three replicate outdoor ponds and in the duplicate indoor LED ponds was only 2% and 0.06% relative standard deviation (σ/mean), respectively, indicating excellent reproducibility of the culture experiments. Furthermore, the average OD_{750} versus time slope for indoor LED pond cultures was only 7.7% greater than for the outdoor Arizona cultures. Similarly, the variability of the AFDW versus time slopes ($R^2 \geq 0.98$ in all cases, Figure 5.9b) in the three replicate outdoor ponds and in the duplicate

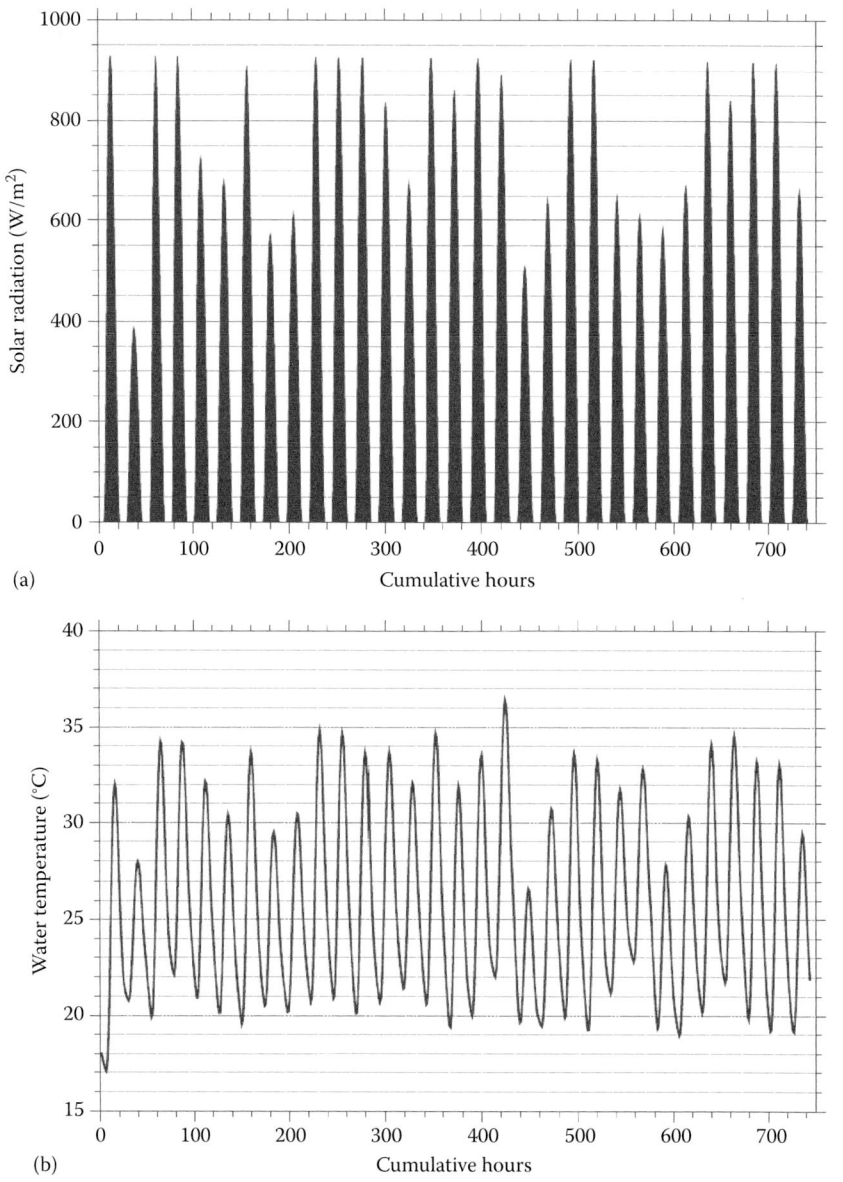

FIGURE 5.6   Estimated incoming hourly solar radiation (a) and water temperature (b) in July for a 25 cm deep pond located in Phoenix, Arizona.

indoor LED ponds was only 1.8% and 0.4% relative standard deviation, respectively, and the average AFDW versus time slope for indoor LED pond cultures was only 9.5% greater than for the outdoor Arizona cultures. The slightly reduced biomass productivities (~slopes) in outdoor pond cultures compared to climate-simulation ponds could be due to a greater abundance of invasive grazers or more UV radiation, factors that could cause suboptimal biomass productivities (Andreasson and Wangberg 2007; Forján et al. 2011). Also, there was about a 1-day-longer lag period for the outdoor pond cultures relative to the indoor ones, suggesting that the former were more stressed.

In summary, the growth performance of outdoor Arizona pond cultures was successfully replicated in indoor LED-lighted and temperature-controlled ponds

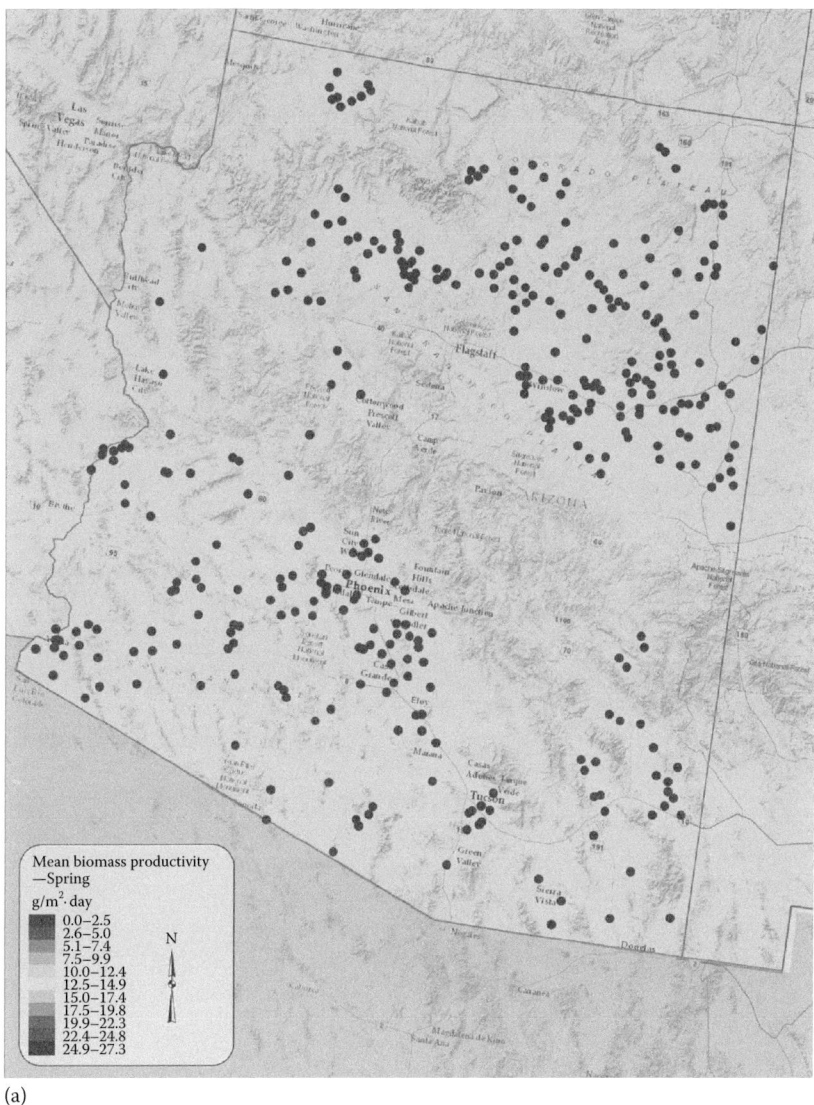

(a)

FIGURE 5.7 **(See color insert.)** Long-term average seasonal biomass productivity of *Chlorella sorokiniana* (DOE 1412) using the BAT pond temperature and biomass growth models for potential pond sites in Arizona for (a) spring. Winter values, as for spring, were in the range 0.0–2.5 g/m²·day at all sites (dark blue), and therefore are not shown. (*Continued*)

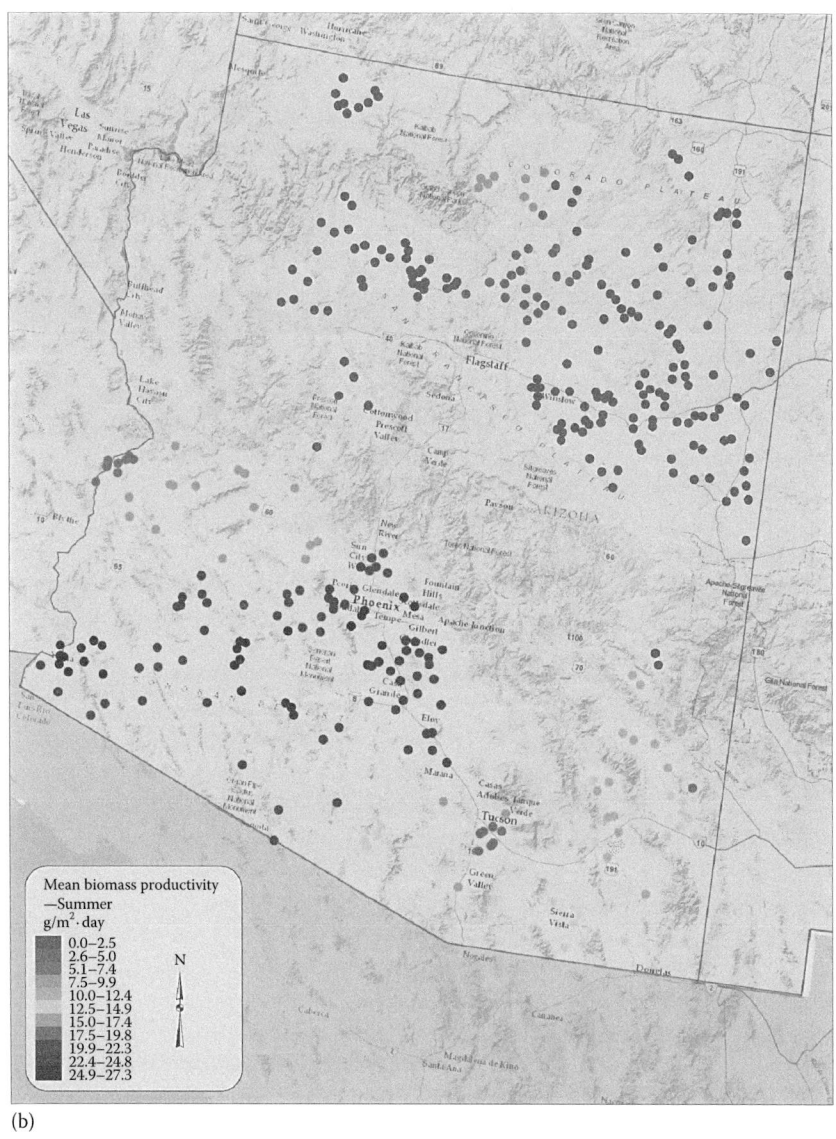

(b)

FIGURE 5.7 (*Continued*)   (See color insert.) Long-term average seasonal biomass productivity of *Chlorella sorokiniana* (DOE 1412) using the BAT pond temperature and biomass growth models for potential pond sites in Arizona for (b) summer.   (*Continued*)

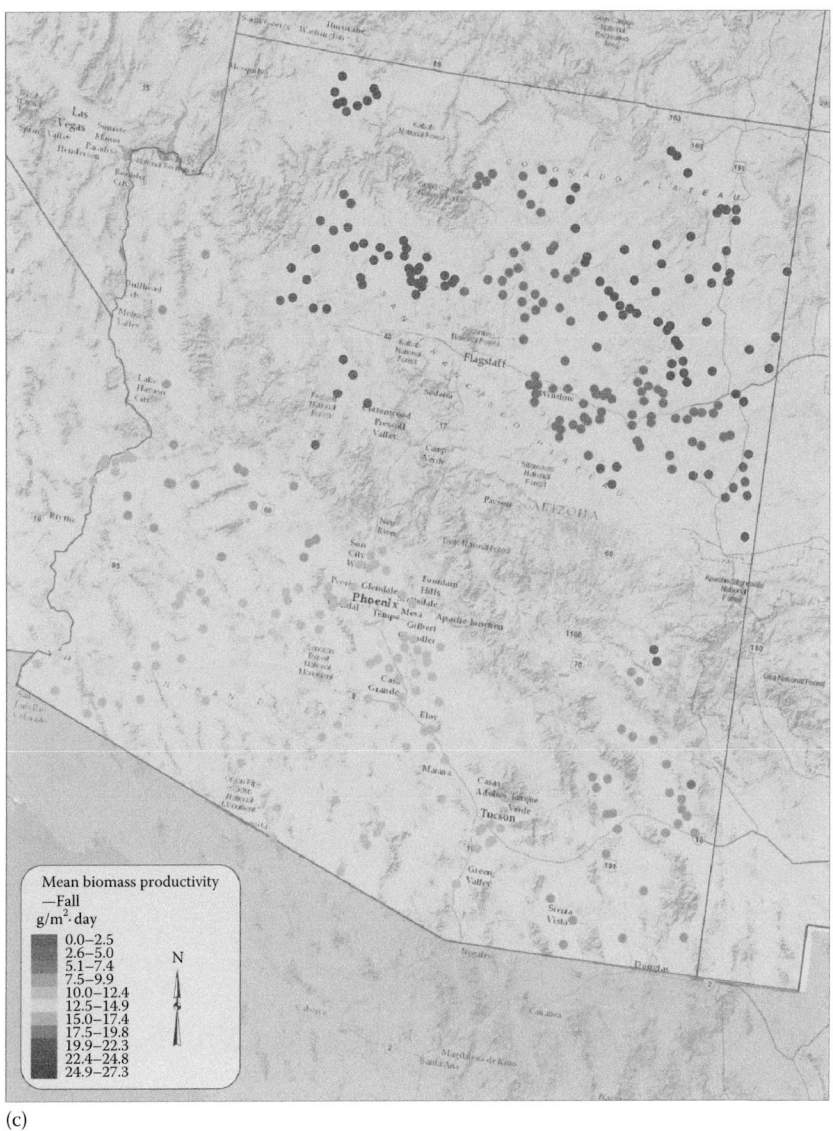

(c)

FIGURE 5.7 (*Continued*)   (See color insert.) Long-term average seasonal biomass productivity of *Chlorella sorokiniana* (DOE 1412) using the BAT pond temperature and biomass growth models for potential pond sites in Arizona for (c) fall.

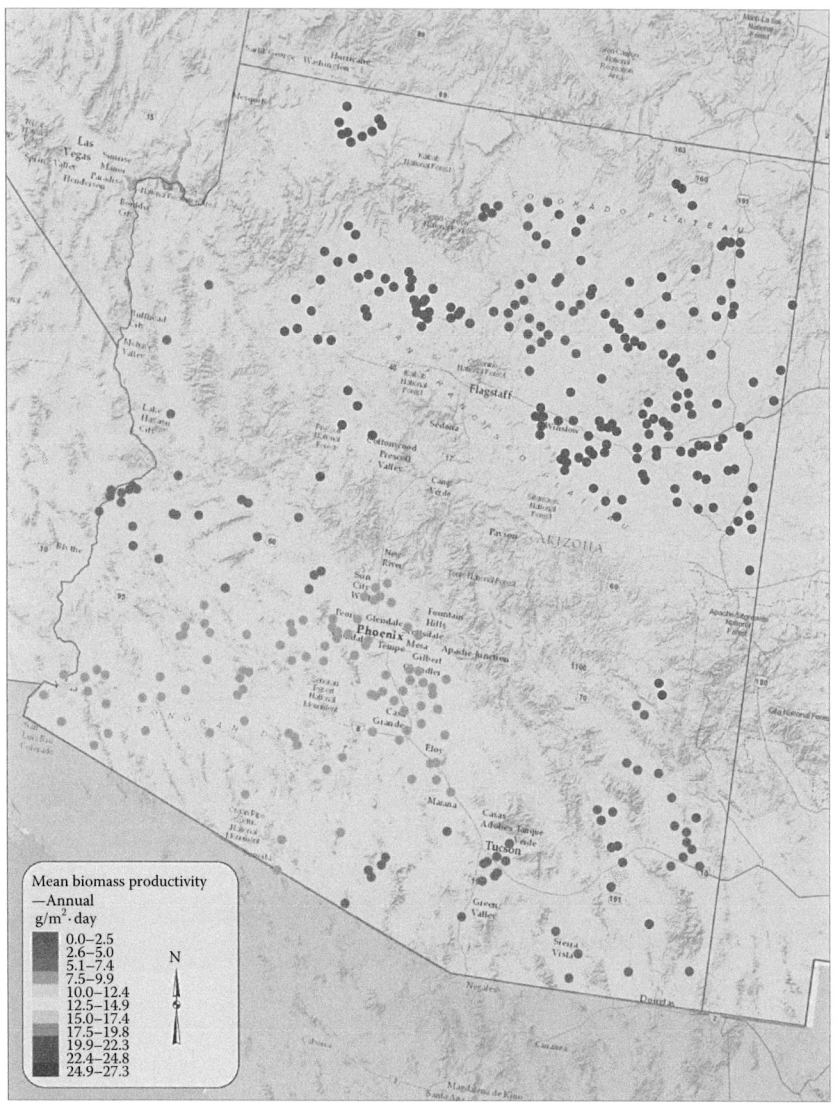

FIGURE 5.8 **(See color insert.)** Average annual biomass productivity for *Chlorella soroki-niana* (DOE 1412) using the BAT pond temperature and biomass growth models for potential pond sites in Arizona.

operated under climate-simulated conditions. Assuming additional, successful vali-dation experiments, these climate-simulation ponds would be useful, with light and temperature scripts provided by the Biomass Assessment Tool, for quantifying the biomass productivity of specific microalgae strains cultured at any hypothetical geo-graphic outdoor pond location of choice.

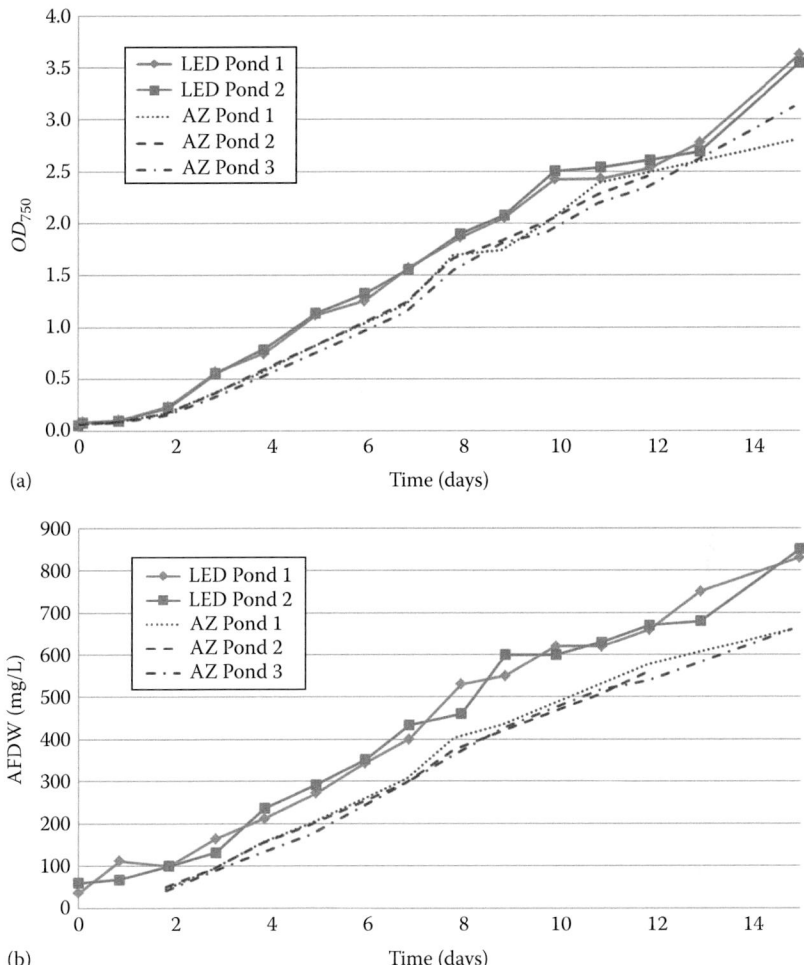

(a)

(b)

FIGURE 5.9   Biomass concentration as a function of time in two replicate indoor climate-simulation ponds and three replicate outdoor ponds in Arizona using *Chlorella sorokiniana* (DOE 1412). (a) OD_{750} versus time; (b) AFDW versus time.

5.4   CONCLUSIONS

A four-step integrated strategy, consisting of strain characterization, biomass growth modeling, productivity mapping, and climate-simulating culturing (Figure 5.1), was developed and successfully validated to determine the maximum achievable monthly and annual biomass productivity of selected microalgae cultured in outdoor ponds at hypothetical geographic locations in the United States. This integrated strategy is an efficient and low-risk approach to screen promising strains for their potential to exhibit high biomass productivities and to validate their model-predicted performance in climate-simulation ponds, before transitioning to cultivation in outdoor raceways.

5A APPENDIX

5A.1 MICROORGANISMS AND MEDIA

C. sorokiniana (DOE 1412) was isolated from surface water in Texas by Dr. Juergen Polle (Brooklyn College) and grown in a BG-11 medium at pH 7 as described in Huesemann et al. (2013).

5A.2 MEASUREMENT OF BIOMASS CONCENTRATION

Biomass concentration was measured as optical density at 750 nm (OD_{750}) and as ash-free dry weight (AFDW; g/L), as described in Van Wagenen et al. (2012).

5A.3 MEASUREMENT OF THE SCATTER-CORRECTED BIOMASS LIGHT ABSORPTION COEFFICIENT (K_{SCA})

C. sorokiniana (DOE 1412) culture was placed in a white translucent container (30 cm diameter, 42 cm height) mixed from below with a magnetic stirrer and illuminated from above with a multicolor LED panel simulating sunlight at ca. 2000 μmol/m²·s, and the light intensity was measured with a leveled LI-COR underwater light sensor (LI-192SA) connected to a LI-COR light meter (model LI-250A) as a function of light penetration distance in 1 cm increments. Light attenuation profiles were measured twice (i.e., stepwise with increasing and decreasing depth) for 10 different biomass concentrations, ranging from an OD_{750} of 0.2 to 3.6. The biomass light absorption coefficient k_a (see Equations 5.1 and 5.2) was separately determined in triplicate for dilute culture samples ($OD_{750} < 0.3$) in a 1 cm path length cuvette by measuring light absorption relative to a medium blank with a quantum sensor (PAR, LI-190) as described in Huesemann et al. (2013). For each light attenuation profile, the Excel solver function was used to obtain the pair of K_B and K_z values (see Equation 5.2) that gave the best curve fit via minimizing the sum of squared residuals. Since K_B and K_z were a function of biomass concentration, interpolated values were used to predict the light attenuation profiles for the numerous incremental OD_{750} values during the modeling exercise.

5A.4 MEASUREMENT OF THE SPECIFIC GROWTH RATE (μ) AS A FUNCTION OF INCIDENT LIGHT INTENSITY AND TEMPERATURE

The maximum specific growth rate of *C. sorokiniana* (DOE 1412) was determined in dilute, light-sufficient, exponentially growing shaker flask batch cultures ($OD_{750} < 0.3$) in a BG-11 medium maintained at pH 7 by sparging with CO_2-enriched air as described in Huesemann et al. (2013). To determine the effects of light intensity, the dilute cultures were subjected to light intensities ranging from 44 to 1950 $\mu mol/m^2 \cdot s$, using a multicolored LED panel simulating the sunlight spectrum. To determine the effects of temperature, culture flasks were placed on an equilibrated thermal gradient incubator (eight temperatures, ranging from 14°C to 41°C) as described in Van Wagenen et al. (2012).

5A.5   MEASUREMENT OF THE BIOMASS LOSS RATE (μ_{dark}) IN THE DARK AS A FUNCTION OF TEMPERATURE AND AVERAGE LIGHT INTENSITY

C. sorokiniana (DOE 1412) was grown in well-mixed 1 L Roux bottles at 24°C containing BG-11 medium sparged continuously with CO_2-enriched air to maintain a pH of 7. The cultures were illuminated with white LED lighting at an average surface light intensity of 500 $\mu mol/m^2 \cdot s$. Following inoculation, the cultures were allowed to grow for 14 hours under continuous illumination, split into four aliquots, which were then incubated in 125 mL shaker flasks for 10 hours in the dark on a thermal gradient incubator at 10.9°C, 17.3°C, 24.4°C, and 30°C, respectively. Biomass concentrations (duplicate AFDW) were measured at the beginning and end of the dark period and used to calculate specific dark loss rates (μ_{dark}). Experiments were carried out repeatedly ensuring reproducibility at three initial optical densities (i.e., OD_{750} of ca. 0.4, 1.0, and 3.0) to simulate different growth stages (i.e., exponential, linear, and stationary, respectively) during the light period prior to the dark incubations. The biomass light absorption coefficient k_a of culture samples was measured as described earlier and used to calculate the average light intensity (I_{avg}) "experienced" by the cells during the light period (Benson and Rusch 2006):

$$I_{avg} = I_0 \cdot \frac{1 - e^{-k_a \cdot OD_{750} \cdot d}}{k_a \cdot OD_{750} \cdot d} \tag{5.6}$$

where d is the shaker flask culture depth ($d = 0.045$ m). The average light intensity was determined using the average OD_{750} during the light period for each of the three growth stages, that is, average $OD_{750} = \frac{1}{2}$(final OD_{750} − initial OD_{750}).

5A.6   OUTDOOR RACEWAY POND CULTURE EXPERIMENTS

C. sorokiniana (DOE 1412) was cultured (depth = 25 cm) in a nutrient-sufficient BG-11 medium maintained at pH 7 by intermittent CO_2 sparging via feedback control in triplicate outdoor raceway ponds, similar to those described by Crowe et al. (2012), in Rimrock, Arizona, from June to August, 2012. During the batch culture phase, which lasted from June 10 to July 10 (i.e., day of year [DOY] 162–192), biomass concentrations (OD_{750} and AFDW) were determined daily. Pond water temperatures and incident photosynthetically active solar radiation (LI-COR quantum sensor, LI-190 SA) were measured every 5 minutes. These temperature and light

FIGURE 5.10   (See color insert.) LED-lighted and temperature-controlled raceway pond for culturing of microalgae under climate-simulated conditions.

data served as inputs to the biomass growth model and were also used to operate the indoor LED-lighted and temperature-controlled ponds under conditions simulating the outdoor ponds in Arizona.

5A.7   INDOOR RACEWAY POND CULTURE EXPERIMENTS

Two indoor 800 L temperature-controlled, via Labview™ software, fiberglass raceway ponds, of similar design as the outdoor ponds, each illuminated with a panel containing 4500 multicolored computer-dimmable LEDs (Figure 5.10) were used to perform culturing experiments in a BG-11 medium maintained at pH 7 via feedback-controlled intermittent CO_2 sparging. The incident light intensity and pond water temperature data collected during the outdoor cultivation study were loaded into the LED and temperature controllers to simulate the daily light and temperature fluctuations observed in the outdoor pond cultures at the Rimrock (Arizona) site.

REFERENCES

Andreasson, K. I. M. and S. A. Wangberg. 2007. Reduction in growth rate in *Phaeodactylum tricornutum* (Bacillarlophyceae) and *Dunaliella tertiolecta* (Chlorophyceae) induced by UV-B radiation. *J. Photochem. Photobiol. B* 86:227–233.

Bechet, Q., A. Shilton, and B. Guieysse. 2013. Modeling the effects of light and temperature on algae growth: State of the art and critical assessment for productivity prediction during outdoor cultivation. *Biotechnol. Adv.* 31:1648–1663.

Behrenfeld, M. J. and P. G. Falkowski. 1997. A consumer's guide to phytoplankton primary productivity models. *Limnol. Oceanogr.* 42:1479–1491.

Benson, B. C. and K. A. Rusch. 2006. Investigation of the light dynamics and their impact on algal growth rate in a hydraulically integrated serial turbidostat algal reactor (HISTAR). *Aquacult. Eng.* 35:122–134.

Bosma, R., E. van Zessen, J. H. Reith, J. Tramper, and R. H. Wijffels. 2007. Prediction of volumetric productivity in an outdoor photobioreactor. *Biotechnol. Bioeng.* 97:1108–1120.

Boyd, P. W., T. A. Rynearson, E. A. Armstrong et al. 2013. Phytoplankton temperature versus growth responses from polar to tropical waters—Outcome of a scientific community-wide study. *PLoS One* 8:1–17.

Chalup, M. S. and E. A. Laws. 1990. A test of the assumptions and predictions of recent microalgal growth models with the marine phytoplankter *Pavlova lutheri*. *Limnol. Oceanogr.* 35:583–596.

Coleman, A. M., J. M. Abodeely, R. L. Skaggs, W. A. Moeglein, D. T. Newby, E. R. Venteris, and M. S. Wigmosta. 2014. An integrated assessment of location-dependent scaling for microalgae biofuel production facilities. *Algal Res.* 5:79–94.

Coles, F. F. and R. C. Jones. 2000. Effect of temperature on photosynthesis-light response and growth of four phytoplankton species isolated from a tidal freshwater river. *J. Phycol.* 36:7–16.

Cornet, J. F. and C. G. Dussap. 2009. A simple and reliable formula for assessment of maximum volumetric productivities in photobioreactors. *Biotechnol. Prog.* 25:424–435.

Cornet, J. F., C. G. Dussap, and G. Dubertret. 1992. A structured model for simulation of cultures of the cyanobacterium *Spirulina platensis* in photobioreactors: I. Coupling between light transfer and growth kinetics. *Biotechnol. Bioeng.* 40:817–825.

Crowe, B., S. Attalah, S. Agrawal et al. 2012. A comparison of *Nannochloropsis salina* growth performance in two outdoor pond designs: Conventional versus the ARID raceway with superior temperature management. *Int. J. Chem. Eng.* 2012: Article ID 920608.

Eppley, R. W. 1972. Temperature and phytoplankton growth in the sea. *Fish. Bull.* 70:1063–1085.

Falkowski, P. G., Z. Dubinsky, and K. Wyman. 1985. Growth-irradiance relationships in phytoplankton. *Limnol. Oceanogr.* 30:311–321.

Fernandez, F. G., F. G. Camacho, J. A. Perez, J. M. Sevilla, and E. M. Grima. 1997. A model for light distribution and average solar irradiance inside outdoor tubular photobioreactors for the microalgal mass culture. *Biotechnol. Bioeng.* 55:701–714.

Forján, E., I. Garbayo, M. Henriques et al. 2011. UV-A mediated modulation of photosynthetic efficiency, xanthophyll cycle, and fatty acid production in *Nannochloropsis*. *Mar. Biotechnol.* 13:366–375.

Geider, R. J. and B. A. Osborne. 1989. Respiration and microalgal growth: A review of the quantitative relationship between dark respiration and growth. *New Phytol.* 112:327–341.

Geider, R. J., B. A. Osborne, and J. A. Raven. 1985. Light dependence of growth and photosynthesis in *Phaeodactylum tricornutum*. *J. Phycol.* 21:609–619.

Goldman, J. C. 1979. Outdoor algal mass cultures. II. Photosynthetic yield limitations. *Water Res.* 13:119–136.

Goldman, J. C. and E. J. Carpenter. 1974. A kinetic approach to the effect of temperature on algal growth. *Limnol. Oceanogr.* 19:756–766.

Grobbelaar, J. U. and C. J. Soeder. 1985. Respiration losses in planktonic green algae cultivated in raceway ponds. *J. Plankton Res.* 7:497–506.

Grima, E. M., J. A. S. Perez, F. G. Camacho, J. M. F. Sevilla, and F. G. A. Fernandez. 1996. Productivity analysis of outdoor chemostat culture in tubular air-lift photobioreactors. *J. Appl. Phycol.* 8:369–380.

Huesemann, M. H., T. S. Hausmann, R. Bartha et al. 2009. Biomass productivities in wild type and a new pigment mutant of *Cyclotella* sp. (Diatom). *Appl. Biochem. Biotechnol.* 157:507–526.

Huesemann, M. H., J. Van Wagenen, T. Miller et al. 2013. A screening model to predict microalgae biomass growth in photobioreactors and ponds. *Biotechnol. Bioeng.* 111:1583–1594.

Ingle, J. D. J. and S. R. Crouch. 1988. *Spectrochemical Analysis*. Englewood Cliffs, NJ: Prentice Hall.

James, S. C., V. Janardhanam, and D. T. Hanson. 2013. Simulating pH effects in an algal growth hydrodynamic model. *J. Phycol.* 49:608–615.

Jensen, S. and G. Knutsen. 1993. Influence of light and temperature on photoinhibition of photosynthesis in *Spirulina platensis*. *J. Appl. Phycol.* 5:495–504.

Meyer, C. R., C. S. Renschler, and R. C. Vining. 2008. Implementing quality control on a random number stream to improve a stochastic weather generator. *Hydrol. Process.* 22:1069–1079.

Montagnes, D. J. S. and D. J. Franklin. 2001. Effect of temperature on diatom volume, growth rate, and carbon and nitrogen content: Reconsidering some paradigms. *Limnol. Oceanogr.* 46:2008–2018.

Nicks, A. D. and L. J. Lane. 1989. Weather generator. In *Profile Model Documentation: USDA Water Erosion Prediction Project, Hillslope Version, Rep. 2*, eds. L. J. Lane and M. A. Nearing, pp. 2.1–2.19. West Lafayette, IN: National Soil Erosion Research Laboratory, Agricultural Research Service, U.S. Department of Agriculture.

Ogbonna, J. C. and H. Tanaka. 1996. Night biomass loss and changes in biochemical composition of cells during light/dark cyclic culture of *Chlorella pyrenoidosa*. *J. Ferment. Bioeng.* 82:558–564.

Packer, A., Y. Li, T. Andersen, Q. Hu, Y. Kuang, and M. Sommerfeld. 2011. Growth and neutral lipid synthesis in green microalgae: A mathematical model. *Bioresour. Technol.* 102:111–117.

Perkins, W. A. and M. C. Richmond. 2007. MASS2, modular aquatic simulation system in two dimensions: Theory and numerical methods, Report PNNL-14820-1. Richland, WA: Pacific Northwest National Laboratory. http://www.pnl.gov/publications/ (accessed December 15, 2015).

Platt, T. and A. D. Jassby. 1976. The relationship between photosynthesis and light for natural assemblages of coastal marine phytoplankton. *J. Phycol.* 12:421–430.

Pottier, L., J. Pruvost, J. Deremetz, J. F. Cornet, J. Legrand, and C. G. Dussap. 2005. A fully predictive model for one-dimensional light attenuation by *Chlamydomonas reinhardtii* in a torus reactor. *Biotechnol. Bioeng.* 91:569–582.

Quinn, J., L. de Winter, and T. Bradley. 2011. Microalgae bulk growth model with application to industrial scale systems. *Bioresour. Technol.* 102:5083–5092.

Ryther, J. H. and R. R. L. Guillard. 1962. Studies of marine planktonic diatoms: III. Some aspects of temperature on respiration of five species. *Can. J. Microbiol.* 8:447–453.

Suh, I. S. and S. B. Lee. 2003. A light distribution model for an internally radiating photobioreactor. *Biotechnol. Bioeng.* 82:180–189.

Torzillo, G., A. Sacchi, R. Materassi, and A. Richmond. 1991. Effect of temperature on yield and night biomass loss in *Spirulina platensis* grown outdoors in tubular photobioreactors. *J. Appl. Phycol.* 3:103–109.

U.S. Department of Energy (DOE), Office of Energy Efficiency and Renewable Energy. 2010. National algal biofuels technology roadmap. http://www1.eere.energy.gov/bioenergy/ pdfs/algal_biofuels_roadmap.pdf (accessed December 15, 2015).

U.S. Department of Energy (DOE), Office of Energy Efficiency and Renewable Energy. 2012. Biomass multi-year program plan, November 2012. http://www1.eere.energy.gov/ bioenergy/pdfs/mypp_november_2012.pdf (accessed December 15, 2015).

Van Wagenen, J., T. W. Miller, S. Hobbs, P. Hook, B. Crowe, and M. H. Huesemann. 2012. Effects of light intensity and temperature on fatty acid composition in *Nannochloropsis salina*. *Energies* 5:731–740.

Venteris, E. R., R. Skaggs, A. M. Coleman, and M. S. Wigmosta. 2013a. A GIS cost model to assess the availability of freshwater, seawater, and saline groundwater for algal biofuel production in the United States. *Environ. Sci. Technol.* 47:4840–4849.

Venteris, E. R., R. L. Skaggs, M. S. Wigmosta, and A. M. Coleman. 2013b. A national-scale comparison of resource and nutrient demands for algae-based biofuel production by lipid extraction and hydrothermal liquefaction. *Biomass Bioenerg.* 64:276–290.

Vonshak, A., G. Torzillo, J. Masojidek, and S. Boussiba. 2001. Sub-optimal morning temperature induces photoinhibition in dense outdoor cultures of the alga *Monodus subterraneus* (Eustigmatophyta). *Plant Cell Environ.* 24:1113–1118.

Waichler, S. R. and M. S. Wigmosta. 2003. Development of hourly meteorological values from daily data and significance to hydrological modeling at the H. J. Andrews Experimental Forest. *J. Hydrometeor.* 4:251–263.

Wigmosta, M. S., A. M. Coleman, R. J. Skaggs, M. H. Huesemann, and L. J. Lane. 2011. National microalgae biofuels production potential and resource demand. *Water Resour. Res.* 47:1–13.

Yoder, J. A. 1979. Effect of temperature on light-limited growth and chemical composition of *Skeletonema costatum*. *J. Phycol.* 15:362–370.

Yun, Y. S. and J. M. Park. 2003. Kinetic modeling of the light-dependent photosynthetic activity of the green microalga *Chlorella vulgaris*. *Biotechnol. Bioeng.* 83:303–311.

6 Genetic Engineering of Microalgae
Current Status and Future Prospects

Andrew Spicer and Saul Purton

CONTENTS

6.1 INTRODUCTION

Microalgae represent a remarkably diverse, yet largely unexploited resource. Currently, the only commercially established products from algae are whole algal extracts from species such as *Chlorella* that are marketed as a health food and as feed for aquaculture, or a few high-value biochemicals purified from microalgae such as β-carotene from *Dunaliella salina*, astaxanthin from *Haematococcus pluvialis*, and the long-chain polyunsaturated fatty acid docosahexaenoic acid from the heterotrophic alga *Crypthecodinium cohnii* (Borowitzka 2013; Leu and Boussiba 2014). These species are essentially wild isolates and, as such, are not adapted to intensive cultivation as monocultures and to the high productivities required for their exploitation in industrial biotechnology. In the absence of a controllable sexual cycle for most microalgal species that would allow the combining of desirable traits and the removal of unwanted alleles (as used in the breeding of crop varieties and livestock), "algal domestication" has simply relied on bioprospecting for superior natural isolates or the use of mutagenesis to select for improved phenotypes (e.g., Jin et al. 2003; Chekanov et al. 2014). Both strategies are somewhat limited in that they offer only incremental improvements, and the latter approach is confounded by the accumulation of additional, deleterious mutations within the selected strain (Bonente et al. 2011).

What is required to bring about a disruptive change in the field of algal biotechnology are advanced genetic engineering methods that allow significant designed improvements to be made to key industrial species, as detailed in Table 6.1. Such improvements could, for example, be in (1) growth performance in photobioreactors where biomass production is affected by factors such as light and CO_2 utilization, tolerance to shear forces, biofouling of surfaces, and resistance to invasion by contaminating organisms; (2) increased product synthesis through metabolic engineering of key biochemical pathways or chemical modifications to the product through the introduction of novel enzymes; (3) changes to the physiology of the alga that reduce cultivation costs, such as conversion of a strict phototroph to one able to utilize sugars as an alternative carbon and energy source (Zaslavskaia et al. 2001) or conversion of a vitamin B_{12} auxotroph to one that no longer requires this expensive addition to the medium (Helliwell et al. 2011); and (4) changes to cell morphology, cell wall composition, and so on, that aid downstream processing steps such as biomass harvesting and cell breakage.

Genetic modification (GM) technologies also provide the potential to exploit microalgae as light-driven biotechnology platforms for the synthesis of completely novel recombinant products. While these algal cell platforms would have to compete with well-established heterotrophic systems such as bacteria, yeasts, and mammalian cells, there are several areas where an algal platform might have an advantage. First, several chlorophyte species such as *Chlorella pyrenoidosa* (now renamed *Chlorella sorokiniana*), *Chlorella vulgaris*, and *H. pluvialis* have GRAS (generally recognized as safe) status and have long-established uses as health food and animal feed. As such, whole algal extracts containing a recombinant product could potentially be blended into cosmetics, skin creams, and so on, or could be used as oral vaccines for livestock or aquaculture where the addition of the dried algae into the feed could provide a low-cost vaccination strategy against a range of pathogens. Second, algae could be engineered to produce high-value bioactive compounds derived from terpenoids, fatty acids, or tetrapyrroles since, like plants, they already possess the basic biosynthetic pathways for these compounds and therefore require the addition of only a few extra biosynthetic genes. Third, algae possess chloroplasts with their own small genetic system. This cellular compartment represents an attractive "prokaryotic-like" platform for the synthesis of recombinant proteins,

TABLE 6.1
Examples of Possible Genetic Engineering Approaches to Algal Domestication

Stage in the Production Pipeline	Example of the Engineering Step	References Illustrating Feasibility
High-Density Industrial Cultivation		
1. Better light penetration	Reduction in size or abundance of light-harvesting complexes within the chloroplast thylakoids	Bonente et al. (2011) Cazzaniga et al. (2014)
2. Modification to trophic growth requirements	Heterotrophic or mixotrophic growth using cheap sugars or other cheap feedstocks	Zaslavskaia et al. (2001) Blifernez-Klassen et al. 2012)
	Removal of vitamin B_{12} auxotrophy	Helliwell et al. (2011)
3. Improved energy to biomass conversion	Improvements to the photosynthetic process and introduction of novel carbon-fixation pathways	Stephenson et al. (2011) Hanson et al. (2013) Bar-Even et al. (2010)
4. Improved tolerance to abiotic stress	Improved tolerance to high light and oxidative stress	Förster et al. (2005)
	Improved salt tolerance	Perrineau et al. (2014)
Downstream Processing		
1. Low-cost cell harvesting	Expression of cell wall structures to induce autoflocculation	Vandamme et al. (2013)
2. Easier cell breakage	Cell-wall weakening through induced expression of viral genes encoding lytic enzymes or induced inhibition of cell wall synthesis (e.g., using a temperature shift)	Cheng et al. (2013) Loppes and Deltour (1978)
3. Product secretion from cells	Introduction of secretory/efflux pathways or use of native pathways for recombinant protein secretion	Ramachandra et al. (2009) Lauersen et al. (2013)
Improved Productivity or Synthesis of Novel Products		
1. Increased lipid productivity	Manipulation of triacylglycerol (TAG) biosynthesis pathways or downregulation of TAG catabolism	Hsieh et al. (2012) Liang and Jiang (2013)
	Redirecting carbon flux from starch to TAGs by blocking starch biosynthesis	de Jaeger et al. (2014)
2. Novel fatty acids	Introduction of thioesterases, elongases, and desaturases to create designer oils	Blatti et al. (2012) Hamilton et al. (2014)
3. Novel pigments	Introduction of carotenoid biosynthesis enzymes	Cordero et al. (2011)
4. Novel terpenoids	Introduction of plant-derived isoprenoid biosynthesis enzymes	Zedler et al. (2015) Kempinski et al. (2015)
5. Recombinant proteins	Proteins for industrial, nutritional, and medical applications	Rasala and Mayfield (2015)

particularly where protein folding and disulfide bond formation has proved challenging using bacterial hosts or where the nature of eukaryotic toxins preclude their
synthesis in eukaryotic hosts such as yeasts or mammalian cell lines. One example
that encompasses both of these issues is the synthesis of immunotoxins; Mayfield
and colleagues have demonstrated recently how the chloroplast of *C. reinhardtii* can
be used to make such complex proteins, with the antibody moiety folding correctly
and the cytotoxin completely contained with the organelle (Tran et al. 2013).

The last 10 years or so have witnessed significant advances in the field of algal
transgenics, initially driven by the interest in the exploitation of microalgae as feedstock for biofuels and more lately by their potential as platforms for high-value
products as discussed earlier. In this chapter, we review these advances and future
directions for the technology. However, such advances in genetic engineering must
be accompanied by careful assessment and mitigation of risk from GM microalgae
with clear regulatory frameworks and public engagement in order to fully realize the
potential of algal biotechnology (Flynn et al. 2012; Henley et al. 2013).

6.2   BASIC REQUIREMENTS FOR GENETIC MANIPULATION

6.2.1   DNA DELIVERY METHODS

DNA delivery into microalgal cells has been achieved through a number of different approaches, each of which has merits for particular species. At present, there
is no "one-size-fits-all" approach to nuclear transformation of microalgae, in large
part due to the diverse biology of this polyphyletic group of eukaryotes including
significant differences in cell size, cell wall composition, and physiology. These
all serve to influence the efficiency in which DNA can be delivered into the cell
interior and thence into the nuclear genome. Transformation efficiencies for microalgae are orders of magnitude lower than those observed in other microorganisms
such as yeasts, with efficiencies being as low as 10^{-8} (i.e., 1 in 100 million cells
stably transformed) for some of the most difficult to transform species, and as high
as 10^{-3} in the most routinely transformed species such as *C. reinhardtii*. DNA
delivery methods that have proven successful for microalgae are summarized in
Table 6.2, with the most established being electroporation and bombardment with
DNA-coated microparticles (so-called biolistic transformation). Another promising method that appears to be applicable to a wide range of species involves the
use of the plant pathogen *Agrobacterium tumefaciens*. Hypervirulent strains of this
bacterium are sufficiently promiscuous to transfer single-stranded DNA carrying
transgenes into the cells of non-plant eukaryotic organisms including freshwater
and marine algae (e.g., Anila et al. 2011; Pratheesh et al. 2014). The mechanism
of DNA transfer is akin to bacterial conjugation, and, indeed, a recent report has
shown that *Escherichia coli* conjugation can be used to transfer plasmid DNA into
two diatom species where it replicates stably in the nucleus as an episome (Karas
et al. 2015). This opens up the possibility of artificial chromosome transfer into
diatoms and other microalgal groups and would represent a substantial advance
to the field as it would enable the one-step transfer of whole metabolic pathways

TABLE 6.2
DNA Delivery Methods Used for the Transformation of Microalgal Species

DNA Delivery Method	Comments	Algal Examples	References
Electroporation	Generic method, but challenging if cells have thick cell wall or where multiple membranes must be breached for organelle transformation	*Chlamydomonas reinhardtii* *Phaeodactylum tricornutum* *P. tricornutum* (chloroplast) *Nannochloropsis oceanica* *Chlorella zofingiensis*	Yamano et al. (2013) Zhang and Hu (2014) Xie et al. (2014) Vieler et al. (2012) Liu et al. (2014)
Agitation with glass beads	Simple, low-cost method but requires removal of cell wall	*C. reinhardtii* *C. reinhardtii* (chloroplast)	Neupert et al. (2012) Economou et al. (2014)
Agitation with silicon carbide whiskers	Much more effective with walled strains than glass beads, but inhalation of whiskers poses a serious health risk	*C. reinhardtii* *Amphidinium* sp. and *Symbiodinium microadriaticum*	Dunahay et al. (1997) ten Lohuis and Miller (1998)
Agitation with positively-charged nanoparticles	Low-cost method that works with walled strains (to date, only one report using a single species)	*C. reinhardtii*	Kim et al. (2014)
Microparticle bombardment (biolistics)	General method allowing DNA transfer across cell walls and multiple membranes	*C. reinhardtii* *C. reinhardtii* (chloroplast) *P. tricornutum* *Platymonas subcordiformis* (chloroplast) *Dunaliella salina* *Pseudochoricystis ellipsoidea*	Neupert et al. (2012) Purton (2007) Apt et al. (1996) Cui et al. (2014) Tan et al. (2005) Imamura et al. (2012)
Natural DNA uptake following weakening of cell wall	Cell wall weakened by treatment with 2-deoxy-ᴅ-glucose and cellulose (to date, only one report using a single species)	*Chlorella ellipsoidea*	Liu et al. (2013)
Agrobacterium conjugation	Numerous reports of successful DNA transfer into different algal species including marine species	*C. reinhardtii* *Nannochloropsis* sp. *Penium margaritaceum*	Pratheesh et al. (2014) Cha et al. (2011) Sørensen et al. (2014)
Escherichia coli conjugation	Efficient episome transfer into walled diatoms	*P. tricornutum* and *Thalassiosira pseudonana*	Karas et al. (2015)

into microalgae via gene assemblies incorporated into artificial chromosomes. Two other simple methods of DNA delivery have recently been described that show promise, although these have yet to be applied to more than one species. The first involves agitation, either by vortexing or sonication, of *C. reinhardtii* cells with positively charged nanoparticles onto which DNA has been bound (Kim et al. 2014). The second involves natural uptake of naked DNA following weakening of the cell wall of *Chlorella ellipsoidea* (Liu et al. 2013).

For DNA delivery into the chloroplast and mitochondrial organelles, the established method involves biolistics, principally because this bombardment technique allows DNA transfer across multiple membranes. However, chloroplast transformation of *C. reinhardtii* can also be achieved, albeit at low efficiency, by agitation with glass beads (Economou et al. 2014), and there is a single report of chloroplast transformation of a diatom using electroporation (Cui et al. 2014).

6.2.2   DNA CIS ELEMENTS REQUIRED FOR TRANSGENE EXPRESSION

Typically, genetic engineering involves as a first step the assembly of a chimeric expression cassette comprising different DNA elements that allow the stable (and possibly tuneable) expression of a transgene in the chosen microalgal host. Thus, a typical configuration of elements would be a promoter, 5′ untranslated region (UTR) including transcription start site, the coding region of the desired protein, and 3′ UTR including a transcription terminator. In most cases, the promoter and UTRs are derived from endogenous genes of the target microalga and are chosen because of the robust expression of that gene. For *C. reinhardtii*, a popular promoter/5′ UTR element is that from *RBCS2*, and this is often used in combination with the *HSP70A* promoter to provide robust, constitutive expression. Many other endogenous promoters have also proved successful for transgene expression in *C. reinhardtii*, and in a recent review, Mussgnug (2015) provides an extensive list including various inducible promoters that allow regulated expression of the transgene. Similarly, in *P. tricornutum* the light-regulated *fcpA* and *fcpB* promoter/5′ UTR elements are commonly used (Kroth 2007), although *EF2* that encodes elongation factor 2 represents a promising new constitutive promoter (Seo et al. 2015). However, for many microalgal species of note, such endogenous promoters have not been mapped and characterized, and researchers have relied on the potential activity of heterologous promoters from related species or plant viral promoters such as the 35S cauliflower mosaic virus promoter (Walker et al. 2005).

Other elements can also be included in the expression cassette, such as intron sequences containing transcriptional enhancers or sequences conferring enhanced mRNA stability or, conversely, enhanced mRNA turnover. In *C. reinhardtii*, intron 1 of *RBCS2* is widely used as an enhancer sequence for nuclear transgene expression (Walker et al. 2005), and the use of sequences that act as matrix attachment regions (MARs), which are thought to act as intergenic spaces, or genetic insulators, has been explored in *D. salina* (Wang et al. 2007). In addition, the design of the coding sequence represents an important consideration since pronounced codon preference can be seen in algal genomes: for example, *C. reinhardtii* nuclear genes show a strong bias for GC-rich codons, whereas chloroplast genes show a pronounced bias for AT-rich codons. The optimization of transgenic coding sequence to match this bias has been shown to improve translational efficiency (Rasala and Mayfield 2015), and the use of synthetic DNA sequences designed using codon-optimization algorithms is now generally preferred over the use of native genes or cDNAs as the source of the coding sequence. Finally, we are now seeing more ambitious construction methods that allow single-step introduction of multiple transgenes (Noor-Mohammadi et al. 2014) and the use of viral 2A peptide sequence elements for cotranslation of multiple genes (Rasala et al. 2012).

6.2.3 Selectable Markers

Numerous marker genes have been used for selection of microalgal transformants (Walker et al. 2005). In most cases, these are based on bacterial antibiotic-resistance genes and include markers conferring resistance to zeomycin (the *ble* gene), paromomycin (*aphAVIII*), hygromycin B (*hpt*), G418 (*nptII*), spectinomycin (*aadA*), chloramphenicol (*cat*), and nourseothricin (*nat*). The choice of marker is determined by the particular antibiotic sensitivities of the target species. However, if we consider the regulatory landscape in relation to genetically modified plants, there is concern about the horizontal transfer of such marker genes to pathogenic bacteria. As such, only two commonly used markers are authorized for widespread outdoor application of transgenic crop plants: these are *nptII* and *hpt*. It is, therefore, likely that for commercial applications, the use of other marker genes in transgenic microalgae would not be granted regulatory approval, and indeed the presence of *nptII* or *hpt* in strains to be cultivated outdoors (whether in open ponds or closed photobioreactors) would need to be strongly justified, regulated and closely monitored. One strategy to circumvent the use of bacterial antibiotic-resistance genes is to use dominant alleles of endogenous algal genes as selectable markers (so-called self-cloning). Examples of such markers include alleles of the *RPS14* or *RPL41* genes that encode 80S ribosomal subunits that confer resistance to emetine and cycloheximide, respectively, or those of the phytoene desaturase gene *PDS* that confer resistance to norflurazon (Walker et al. 2005).

Moreover, it is likely that emerging technologies that allow the precise manipulation of native and engineered sequences (see Section 6.4.2) will allow the creation of transgenic lines free of any selectable marker. This would clearly be a more satisfactory approach to the deployment of any modified strains in an outdoor cultivation setting.

For chloroplast transformation, the most widely used selectable marker is *aadA*, which confers resistance to spectinomycin and streptomycin. In addition, dominant alleles of endogenous genes such as herbicide-resistance alleles of *psbA* can be used to circumvent concerns about bacterial antibiotic-resistance genes in the polyploid organelle genome (Purton 2007). In addition, strategies for marker-free generation of chloroplast transformants of *C. reinhardtii* have already been established (Economou et al. 2014).

6.3 CURRENT STATUS OF THE FIELD AND RECENT DEVELOPMENTS

6.3.1 RECENT ADVANCES IN ALGAL OMICS

With the availability of inexpensive next-generation sequencing and ever-increasing improvements in computer algorithms for sequence assembly and annotation, we are witnessing a rapid rise in the number of algal genomes being sequenced (Tirichine and Bowler 2011). A quick Internet survey identifies nearly 30 complete microalgal genomes at the time of writing, although many more are known to be in progress or have not been made publicly available. For several of the model species such as *C. reinhardtii* and *P. tricornutum*, the genomic data continue to be refined and improved (Blaby et al. 2014) and are being complemented with extensive transcript, proteomic and metabolite data gathered under a range of physiological conditions. To make these big data accessible to researchers, online resources such as the Algal Functional Annotation Tool (pathways.mcdb.ucla.edu), DiatomCyc (www.diatomcyc.org), and the *Nannochloropsis oceanica* genome browser (benning-linux.bch.msu.edu) have been established. Importantly, the integration of the omics data, together with flux balance analysis, is allowing the development of dynamic models of algal cell biology (e.g., Boyle and Morgan 2009; Dal'Molin et al. 2011; Fabris et al. 2014). Such models serve to underpin and guide genetic engineering strategies and are catalyzing efforts to further develop and refine the transformation tools (Schmidt et al. 2010).

6.3.2 RECENT PROGRESS IN NUCLEAR TRANSFORMATION OF MICROALGAE

Nuclear transformation methodology is well established for a handful of chlorophyte and diatom species (Walker et al. 2005; Kroth 2007; Siaut et al. 2007; Gong et al. 2011), and although *C. reinhardtii* and *P. tricornutum* remain the most advanced models for each group, there has been steady progress in the development of methods for other industrially relevant species including key green algal genera (e.g., *Chlorella, Dunaliella, Haematococcus*, and *Scenedesmus*) and new diatoms (e.g., *Thalassiosira, Fistulifera*, and *Chaetoceros*). In addition, promising genetic engineering technologies have been developed for representatives of other algal groups including the eustigmatophyte genus *Nannochloropsis* and the cyanidiale *Cyanidoschyzon merolae*. Tables 6.2 and 6.3 provide examples of recent success in the transformation of these species, and we discuss recent advances for *C. reinhardtii* and *P. tricornutum* in the following text.

As mentioned in Section 6.2.1, new technologies for *C. reinhardtii* have included novel DNA delivery methods involving advanced electroporators (Yamano et al. 2013), positively charged nanoparticles (Kim et al. 2014), or improvements in *Agrobacterium*-mediated delivery (Pratheesh et al. 2014) that increase the transformation efficiency. This efficiency has allowed the generation of libraries of insertional mutants containing many thousands of members (e.g., Zhang et al. 2014). The issue of weak or unstable transgene expression in *C. reinhardtii* has been addressed in part by the selection of new mutant strains as expression hosts (Neupert et al. 2009) and the

TABLE 6.3

Recent Reports of Nuclear Transformation of Additional Microalgal Species

Species (Phylum)	Transformation Method	Selectable Marker	Significance	References
Dunaliella bardawil (Chlorophyta)	Agrobacterium	*hpt*	Strain already commercially exploited due to hyperaccumulation of β-carotene and ability to grow in hypersaline conditions.	Anila et al. (2011)
Lobosphaera incisa (Chlorophyta)	Electroporation	*ble*	Promising candidate for production of arachidonic acid, an essential ω-6 PUFA.	Zorin et al. (2014)
Pandorina morum (Chlorophyta)	Particle gun	*aphVIII*	Model alga for early emergence of multicellularity.	Lerche and Hallman (2014)
Parachlorella kessleri (Chlorophyta)	Agrobacterium	*hpt*	Marine variant of this species with potential as biomass feedstock and in bioremediation.	Rathod et al. (2013)
Ostreococcus tauri (Chlorophyta)	Electroporation	*nptII* or *nat*	Species represents the smallest known free-living eukaryote; gene-targeting possible.	Lozano et al. (2014)
Pseudochoricystis ellipsoidea (Chlorophyta)	Particle gun	*nptII*	This species accumulates significant amounts of aliphatic hydrocarbons.	Imamura et al. (2012)
Closterium psl complex (Chlorophyta)	Particle gun	*ble*	First report for a Charophycean alga, and only second example of a transformable alga that has a controllable sexual cycle.	Abe et al. (2011)
Scenedesmus obliquus (Chlorophyta)	Electroporation	*cat*	Commercially important species for production of biofuels and other lipid products.	Guo et al. (2013)
Cyanidioschyzon merolae (Rhodophyta)	Polyethylene-glycol (PEG)	*URA$_{Cm-Cm}$*	Extremophile found in sulfate-rich hot springs; gene-targeting possible.	Fujiwara et al. (2013)
Fistulifera sp. (Heterokonta)	Particle gun	*nptII*	Oil-rich marine diatom with potential as feedstock for biofuels.	Muto et al. (2013)
Chaetoceros gracilis (Heterokonta)	Electroporation	*nat*	Commercially grown as feed for fisheries.	Ifuku et al. (2015)

development of "2A peptide" technology that allows the efficient cotranslation of a transgene with the selectable marker (Rasala et al. 2012). A wide range of fluorescent reporter genes is now available for gene expression and protein localization studies (Rasala et al. 2013), together with new selectable markers (Meslet-Cladière and Vallon 2011) and new regulated promoters such as the *METE* promoter that is repressed by vitamin B_{12} (Helliwell et al. 2014). Molecular tools for targeting of recombinant proteins into subcellular compartments have been developed (Lauersen et al. 2015), as have methods for efficient secretion of recombinant proteins into the growth medium that will facilitate protein recovery and purification (Lauersen et al. 2013; Rasala et al. 2013). There has also been significant progress in the development of reliable reverse-genetic tools for the targeted knockdown of endogenous genes in *C. reinhardtii* using RNA silencing approaches (e.g., van Dijk and Sarkar 2011; Hu et al. 2014), and a search of the PubMed database (www.pubmed.com) reveals over 100 publications reporting the successful knockdown of specific genes in *C. reinhardtii*. In contrast, the development of genome targeting methods for the knockout/modification of any chosen endogenous gene or the precise insertion of transgenes into target loci has proved particularly challenging. An increase in the frequency of rare homologous recombination events can be achieved by using single-stranded DNA for transformation (Zorin et al. 2009), and attempts to induce double-strand breaks (DSB) at target loci using an engineered zinc-finger nuclease (ZFN) or the Cas9 nuclease have been reported (Sizova et al. 2013; Jiang et al. 2014). However, as discussed in Section 6.4.2, we still await a reliable method of nuclear genome editing for *C. reinhardtii*.

For the diatom *P. tricornutum*, similar advances are being made in transformation efficiency through refinements in electroporation (Miyahara et al. 2013; Zhang and Hu 2014). New expression vectors and engineering strategies allow efficient synthesis of recombinant proteins (Hempel et al. 2011b), and also their secretion into the growth medium (Hempel and Maier 2012). Techniques for gene knockdowns via RNA interference have been established (Lavaud et al. 2012), and, importantly, genome editing using meganucleases and transcription activator-like effector nucleases (TALENs) is now feasible (see Section 6.4.2.2) allowing the knockout of target genes (Daboussi et al. 2014). It is now routine to express multiple transgenes in *P. tricornutum*, and this species is rapidly emerging as an attractive industrial biotechnology platform both for recombinant protein production (Hempel and Maier 2012) and for the synthesis of valuable metabolites such as biodegradable plastics and long-chain fatty acids (Hempel et al. 2011a; Hamilton et al. 2014).

6.3.3 RECENT PROGRESS IN ORGANELLAR TRANSFORMATION OF ALGAE

The full exploitation of an algal species as a biotechnological platform requires the development of routine tools for engineering not only the nuclear genome but also the genetic systems located in the chloroplast and mitochondrial organelles. The chloroplast is particularly attractive as a compartment both for recombinant protein synthesis and for metabolic engineering strategies given that the organelle appears to be capable of accumulating very high levels of foreign proteins in a stable, folded form (Bock 2014) and is the site of numerous biosynthetic pathways including those for carbohydrates, fatty acids, terpenoids, and tetrapyrroles. Chloroplast transformation of *C. reinhardtii* was first reported by Boynton et al. (1988), and the tools and technologies for genetic manipulation using this model species have improved significantly over the years. Several detailed reviews of these developments have been published (see Purton 2007); however, important recent advances include (1) the establishment of a simple method of DNA delivery based on agitation with glass beads that avoids the use of expensive biolistic equipment (Economou et al. 2014); (2) new selectable markers, including those based on the rescue of phototrophic or auxotrophic mutants that avoid the use of antibiotic-resistance markers (Purton et al. 2013); (3) new reporter genes and a negative selectable marker that can be used to study chloroplast gene regulation and optimization of transgene expression (Young and Purton 2014); (4) technologies for the regulated translation of transgenes when linked to the 5′ UTR of the endogenous *psbD* gene, such that the synthesis of toxic products in the chloroplast can be induced following biomass production by removal of copper or vitamins from the medium (Rochaix et al. 2014); and (5) the implementation of synthetic biology strategies for the design, synthesis, and assembly of transgenes in chloroplast expression vectors that allow a rapid pipeline from conception to creation of transgenic lines (Purton et al. 2013).

The last few years have seen the increasing development of the *C. reinhardtii* chloroplast as a host for the synthesis of therapeutic proteins with over 25 different such proteins reported (reviewed in Rasala and Mayfield 2015). Although the commercial exploitation of *C. reinhardtii* must compete against long-established bacterial and yeast hosts, the alga offers various advantages including its GRAS status as an edible organism, the low cost of cultivation, the ability of the chloroplast to correctly fold proteins including forming disulfide bonds, and the suitability of the chloroplast as a site of synthesis for proteins that are cytotoxic to eukaryotic cells (Rasala and Mayfield 2015). Recently, the technology has started to move out of the lab with the establishment of companies such as Solarvest (www.solarvest.ca) and Triton Algae Innovations (www.tritonhn.com), and the demonstration of pilot scale cultivation of transgenic lines (Gimpel et al. 2014).

However, *C. reinhardtii* is not necessarily the best choice for an industrial platform, and other chlorophyte algae such as species of *Chlorella* or *Dunaliella* offer various advantages such as superior growth rates and improved tolerances to abiotic stresses during cultivation (Rasala and Mayfield 2015). In principle, it should be relatively straightforward to transfer the chloroplast engineering technology developed for *C. reinhardtii* to other unicellular algae. However, different species present particular challenges such as the delivery of DNA across thick cell walls and multiple chloroplast membranes, identification of suitable markers for selection, and identification of neutral regions within the chloroplast genome suitable for DNA insertion. As detailed in Table 6.4, stable chloroplast transformation has now been reported in seven different species including four chlorophyte species. One surprising report details the transformation of the chloroplast in the marine diatom *P. tricornutum* using electroporation (Xie et al. 2014). Given that the DNA needs to cross five membranes (the cell membrane and the four membranes surrounding the secondary chloroplast of this heterokont), then it is indeed impressive that an electrical pulse can deliver DNA into the organelle and opens the way for chloroplast transformation of other commercially important heterokonts such as *Nannochloropsis*.

TABLE 6.4

Microalgal Species for Which Chloroplast Transformation Has Been Reported

Species (Phylum)	Transformation Method	Selectable Marker	Selection	References
Chlamydomonas reinhardtii (Chlorophyta)	Particle gun	atpB	Photoautotrophy	Boynton et al. (1988)
		aadA	Spectinomycin, streptomycin	Goldschmidt-Clermont (1991)
		psbA	Metribuzin, DCMU	Newman et al. (1992)
		aphA-6	Kanamycin, amikacin	Bateman and Purton (2000)
		16S or 23S rDNA	Spectinomycin, streptomycin, erythromycin	Newman et al. (1990)
		ARG9	Arginine auxotrophy	Remacle et al. (2009)
	Glass beads	tscA, psbH	Photoautotrophy	Kindle et al. (1991)
				Economou et al. (2014)
		codA	Negative selection for 5-fluorocytosine	Young and Purton (2014)
Haematococcus pluvialis (Chlorophyta)	Particle gun	aadA	Spectinomycin	Gutiérrez et al. (2012)
Dunaliella tertiolecta (Chlorophyta)	Particle gun	ereB	Erythromycin	Georgianna and Mayfield (2012)
Platymonas subcordiformis (Chlorophyta)	Particle gun	bar	Basta	Cui et al. (2014)
Porphyridium sp. (Rhodophyta)	Particle gun	AHAS (W492S)	Sulfometuron methyl	Lapidot et al. (2002)
Euglena gracilis (Euglenozoa)	Particle gun	aadA	Streptomycin and spectinomycin	Doetsch et al. (2001)
Phaeodactylum tricornutum (Heterokontophyta)	Electroporation	cat	Chloramphenicol	Xie et al. (2014)

Genetic engineering of the algal mitochondrial genome is still very much in its infancy, with only a few reports, and all confined to *C. reinhardtii*. Early work demonstrated that respiratory mutants carrying deletions in the cytochrome *b* gene (*cob*) on the mitochondrial genome could be rescued by transformation with DNA carrying the wild-type *cob* (reviewed in Larosa and Remacle 2013). Two promising reports have described the introduction of marker genes into the mitochondrial genome (Hu et al. 2011, 2012), although the engineering of mitochondrial metabolic pathways remains a technical challenge for the future.

6.4 CURRENT CHALLENGES

6.4.1 SYNTHETIC BIOLOGY APPROACHES TO EFFICIENT AND PREDICTABLE ENGINEERING

A UK Synthetic Biology Roadmap recently defined synthetic biology as follows: "Synthetic biology is the design and engineering of biologically based parts, novel devices and systems as well as the redesign of existing, natural biological systems." As such, improvement of existing microalgal strains for enhanced or altered productivity or, alternately, genetic modifications that result in the production of novel metabolites by a microalgal strain would equally fall under the umbrella of synthetic biology. Indeed, under these terms, most of what has been previously described in this chapter falls under this definition as it applies to microalgae. In more specific terms, the principles of synthetic biology include the use of validated genetic parts, applied with precision and predictability to produce end products for purpose. Synthetic biology attempts to apply engineering principles to biology.

6.4.2 REPRODUCIBLE AND RELIABLE NUCLEAR GENE TARGETING

In microalgal nuclei, typical transgene insertion overwhelmingly occurs through random, non-homologous integration. Site of integration effects can impact on transgene expression and stability. Nuclear transgene expression also requires validated promoters, enhancers, and terminators for a given strain. This can be a limiting factor when attempting transgenesis experiments with new microalgal strains. Random transgene insertion is also problematic when comparing strains as it is unlikely that a transgene will integrate into the same genomic site in independently derived strains. Recent advances in strain engineering approaches are offering alternate, more precise approaches to nuclear transgenesis in microalgae.

6.4.2.1 Algenious® Gene Trapping System

One of these novel approaches relies upon a gene trapping, genome interrogation platform called Algenious to identify nuclear "safe harbors" for transgene integration (A. Spicer, unpublished data). This system uses selectable marker gene cassettes flanked by intron splice donor and/or consensus sequences to trap actively expressed sequences within microalgal genomes (Figure 6.1). A safe harbor within this context is an actively expressed native gene, the disruption of which does not act as a lethal mutation, and whose regulatory sequences are capable of driving robust expression of an inserted transgene, integrated downstream of the respective transcription start site. This gene trapping approach has worked well for all eukaryotic microalgal species tested to date, including chlorophytes, eustigmatophytes, and diatoms. It is not

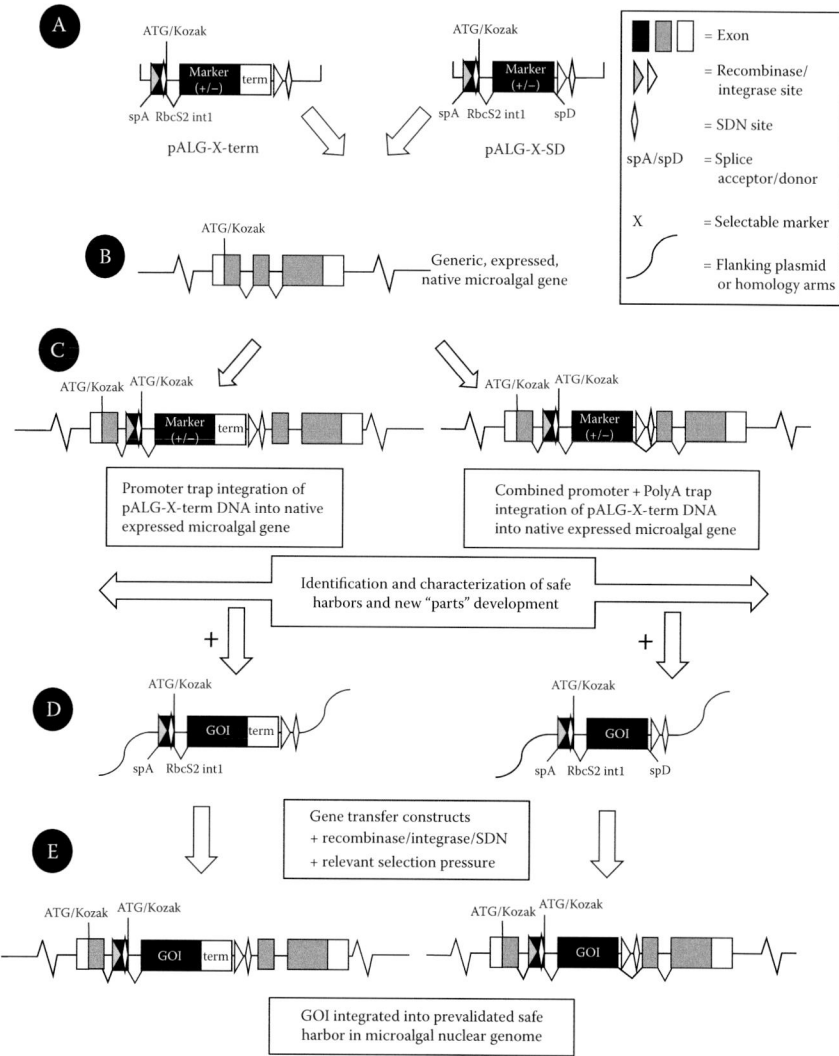

FIGURE 6.1 An Algenious system for rapid parts development, identification of ranked and validated safe harbor sites, and targeted gene insertion into microalgal nuclear genomes. Two alternative configurations of gene trapping vectors can be used, one is referred to as pALG-X-term and one as pALG-X-SD, representing vectors where X is a selectable marker cassette of choice and term represents a 3′ UTR including a termination/polyadenylation signal, whereas SD represents a consensus exon–intron splice donor sequence. A common eukaryotic microalgal intron sequence, in this instance *Chlamydomonas reinhardtii* RbcS2 intron 1 is incorporated into the constructs. The X-term constructs rely upon trapping of a functional promoter and transcription start site, whereas the X-SD constructs rely upon trapping both a functional promoter and transcription start site as well as a functional terminator/polyA signal. Constructs can be configured to include recognition sequences for site-specific recombinases, integrases, or site-directed nucleases such that trapped sites can be retargeted afterward through single or multiple rounds of insertion. *(Continued)*

dependent upon having validated promoters or other "parts" for a given species. Rather, it effectively identifies promoters and terminators that can be tested rapidly for their utility in regulating transgene expression. Not only does this approach identify safe harbors where transgene expression is tolerated and stably maintained even in the absence of selective pressure (A. Spicer, unpublished data), but it also acts to create insertional strain banks (that can be extremely useful for phenotype screens) in those microalgae that exist primarily as haploid cells and can be used to integrate so-called landing pad sequences into the safe harbor sites. In contrast to those insertional mutagenesis strategies that utilize conventional transgene DNA constructs incorporating promoter, 5′ UTR, selectable marker ORF, and 3′ UTR/terminator and that can still produce an antibiotic-resistant transgenic strain through random integration irrespective of whether or not a native gene has been disrupted, the Algenious strategy exclusively identifies insertions into actively transcribed genes. In the absence of any validated promoter or other regulatory sequences for a given species, insertional mutagenesis strategies can be applied through use of a common suite of Algenious gene trapping DNAs.

Safe harbor sites can be ranked with respect to each other in regard to the relative levels of expression of a marker transgene or fluorescent protein. A log of safe harbor sites where transgenes are not only tolerated but also faithfully and predictably expressed can be used as a resource to apply genome editing technology as it comes into its own, that is, to target transgenes into nuclear genome sites where expression would be predictable.

Landing pad sites can be used to facilitate subsequent reproducible transgene knockins into the safe harbor sites using either site-specific recombinases, meganucleases, or other site-directed nucleases (SDNs), including transcription activator-like effector nucleases (TALENs), and clustered regularly interspaced short palindromic repeats (CRISPR)/Cas9 (see later). A validated strain bank can be derived with preconfigured safe harbor transgene landing pads into which DNA sequences can be integrated with precision to create reliable expression strains. This can include replacement of selectable marker sequences with a gene of interest sequence. As described earlier, a safe harbor inventory for a given microalgal species can also be derived and subsequently used within a TALEN, CRISPR/Cas9, or other SDN strategy to integrate transgenes of interest into prevalidated sites within a given microalgal genome.

FIGURE 6.1 (*Continued*)   Selectable markers are those that can be employed in a positive manner (such as *ble*/zeocin, *hpt*/hygromycin B) or both positive and negative (such as fusions between *ble* and *codA*, cytosine deaminase, or ble and a fluorescent protein). Trapped sequences can be used to rapidly identify useful "parts" including new promoters and terminators as well as potential enhancer sequences. Microalgal cells that have stably integrated DNAs are identified by their growth on the appropriate selectable agent with respective colonies representing a library of insertion sites for a given microalgal strain. Trapped strains can also be used to monitor the relative strength and/or stability of an insertion site with regard to transgene expression levels. This can lead to the identification of useful parts for further application within transgenic microalgal strains as well as a ranked set of validated transgene insertion sites where transgene expression levels are predicted to be stable and reproducible.

6.4.2.2 Site-Directed Nucleases and Other Genome Editing Tools

Genome editing refers to the application of a range of molecular tools to target a precise site within a genome, generally introducing double-strand breaks (DSBs) into that site that are then repaired to result in a modest deletion of a few nucleotides in the native gene. Using these tools, mutations can be created within native genes without the need for insertional inactivation by incorporation of a foreign gene encoding a selectable marker. As such, the various regulatory bodies that oversee the application of genetically modified organisms (GMOs) including modified plants (including the European Food Standards Agency [EFSA] and the German Biosafety Commission [ZKBS]) have considered that organisms created using similar approaches where there is no foreign transgene incorporated into the genome and where the change that has been effected is small (more similar to a naturally occurring mutation) should not be subject to the regulations governing the use of GMOs (Podevin et al. 2013). Of course, this is a subject of formal regulatory review with the science being ahead of the regulatory framework and presenting a significant challenge to the current definitions of what constitutes a GM organism, particularly under current EU definitions.

Genome editing tools include site-directed nucleases (SDNs) such as zinc-finger nucleases (ZFNs), meganucleases, TALENs and the more recently described compact TALENs (Beurdeley et al. 2013), and CRISPR/Cas9 (clustered regularly interspaced short palindromic repeats) (for review, see Hsu et al. 2014; Kim and Kim 2014). All can be used to create mutations within native genes (knockouts) or also to insert DNA sequences with precision into specific sites within the genome (knockins). Cellectis (France) has recently reported the successful application of TALENs in *P. tricornutum* (Weyman et al. 2014), while a transient CRISPR/Cas9 approach has been reported in *C. reinhardtii* (Jiang et al. 2014), and there has been one meeting report made to date where CRISPR/Cas9 has been reported to work in *N. oceanica*. It is likely that CRISPR/Cas9 as well as other genome editing approaches will be successfully developed and applied within microalgae as advances in this area are being made at an ever-increasing pace and as microalgae are the subject of increasing interest as production platforms for natural bioproducts as well as newly (engineered in) functionalities. One key challenge that remains in microalgae is the extremely low (in comparative terms) transformation efficiencies, which make the application of a robust selectable phenotype strategy of paramount importance in the identification of those rare cells within which the desired genetic change has been incorporated.

6.4.3 "Marker-Free" or "Marker-Benign" Technologies

The importance of "marker-free" or "marker-benign" technologies—where no foreign selectable marker gene has been employed at all or where the marker gene has been removed from the modified strain after the strain has been established—has been described previously and is expected to have increasing relevance to the algal biotechnology industry, as more optimized strains are applied within a commercial setting. In the chloroplast, those microalgal strains that are able to be grown in either a heterotrophic or phototrophic modes are amenable to the development and application of marker-free technologies. Here, homologous recombination of a selectable marker gene into a chloroplast gene encoding a necessary component of photosystem II (PSII) renders cells only able to grow on simple carbon sources and unable to grow photosynthetically. This is then rescued by replacement of the selectable marker gene with a desired gene of interest, concomitant with restoration of the PSII gene and phototrophic growth (Economou et al. 2014). Other strategies that can be employed to facilitate marker-free strain derivation include the use of positive–negative selection (PNS) strategies that have been widely applied in other eukaryotic organisms. In PNS, the marker gene confers a positive attribute, such as ability to grow in the presence of an antibiotic, but also can be selected against by virtue of the encoded marker gene's activity to produce a toxic or lethal (negative) effect in the presence of another exogenous factor, such as a chemical that is converted to a toxic by-product through enzymatic action. In some instances, PNS is achieved by fusing of sequences encoding the positive selectable marker to sequences conferring negative selection. An example of this would be fusion of the *ble* gene (conferring zeomycin resistance) coding sequence to a cytosine deaminase (*codA*) coding sequence, conferring sensitivity to 5-fluorocytosine.

6.5 CONCLUSIONS

It is clear that microalgal transgenic technology has received a resurgence in activity in recent years. Renewed interest in algae as a potential biofuel source has driven increased funding and, subsequently, numbers of researchers, as well as companies (see Table 6.5), into the area. This, in addition to the maturation of genomic technologies and the development of more sophisticated genetic manipulation strategies, is allowing more microalgal strains to be explored in relation to an increased understanding of the fundamental biology of these diverse organisms, but also in relation to their application as commercial platforms for the production of industrially relevant products. The range of improvements that has been reported continues to increase and impress with regard to the diversity of innovation that is being applied. The continued application of sophisticated strain improvement strategies is expected to be of increasing relevance to the algal biotechnology products market within the next 5–10 years and may have a direct impact on the future economically realistic application of algal biomass in liquid biofuel production. It is of note that there are a number of companies that are emerging as new business start-ups or existing companies that have chosen to adopt a tool and service approach to this marketplace. Significantly, these businesses fall strongly within the synthetic biology technology development sphere, suggesting that microalgae are being recognized as a strong opportunity for development and commercial relevance in this area. Continued commercial buy-in as well as public engagement is expected to have a strong influence upon the growth, success, and scale of the impact that algal biotechnology will play within society in future years.

TABLE 6.5

Companies Developing Genetic Engineering Technologies for Algae

Company	Location	Website	Research Areas
AB Seeds, Monsanto (formerly Rosetta Green)	Rehovot, Israel	www.abseeds.co.il	Production of human proteins in microalgae; microRNA technologies for engineering microalgae
Algenics[a]	Saint-Herblain, France		Microalgae-based technology to produce recombinant therapeutics for animal and human health
Algenuity	Stewartby, United Kingdom	www.algenuity.com	Advanced GM technologies for a range of microalgal species; site-specific nuclear genome modifications including the Algenious platform; Chloroplast engineering
Aurora Algae[a]	Hayward, California Perth, Western Australia		Strain engineering technologies to underpin development of high-value oils and protein products from microalgae
Cellectis	Paris, France	www.cellectis.com	Diatom nuclear engineering using site-specific endonucleases (meganucleases and TALENs/cTALENs)
Kuehnle AgroSystems	Honolulu, Hawaii	www.kuehnleagro.com	Synthetic biology tools for chloroplast and nucleus engineering
Life Technologies[b]	Carlsbad, California	www.thermofisher.com	Microalgal engineering kits; genome editing tools and custom vector production
PlanktOMICS	Laramie, Wyoming	www.planktomics.com	Microalgal strain engineering services
Sapphire Energy	San Diego, California	www.sapphireenergy.com	Strain engineering technologies to underpin development of biofuels, animal feed, and food from microalgae
Solarvest Bioenergy, Inc.	Summerville, Canada	www.solarvest.ca	Microalgal strain engineering for recombinant proteins and biohydrogen production
Solazyme	San Francisco, California	www.solazyme.com	Metabolic engineering of microalgae to produce tailored oils for various sectors including food and personal care
Synthetic Genomics	La Jolla, California	www.syntheticgenomics.com	Synthetic biology approaches to make designer microalgae for biofuel production and other applications
TransAlgae	Rehovot, Israel	www.transalgae.com	Microalgae as oral delivery platforms for recombinant vaccines and other products in farming and aquaculture
Triton Algae Innovations	San Diego, California	www.tritonai.com	Therapeutic proteins for livestock intestinal diseases produced in microalgal chloroplasts

[a] Operations halted in 2015.

[b] Now Thermofisher Scientific, as of 2015.

ACKNOWLEDGMENTS

Algal research in the Purton Laboratory is funded by Grants #BB/I007660/1 and #BB/L002957/1 from the UK Biotechnology and Biological Sciences Research Council and by the EU Framework Program 7 Projects SUNBIOPATH (KBBE 2009-3: GA245070) and GIAVAP (KBBE 2010-3: GA266401).

REFERENCES

Abe, J., S. Hori, Y. Tsuchikane, N. Kitao, M. Kato, and H. Sekimoto. 2011. Stable nuclear transformation of the *Closterium peracerosum-strigosum-littorale* complex. *Plant Cell Physiol.* 52:1676–1685.

Anila, N., A. Chandrashek, G. A. Ravishankar, and R. Sarada. 2011. Establishment of *Agrobacterium tumefaciens*-mediated genetic transformation in *Dunaliella bardawil*. *Eur. J. Phycol.* 46:36–44.

Apt, K. E., P. G. Kroth-Pancic, and A. R. Grossman. 1996. Stable nuclear transformation of the diatom *Phaeodactylum tricornutum*. *Mol. Gen. Genet.* 252:572–579.

Bar-Even, A., E. Noor, N. E. Lewis, and R. Milo 2010. Design and analysis of synthetic carbon fixation pathways. *Proc. Natl. Acad. Sci. USA* 107:8889–8894.

Bateman, J. M. and S. Purton. 2000. Tools for chloroplast transformation in *Chlamydomonas*: Expression vectors and a new dominant selectable marker. *Mol. Gen. Genet.* 263:404–410.

Beurdeley, M., F. Bietz, J. Li et al. 2013. Compact designer TALENs for efficient genome engineering. *Nat. Commun.* 4:1762.

Blaby, I. K., C. E. Blaby-Haas, N. Tourasse et al. 2014. The *Chlamydomonas* genome project: A decade on. *Trends Plant Sci.* 19:672–680.

Blatti, J. L., J. Beld, C. A. Behnke, M. Mendez, S. P. Mayfield, and M. D. Burkart. 2012. Manipulating fatty acid biosynthesis in microalgae for biofuel through protein-protein interactions. *PLoS One* 7:e42949.

Blifernez-Klassen, O., V. Klassen, A. Doebbe et al. 2012. Cellulose degradation and assimilation by the unicellular phototrophic eukaryote *Chlamydomonas reinhardtii*. *Nat. Commun.* 3:1214.

Bock, R. 2014. Engineering chloroplasts for high-level foreign protein expression. *Methods Mol. Biol.* 1132:93–106.

Bonente, G., C. Formighieri, M. Mantelli et al. 2011. Mutagenesis and phenotypic selection as a strategy toward domestication of *Chlamydomonas reinhardtii* strains for improved performance in photobioreactors. *Photosynth. Res.* 108:107–120.

Borowitzka, M. A. 2013. High-value products from microalgae—Their development and commercialization. *J. Appl. Phycol.* 25:743–756.

Boyle, N. R. and J. A. Morgan. 2009. Flux balance analysis of primary metabolism in *Chlamydomonas reinhardtii*. *BMC Syst. Biol.* 3:4.

Boynton, J. E., N. W. Gillham, E. H. Harris et al. 1988. Chloroplast transformation in *Chlamydomonas* with high velocity microprojectiles. *Science* 240:1534–1538.

Cazzaniga, S., L. Dall'Osto, J. Szaub et al. 2014. Domestication of the green alga *Chlorella sorokiniana*: Reduction of antenna size improves light-use efficiency in a photobioreactor. *Biotechnol. Biofuels* 7:157.

Cha, T. S., C. F. Chen, W. Yee, A. Aziz, and S. H. Loh. 2011. Cinnamic acid, coumarin and vanillin: Alternative phenolic compounds for efficient *Agrobacterium*-mediated transformation of the unicellular green alga, *Nannochloropsis* sp. *J. Microbiol. Met.* 84:430–434.

Chekanov, K., E. Lobakova, I. Selyakh, L. Semenova, R. Sidorov, R., and A. Solovchenko. 2014. Accumulation of astaxanthin by a new *Haematococcus pluvialis* strain BM1 from the white sea coastal rocks (Russia). *Mar. Drugs* 12:4504–4520.

Cheng, Y. S., Y. Zheng, J. M. Labavitch, and J. S. VanderGheynst. 2013. Virus infection of *Chlorella variabilis* and enzymatic saccharification of algal biomass for bioethanol production. *Bioresour. Technol.* 137:326–331.

Cordero, B. F., I. Couso, R. León, H. Rodríguez, and M. Á. Vargas. 2011. Enhancement of carotenoids biosynthesis in *Chlamydomonas reinhardtii* by nuclear transformation using a phytoene synthase gene isolated from *Chlorella zofingiensis*. *Appl. Microbiol. Biotechnol.* 91:341–351.

Cui, Y., S. Qin, and P. Jiang. 2014. Chloroplast transformation of *Platymonas* (*Tetraselmis*) *subcordiformis* with the *bar* gene as selectable marker. *PLoS One* 9:e98607.

Daboussi, F., S. Leduc, A. Maréchal et al. 2014. Genome engineering empowers the diatom *Phaeodactylum tricornutum* for biotechnology. *Nat. Commun.* 5:3831.

Dal'Molin, C. G., L. E. Quek, R. W. Palfreyman, and L. K. Nielsen. 2011. AlgaGEM—A genome-scale metabolic reconstruction of algae based on the *Chlamydomonas reinhardtii* genome. *BMC Genomics* 12(Suppl. 4):S5.

de Jaeger, L., R. E. Verbeek, R. B. Draaisma et al. 2014. Superior triacylglycerol (TAG) accumulation in starchless mutants of *Scenedesmus obliquus*: (I) mutant generation and characterization. *Biotechnol. Biofuels* 7:69.

Doetsch, N. A., M. R. Favreau, N. Kuscuoglu, M. D. Thompson, and R. B. Hallick. 2001. Chloroplast transformation in *Euglena gracilis*: Splicing of a group III twintron transcribed from a transgenic *psbK* operon. *Curr. Genet.* 39:49–60.

Dunahay, T. G., S. A. Adler, and J. W. Jarvik. 1997. Transformation of microalgae using silicon carbide whiskers. *Methods Mol. Biol.* 62:503–509.

Economou, C., T. Wannathong, J. Szaub, and S. Purton. 2014. A simple, low cost method for chloroplast transformation of the green alga *Chlamydomonas reinhardtii*. *Methods Mol. Biol.* 1132:401–411.

Fabris, M., M. Matthijs, S. Carbonelle et al. 2014. Tracking the sterol biosynthesis pathway of the diatom *Phaeodactylum tricornutum*. *New Phytol.* 204:521–535.

Flynn, K., A. Mitra, H. Greenwell, and J. Sui. 2012. Monster potential meets potential monster: Pros and cons of deploying genetically modified microalgae for biofuels production. *Interface Focus* 3:20120037.

Förster, B., C. B. Osmond, and B. J. Pogson. 2005. Improved survival of very high light and oxidative stress is conferred by spontaneous gain-of-function mutations in *Chlamydomonas*. *Biochim. Biophys. Acta* 1709:45–57.

Fujiwara, T., M. Ohnuma, M. Yoshida, T. Kuroiwa, and T. Hirano. 2013. Gene targeting in the red alga *Cyanidioschyzon merolae*: Single- and multi-copy insertion using authentic and chimeric selection markers. *PLoS One* 8:e73608.

Georgianna, D. R. and S. P. Mayfield. 2012. Exploiting diversity and synthetic biology for the production of algal biofuels. *Nature* 488:329–335.

Gimpel, J., J. S. Hyun, N. G. Schoepp, and S. P. Mayfield. 2014. Production of recombinant proteins in microalgae at pilot greenhouse scale. *Biotechnol. Bioeng.* 112:339–345.

Goldschmidt-Clermont, M. 1991. Transgenic expression of aminoglycoside adenine transferase in the chloroplast: A selectable marker of site-directed transformation of *Chlamydomonas*. *Nucleic Acids Res.* 19:4083–4089.

Gong, Y., H. Hu, Y. Gao, X. Xu, and H. Gao. 2011. Microalgae as platforms for production of recombinant proteins and valuable compounds: Progress and prospects. *J. Ind. Microbiol. Biotechnol.* 38:1879–1890.

Guo, S. L., X. Q. Zhao, Y. Tang et al. 2013. Establishment of an efficient genetic transformation system in *Scenedesmus obliquus*. *J. Biotechnol.* 163:61–68.

Gutiérrez, C. L., J. Gimpel, C. Escobar, S. H. Marshall, and V. Henríquez. 2012. Chloroplast genetic tool for the green microalgae *Haematococcus pluvialis* (Chlorophyceae, Volvocales). *J. Phycol.* 48:976–983.

Hamilton, M. L., R. P. Haslam, J. A. Napier, and O. Sayanova. 2014. Metabolic engineering of *Phaeodactylum tricornutum* for the enhanced accumulation of omega-3 long chain polyunsaturated fatty acids. *Metab. Eng.* 22:3–9.

Hanson, M. R., B. N. Gray, and B. A. Ahner. 2013. Chloroplast transformation for engineering of photosynthesis. *J. Exp. Bot.* 64:731–742.

Helliwell, K. E., M. Scaife, S. Sasso, A. P. U. Araujo, S. Purton, and A. G. Smith. 2014. Unraveling vitamin B_{12}-responsive gene regulation in algae. *Plant Physiol.* 165:388–397.

Helliwell, K. E., G. W. Wheeler, K. Leptos, R. Goldstein, and A. G. Smith. 2011. Insights into the evolution of vitamin B_{12} auxotrophy from sequenced algal genomes. *Mol. Biol. Evol.* 28:2921–2933.

Hempel, F., A. S. Bozarth, N. Lindenkamp et al. 2011a. Microalgae as bioreactors for bioplastic production. *Microb. Cell Fact.* 10:81.

Hempel, F., J. Lau, A. Klingl, and U. G. Maier. 2011b. Algae as protein factories: Expression of a human antibody and the respective antigen in the diatom *Phaeodactylum tricornutum*. *PLoS One* 6:e28424.

Hempel, F. and U. G. Maier. 2012. An engineered diatom acting like a plasma cell secreting human IgG antibodies with high efficiency. *Microb. Cell Fact.* 11:126.

Henley, W. J., R. W. Litaker, L. Novoveská, C. S. Duke, H. D. Quemada, and R. T. Sayre. 2013. Initial risk assessment of genetically modified (GM) microalgae for commodity-scale biofuel cultivation. *Algal Res.* 2:66–77.

Hsieh, H. J., C. H. Su, and L. J. Chien. 2012. Accumulation of lipid production in *Chlorella minutissima* by triacylglycerol biosynthesis-related genes cloned from *Saccharomyces cerevisiae* and *Yarrowia lipolytica*. *J. Microbiol.* 50:526–534.

Hsu, P. D., E. S. Lander, and F. Zhang. 2014. Development and applications of CRISPR-Cas9 for genome engineering. *Cell* 157:1262–1278.

Hu, J., X. Deng, N. Shao, G. Wang, and K. Huang. 2014. Rapid construction and screening of artificial microRNA systems in *Chlamydomonas reinhardtii*. *Plant J.* 79:1052–1064.

Hu, Z., Z. Fan, Z. Zhao, J. Chen, and J. Li. 2012. Stable expression of antibiotic-resistant gene *ble* from *Streptoalloteichus hindustanus* in the mitochondria of *Chlamydomonas reinhardtii*. *PLoS One* 7:e35542.

Hu, Z., Z. Zhao, Z. Wu et al. 2011. Successful expression of heterologous eGFP gene in the mitochondria of a photosynthetic eukaryote *Chlamydomonas reinhardtii*. *Mitochondrion* 11:716–721.

Ifuku, K., D. Yan, M. Miyahara, N. Inoue-Kashino, Y. Y. Yamamoto, and Y. Kashino. 2015. A stable and efficient nuclear transformation system for the diatom *Chaetoceros gracilis*. *Photosynth. Res.* 123:203–211.

Imamura, S., D. Hagiwara, F. Suzuki, N. Kurano, and S. Harayama. 2012. Genetic transformation of *Pseudochoricystis ellipsoidea*, an aliphatic hydrocarbon-producing green alga. *J. Gen. Appl. Microbiol.* 58:1–10.

Jiang, W., A. J. Brueggeman, K. M. Horken, T. M. Plucinak, and D. P. Weeks. 2014. Successful transient expression of Cas9 and single guide RNA genes in *Chlamydomonas reinhardtii*. *Eukaryot. Cell* 13:1465–1469.

Jin, E., B. Feth, and A. Melis. 2003. A mutant of the green alga *Dunaliella salina* constitutively accumulates zeaxanthin under all growth conditions. *Biotechnol. Bioeng.* 81:115–124.

Karas, B. J., R. E. Diner, S. C. Lefebvre et al. 2015. Designer diatom episomes delivered by bacterial conjugation. *Nat. Commun.* 6:6925.

Kempinski, C., Z. Jiang, S. Bell, and J. Chappell. 2015. Metabolic engineering of higher plants and algae for isoprenoid production. *Adv. Biochem. Eng. Biotechnol.* 148:161–199.

Kim, H. and J.-S. Kim. 2014. A guide to genome engineering with programmable nucleases. *Nat. Rev. Genet.* 15:321–334.

Kim, S., Y. C. Lee, D. H. Cho et al. 2014. A simple and non-invasive method for nuclear transformation of intact-walled *Chlamydomonas reinhardtii*. *PLoS One* 9:e101018.

Kindle, K. L., K. L. Richards, and D. B. Stern. 1991. Engineering the chloroplast genome: Techniques and capabilities for chloroplast transformation in *Chlamydomonas reinhardtii*. *Proc. Natl. Acad. Sci. USA* 88:1721–1725.

Kroth, P. 2007. Molecular biology and the biotechnological potential of diatoms. *Adv. Exp. Med. Biol.* 616:23–33.

Lapidot, M., D. Raveh, A. Sivan, S. M. Arad, and M. Shapira. 2002. Stable chloroplast transformation of the unicellular red alga *Porphyridium* species. *Plant Physiol.* 129:7–12.

Larosa, V. and C. Remacle. 2013. Transformation of the mitochondrial genome. *Int. J. Dev. Biol.* 57:659–665.

Lauersen, K. J., H. Berger, J. H. Mussgnug, and O. Kruse. 2013. Efficient recombinant protein production and secretion from nuclear transgenes in *Chlamydomonas reinhardtii*. *J. Biotechnol.* 167:101–110.

Lauersen, K. J., O. Kruse, and J. H. Mussgnug. 2015. Targeted expression of nuclear transgenes in *Chlamydomonas reinhardtii* with a versatile, modular vector toolkit. *Appl. Microbiol. Biotechnol.* 99:3491–3503.

Lavaud, J., A. C. Materna, S. Sturm, S. Vugrinec, and P. G. Kroth. 2012. Silencing of the violaxanthin de-epoxidase gene in the diatom *Phaeodactylum tricornutum* reduces diatoxanthin synthesis and non-photochemical quenching. *PLoS One* 7:e36806.

Lerche, K. and A. Hallmann. 2014. Stable nuclear transformation of *Pandorina morum*. *BMC Biotechnol.* 14:65.

Leu, S. and S. Boussiba. 2014. Advances in the production of high-value products by microalgae. *Ind. Biotechnol.* 10:169–183.

Liang, M. H. and J. G. Jiang. 2013. Advancing oleaginous microorganisms to produce lipid via metabolic engineering technology. *Prog. Lipid Res.* 52:395–408.

Liu, J., Z. Sun, H. Gerken, J. Huang, Y. Jiang, and F. Chen. 2014. Genetic engineering of the green alga *Chlorella zofingiensis*: A modified norflurazon-resistant phytoene desaturase gene as a dominant selectable marker. *Appl. Microbiol. Biotechnol.* 98:5069–5079.

Liu, L., Y. Wang, Y. Zhang, X. Chen, P. Zhang, and S. Ma. 2013. Development of a new method for genetic transformation of the green alga *Chlorella ellipsoidea*. *Mol. Biotechnol.* 54:211–219.

Loppes, R. and R. Deltour. 1978. A temperature-conditional protoplast of *Chlamydomonas reinhardi*. *Exp. Cell Res.* 117:439–441.

Lozano, J. C., P. Schatt, H. Botebol et al. 2014. Efficient gene targeting and removal of foreign DNA by homologous recombination in the picoeukaryote *Ostreococcus*. *Plant J.* 78:1073–1083.

Meslet-Cladière, L. and O. Vallon. 2011. Novel shuttle markers for nuclear transformation of the green alga *Chlamydomonas reinhardtii*. *Eukaryot. Cell* 10:1670–1678.

Miyahara, M., M. Aoi, N. Inoue-Kashino, Y. Kashino, and K. Ifuku. 2013. Highly efficient transformation of the diatom *Phaeodactylum tricornutum* by multi-pulse electroporation. *Biosci. Biotechnol. Biochem.* 77:874–876.

Mussgnug, J. H. 2015. Genetic tools and techniques for *Chlamydomonas reinhardtii*. *Appl. Microbiol. Biotechnol.* 99:5407–5418.

Muto, M., Y. Fukuda, M. Nemoto, T. Yoshino, T. Matsunaga, and T. Tanaka. 2013. Establishment of a genetic transformation system for the marine pennate diatom *Fistulifera* sp. strain JPCC DA0580—A high triglyceride producer. *Mar. Biotechnol.* 15:48–55.

Neupert, J., D. Karcher, and R. Bock. 2009. Generation of *Chlamydomonas* strains that efficiently express nuclear transgenes. *Plant J.* 57:1140–1150.

Neupert, J., N. Shao, Y. Lu, and R. Bock. 2012. Genetic transformation of the model green alga *Chlamydomonas reinhardtii*. *Methods Mol. Biol.* 847:35–47.

Newman, S. M., J. E. Boynton, N. W. Gillham, B. L. Randolph-Anderson, A. M. Johnson, and E. H. Harris. 1990. Transformation of chloroplast ribosomal RNA genes in *Chlamydomonas*: Molecular and genetic characterization of integration events. *Genetics* 126:875–888.

Newman, S. M., E. H. Harris, A. M. Johnson, J. E. Boynton, and N. W. Gillham. 1992. Nonrandom distribution of chloroplast recombination events in *Chlamydomonas reinhardtii*: Evidence for a hotspot and an adjacent cold region. *Genetics* 132:413–429.

Noor-Mohammadi, S., A. Pourmir, and T. W. Johannes. 2014. Method for assembling and expressing multiple genes in the nucleus of microalgae. *Biotechnol. Lett.* 36:561–566.

Perrineau, M. M., E. Zelzion, J. Gross, D. C. Price, J. Boyd, and D. Bhattacharya. 2014. Evolution of salt tolerance in a laboratory reared population of *Chlamydomonas reinhardtii*. *Environ. Microbiol.* 16:1755–1766.

Podevin, N., H. V. Davies, F. Hartung, F. Nogué, and J. M. Casacuberta. 2013. Site-directed nucleases: A paradigm shift in predictable, knowledge-based plant breeding. *Trends Biotechnol.* 31:375–383.

Pratheesh, P. T., M. Vineetha, and G. M. Kurup. 2014. An efficient protocol for the Agrobacterium-mediated genetic transformation of microalga *Chlamydomonas reinhardtii*. *Mol. Biotechnol.* 56:507–515.

Purton, S. 2007. Tools and techniques for chloroplast transformation of *Chlamydomonas*. *Adv. Exp. Med. Biol.* 616:34–45.

Purton, S., J. B. Szaub, T. Wannathong, R. Young, and C. K. Economou. 2013. Genetic engineering of algal chloroplasts: Progress and prospects. *Russ. J. Plant Physiol.* 60:491–499.

Ramachandra, T. V., D. M. Mahapatra, and B. Karthick. 2009. Milking diatoms for sustainable energy: Biochemical engineering versus gasoline-secreting diatom solar panels. *Ind. Eng. Chem. Res.* 48:8769–8788.

Rasala, B. A., D. J. Barrera, and J. Ng et al. 2013. Expanding the spectral palette of fluorescent proteins for the green microalga *Chlamydomonas reinhardtii*. *Plant J.* 74:545–556.

Rasala, B. A., P. A. Lee, Z. Shen, S. P. Briggs, M. Mendez, and S. P. Mayfield. 2012. Robust expression and secretion of Xylanase1 in *Chlamydomonas reinhardtii* by fusion to a selection gene and processing with the FMDV 2A peptide. *PLoS One* 7:e43349.

Rasala, B. A. and S. P. Mayfield. 2015. Photosynthetic biomanufacturing in green algae; production of recombinant proteins for industrial, nutritional, and medical uses. *Photosynth. Res.* 123:227–239.

Rathod, J. P., G. Prakash, R. Pandit, and A. M. Lali. 2013. *Agrobacterium*-mediated transformation of promising oil-bearing marine algae *Parachlorella kessleri*. *Photosynth. Res.* 118:141–146.

Remacle, C., S. Cline, L. Boutaffala et al. 2009. The *ARG9* gene encodes the plastid-resident *N*-acetyl ornithine aminotransferase in the green alga *Chlamydomonas reinhardtii*. *Eukaryot. Cell* 8:1460–1463.

Rochaix, J. D., R. Surzycki, and S. Ramundo. 2014. Tools for regulated gene expression in the chloroplast of *Chlamydomonas*. *Methods Mol. Biol.* 1132:413–424.

Schmidt, B. J., X. Lin-Schmidt, A. Chamberlin, K. Salehi-Ashtiani, and J. A. Papin. 2010. Metabolic systems analysis to advance algal biotechnology. *Biotechnol. J.* 5:660–670.

Seo, S., H. Jeon, S. Hwang, E. Jin, and K. S. Chang. 2015. Development of a new constitutive expression system for the transformation of the diatom *Phaeodactylum tricornutum*. *Algal Res.* 11:50–54.

Siaut, M., M. Heijde, M. Mangogna et al. 2007. Molecular toolbox for studying diatom biology in *Phaeodactylum tricornutum*. *Gene* 406:23–35.

Sizova, I., A. Greiner, M. Awasthi, S. Kateriya, and P. Hegemann. 2013. Nuclear gene targeting in *Chlamydomonas* using engineered zinc-finger nucleases. *Plant J.* 73:873–882.

Sørensen, I., Z. Fei, A. Andreas, W. G. T. Willats, D. S. Domozych, and J. K. C. Rose. 2014. Stable transformation and reverse genetic analysis of *Penium margaritaceum*: A platform for studies of Charophyte green algae, the immediate ancestors of land plants. *Plant J.* 77:339–495.

Stephenson, P. G., C. M. Moore, M. J. Terry, M. V. Zubkov, and T. S. Bibby. 2011. Improving photosynthesis for algal biofuels: Toward a green revolution. *Trends Biotechnol.* 29:615–623.

Tan, C., S. Qin, Q. Zhang, P. Jiang, and F. Zhao. 2005. Establishment of a micro-particle bombardment transformation system for *Dunaliella salina*. *J. Microbiol.* 43:361–365.

ten Lohuis, M. R. and D. J. Miller. 1998. Genetic transformation of dinoflagellates (*Amphidinium* and *Symbiodinium*): Expression of GUS in microalgae using heterologous promoter constructs. *Plant J.* 13:427–435.

Tirichine, L. and C. Bowler. 2011. Decoding algal genomes: Tracing back the history of photosynthetic life on Earth. *Plant J.* 66:45–57.

Tran, M., R. E. Henry, D. Siefker et al. 2013. Production of anti-cancer immunotoxins in algae: Ribosome inactivating proteins as fusion partners. *Biotechnol. Bioeng.* 110:2826–2835.

van Dijk, K. and N. Sarkar. 2011. Selectable and inheritable gene silencing through RNA interference in the unicellular alga *Chlamydomonas reinhardtii*. *Methods Mol. Biol.* 765:457–476.

Vandamme, D., I. Foubert, and K. Muylaert. 2013. Flocculation as a low-cost method for harvesting microalgae for bulk biomass production. *Trends Biotechnol.* 31:233–239.

Vieler, A., G. Wu, C. H. Tsai et al. 2012. Genome, functional gene annotation, and nuclear transformation of the heterokont oleaginous alga *Nannochloropsis oceanica* CCMP1779. *PLoS Genet.* 8:e1003064.

Walker, T. L., C. Collet, and S. Purton. 2005. Algal transgenics in the genomic era. *J. Phycol.* 41:1077–1093.

Wang, T., L. Xue, W. Hou et al. 2007. Increased expression of transgene in stably transformed cells of *Dunaliella salina* by matrix attachment regions. *Appl. Microbiol. Biotechnol.* 76:651–657.

Weyman, P. D., K. Beeri, S. C. Lefebvre et al. 2014. Inactivation of *Phaeodactylum tricornutum* urease gene using transcription activator-like effector nuclease-based targeted mutagenesis. *Plant Biotechnol. J.* 13:460–470.

Xie, W.-H., C.-C. Zhu, N.-S. Zhang et al. 2014. Construction of novel chloroplast expression vector and development of an efficient transformation system for the diatom *Phaeodactylum tricornutum*. *Mar. Biotechnol.* 16:538–546.

Yamano, T., H. Iguchi, and H. Fukuzawa. 2013. Rapid transformation of *Chlamydomonas reinhardtii* without cell-wall removal. *J. Biosci. Bioeng.* 115:691–694.

Young, R E. B. and S. Purton. 2014. Cytosine deaminase as a negative selectable marker for the microalgal chloroplast: A strategy for the isolation of nuclear mutations that affect chloroplast gene expression. *Plant J.* 80:915–925.

Zaslavskaia, L. A., J. C. Lippmeier, C. Shih, D. Ehrhardt, A. R. Grossman, and K. E. Apt. 2001. Trophic obligate conversion of a photoautotrophic organism through metabolic engineering. *Science* 292:2073–2075.

Zedler, J. A., D. Gangl, B. Hamberger, S. Purton, and C. Robinson. 2015. Stable expression of a bifunctional diterpene synthase in the chloroplast of *Chlamydomonas reinhardtii*. *J. Appl. Phycol.* 27:2271–2277.

Zhang, C. and H. Hu. 2014. High-efficiency nuclear transformation of the diatom *Phaeodactylum tricornutum* by electroporation. *Mar. Genomics* 16:63–66.

Zhang, R., W. Patena, U. Armbruster, S. S. Gang, S. R. Blum, and M. C. Jonikas. 2014. High-throughput genotyping of green algal mutants reveals random distribution of mutagenic insertion sites and endonucleolytic cleavage of transforming DNA. *Plant Cell* 26:1398–1409.

Zorin, B., O. Grundman, I. Khozin-Goldberg et al. 2014. Development of a nuclear transformation system for oleaginous green alga *Lobosphaera* (*Parietochloris*) *incisa* and genetic complementation of a mutant strain, deficient in arachidonic acid biosynthesis. *PLoS One* 9:e105223.

Zorin, B., Y. Lu, I. Sizova, and P. Hegemann. 2009. Nuclear gene targeting in *Chlamydomonas* as exemplified by disruption of the PHOT gene. *Gene* 432:91–96.

7 Crop Protection in Open Ponds

Robert C. McBride, Val H. Smith,
Laura T. Carney, and Todd W. Lane

CONTENTS

7.1 CHALLENGES OF OPEN POND ALGAE CULTIVATION

Algae can be cultivated in open or closed bioreactors. Open bioreactors are defined as any reactor that is exposed to the environment. These reactors can take many different forms, but most conform to one of the following broad categories: shallow lagoons and ponds, inclined cascade systems, circular central pivot ponds, mixed ponds, and raceway ponds (Borowitzka and Moheimani 2013). While these ponds are configured differently in terms of their construction, lining, means of propulsion/ mixing, and intensity of management, they all share the common element of being fully exposed to the external environment. Research on how to successfully cultivate microalgae using open systems was initiated in the late 1940s and early 1950s in the United States, Germany, and Japan (Cook 1950; Gummert et al. 1953; Mituya et al. 1953). While significant progress has been made over the intervening decades, the open pond systems still face serious challenges that stem from being exposed to unpredictable and uncontrollable meteorological conditions, suboptimal mixing within the culture, and exposure to many forms of contamination. These problems limit productivity, nutrient utilization efficiency, and performance stability. Despite these challenges, open ponds continue to be used and developed primarily because they are cheaper and easier to scale, build, and operate when compared to closed photobioreactors (Sheehan et al. 1998; Waltz 2009).

The most prominent limiting factor for obtaining reliable productivity in open systems is the problem of biological contamination (Chaumont 1993). Unwanted organisms inevitably enter into the culture and impact target algae strains (Shimamatsu 2004; Tredici 2004; Richmond 2007; Schenk et al. 2008). If unmanaged, the growth and proliferation of these contaminants inevitably reduce the productivity of the ponds (estimates of impact have been made of 10%–30%) (Richardson et al. 2014) and may destabilize the ponds, causing them either to become dominated by nontarget organisms or to fail entirely.

Open pond algal contaminants can be broadly described as unseeded, invading organisms. A broad range of organisms fit this classification, including viruses, bacteria, fungi, protozoa, rotifers, and crustacean zooplankton, as well as nontarget species of algae (Figure 7.1). Most contaminants have little impact on the quality and stability of production. However, a subset of these contaminants can very negatively impact the quality and stability of biomass production. These negative impacts can result from direct interactions of the contaminants with the algal strain being cultivated, such as cell consumption by invertebrate herbivores or cell death due to infections by virulent pathogens, or from indirect interactions such as interspecific resource competition between the desired algal strain and undesirable algal invaders (*weed* algae). When invading organisms negatively impact open pond algae production, they are considered pests.

Like any other living organisms, algae can be plagued by diseases caused by viruses, bacteria, and fungi, and they also can be attacked by protozoa (Gachon et al. 2010). These microbial parasites thus pose a significant threat to the successful outdoor mass culture of algae (Carney and Lane 2014). For example, viruses are known algal pathogens, but while there are some reports of open pond cultures of

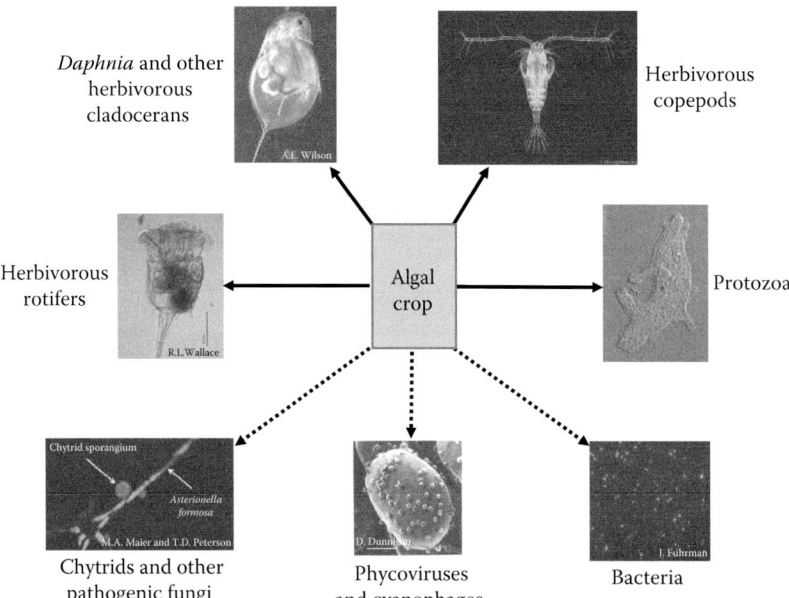

FIGURE 7.1 (See color insert.) Examples of contaminating pest organisms that can impact the productivity of algae grown in open ponds. Herbivorous consumers are denoted by the solid arrows, and potential pathogens are denoted by the dotted lines.

Arthrospira (previously *Spirulina*) collapsing due to phage-like activity (Shimamatsu 2004), viruses do not seem to have presented major problems thus far for commercial algae ponds despite being widely present in the water column (Van Etten et al. 1991; Polson et al. 2011; Jacquet et al. 2013). Bacteria can negatively interact with algae either indirectly through interspecific resource competition for inorganic nutrients such as phosphorus (Liu et al. 2012) or directly as pathogens. For example, there are cases in which predatory bacteria attach to and kill algae via the release of lytic agents (Daft and Stewart 1973), and it has been shown that these bacteria can impact mass cultures of algae (Wen-li et al. 2011). In certain conditions, such as high organic carbon loading, bacteria may bloom and cause the loss of entire cultured populations of strict photoautotrophs such as *Arthrospira* and *Dunaliella* (Richmond 2013). Protozoa also can be a problem: amoebae can be a significant challenge to commercial production of algae and have been noted to cause cell losses, particularly in mass cultures of *Arthrospira* (Post et al. 1983; Sigee et al. 1999). Ciliates have been shown to clear dense outdoor cultures of *Dunaliella salina* within a few days (Moreno-Garrido and Canavate 2001).

Many taxa of basal fungi can cause significant damage to eukaryotic algae grown in open ponds. Some chytrids are host-specific parasitic fungi that can have a considerable impact on algal dynamics (Ibelings et al. 2004). The chytrid *Phlyctidium scenedesmi* has been noted to impact *Scenedesmus* in open ponds (Fott 1967; Ilkov 1975). The infection of chytrid species has been observed in high-rate *Scenedesmus obliquus* sewage oxidation ponds as well as in outdoor mass culture systems (Lukavsky 1971; Abeliovich and Dikbuck 1977). Similarly, a chytrid-like fungus, *Amoeboaphelidium protococcarum,* as well as an undescribed chytrid related to *A. protococcarum* was shown to rapidly decimate cultures of *Scenedesmus* in open ponds (Letcher et al. 2013; McBride et al. 2014). *Pseudosphaerita euglenae*, a fungal parasite of *Euglena*, has been observed in oxidation ponds (Anderson et al. 1995). The parasitic fungi *Rhizophydium* sp. was found to infect and destroy *Haematococcus pluvialis* cells as culture temperatures increased in summer (Hoffman et al. 2002). Diatoms also are highly susceptible to fungal infection (Ibelings et al. 2004).

In addition to potential losses to pathogens and protozoa, macrozooplankton such as rotifers and cladocerans are consistently a challenge in large-scale outdoor cultures of many kinds of algal species (Lincoln et al. 1983) and can reduce algal biomass concentrations to low levels within a matter of days if environmental conditions are optimal (Benemann 2008). Rotifers such as *Brachionus* have been reported to impact *Arthrospira* (Méndez and Uribe 2012). Similarly, invasions by herbivorous copepods led to major productivity losses in outdoor mass cultures of marine microalgae (Loosanoff et al. 1957). Herbivorous insect larvae also have been reported to impact open ponds of *Arthrospira* (Belay 1997).

Contaminating nontarget strains of algae are another significant challenge in open pond algal cultivation systems. Algae are widely distributed in the landscape, and many species are capable of rapid growth in a broad range of environmental conditions (Wang et al. 2013). The DOE Aquatic Species Program screened hundreds of strains of algae and could not maintain dominant single strains for extended periods of time in open culture due to contaminating invasions of green algae (Sheehan et al. 1998). Contamination of *Dunaliella* strains by non-carotenoid producing *Dunaliella* strains has also been reported (Day and Stanley 2012). Outdoor cultures of the cyano-bacterium *Arthrospira* have been known to become contaminated with eukaryotic microalgae such as *Chlorella* and *Oocystis*, as well as by other species of cyanobac-teria (Vonshak et al. 1983; Vonshak and Richmond 1988; Belay 1997; Vonshak 1997).

These examples are not an exhaustive review of the known pests of algae but serve to illustrate the breadth of challenges faced in open pond algal cultivation. It is important to note that in addition to the known pathogens, the nascent field of algae crop protection can expect to see many new and as yet undescribed pathogens aris-ing as the open pond platform continues to expand and as the field of crop protection continues to advance.

Biological contaminants can be introduced to open ponds in a number of ways. Their propagules may be introduced into a pond along with the algal inoculum or may be present in the pond basin prior to inoculation as residual from a previous culture. Microbes can be transported into open ponds via wind, dust, or rain. For microbes to be transported this way, they must first become aerosolized and then must survive transport in the air and be deposited onto the water surface. The pri-mary factors that drive the aerosolization of microbes are storm events, high-wind events, rain, wave action, foam and scum formation, and human activities (Maynard 1968; Schlichting 1974; Burge and Rogers 2000). Once aerosolized, microbes also must survive the harsh conditions of the air where desiccation is a key challenge. It has been noted that the increased humidity generally increases survival of bio-aerosols (Ehresmann and Hatch 1975; Sharma et al. 2007). Deposition of airborne microbes is impacted by many conditions. Humidity tends to increase their settling in general. For small particles between 0.1 and 10 μm, rain is the most important deposition mechanism, and for larger particles (diameter greater than 10 μm), gravity plays a larger role in deposition (Després et al. 2012).

Contaminants can also be introduced to open ponds via vectors such as insects or birds. Insects transport a wide variety of contaminants both in their guts and on their exoskeletons (Parsons et al. 1966; Milliger and Schlichting 1968), and it has been suggested that insect vectors are a major source of undesirable invaders for algae ponds (Mendes and Vermelho 2013). As with insects, contaminants can be transported by birds both in their guts and externally. Studies have illustrated the viable transport of various species of microbes for a number of hours in the intestines of waterfowl, as well as on the feet, beaks, and feathers of waterfowl (Kristiansen 1996; Charalambidou and Santamaría 2002; Leeuwen et al. 2012). Waterfowl can transport aquatic organisms from water body to water body (Green and Figuerola 2005) and are also considered to be the most important transporters of microalgae (Kristiansen 1996).

7.2   EXISTING STRATEGIES FOR CROP PROTECTION

It is generally acknowledged that the early detection of pest challenges typically results in cheaper and more effective management strategies (where problem solu-tions are available). A broad review of molecular detection tools can be found in Chapter 8, which summarizes recent developments and advances in the field of pest detection and identification.

The present section focuses on the tools and procedures that will be necessary to resolve pest challenges once they have been detected. Diverse solutions have been considered and evaluated for the management of algae in open pond systems used for the production of low-cost biomass. These crop protection solutions fall into two broad categories: solutions that seek to solve crop problems when they arise (*reactive strategies*) and solutions that seek to prevent crop problems from arising (*preventive strategies*). The majority of crop protection efforts employed in the algal industry thus far have focused on reacting to contamination and deploying strategies intended to minimize the impact of these contaminants on the production of low-cost algal biomass. In contrast, far less work has been done to create pond environments that either reduce or completely prevent biological contamination from becoming an economically important issue. This section will highlight some of the core work that has been done in developing both of these strategies.

Figure 7.2 shows a simplified fitness surface that illustrates the growth rate responses of a hypothetical alga and its pest nemesis across a landscape of varying environmental gradients. These landscapes are the product of the genotypes of both the algae and the nuisance pest organisms. Over time, the environment can select for these landscapes to shift and result in different solutions to finding the right environment for a pond to be productive. In this example, the environmental space in which the green surface is above the red surface represents the variable space in which the

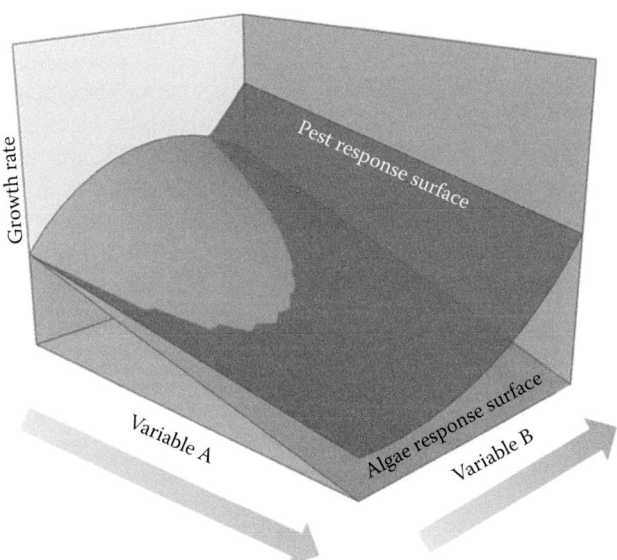

FIGURE 7.2 **(See color insert.)** Hypothetical growth rate landscapes for algae and pests of algae across environmental gradients. This figure illustrates the concept that modifying the cultivation environment can differentially impact the relative growth rates of algae and their pests. Careful manipulations can be designed that drive the pond to a solution space in which the algal growth rate is higher than that of the pest, resulting in productive algal biomass cultivation systems.

growth rate of the algae would be greater than that of the pest, and the pond would be productive. While Figure 7.2 is a simplified example, it is valuable to have as a frame of reference when thinking about crop protection solutions. Solutions that react to challenges seek to find a gradient that shifts the algae growth rate above the pest, while solutions that seek to prevent challenges from arising try to create landscapes in which the green surface is always above the red surface. In addition to being a useful heuristic, this approach can also be used to optimize conditions to minimize pest pressure using response surface methodology (Kim et al. 2012; Aguirre and Bassi 2013).

7.2.1 REACTIVE STRATEGIES: CURRENT SOLUTIONS

Reactive crop protection strategies require a pest challenge to be detected and recognized, and then a carefully chosen solution is triggered that resolves this challenge. Because of the potential for rapid biomass damage, early pest detection is critical to avoid economically significant losses. Many solutions have been proposed in the literature, ranging from mechanical and cultural to chemical and biological. This subsection will review some of these strategies.

In response to an imminent culture crash, the entire culture biomass can be harvested, and the culture then restarted, limiting lost production. However, the economic cost or loss of culture, cleaning the ponds, and restarting the culture would be significant. Additionally, contaminating organisms can be carried over to infect successive cultures if the pond cultivation system is not sufficiently sterilized before reinoculating the pond. Further, this practice thus is not considered to be ideal for the large-scale production of low-cost biomass (Borowitzka and Moheimani 2013).

Mechanical strategies also have been deployed to mitigate the impact of contaminants. For example, filtration with netting or filters with large mesh or pore sizes can be used either during harvest or in the course of normal cultivation to remove larger contaminants and insect larvae (Belay et al. 1994; Borowitzka 2005; Ravikumar 2014). Anecdotal data suggest that pumping the algal culture using a high shear rate pump also can be a successful strategy to remove large zooplankton that are more negatively impacted by the action of the pump than the much smaller algae. Pulsed electric fields have also been shown to have success in controlling protozoa in algal cultures (Rego et al. 2014).

Other practices such as manipulating the pH also have been successfully used to mitigate the impact of some biological contaminants. For example, the pH of the culture water can be drastically lowered by injecting gaseous CO_2 or by adding acid as a means to control the growth of rotifers (Becker 1994). This method has been successfully used to control rotifers in cultures of *Nannochloropsis* (Zmora and Richmond 2008). Similarly, it was reported in wastewater algae cultures that increasing the pH to 9.5 removed undesirable herbivorous cladoceran zooplankton (Lincoln et al. 1983). In addition, the regulation of pH also can be used to help prevent the growth of weedy algal contaminants. For example, in clay-lined ponds, maintaining a pH above 9 favored the dominance of target strain cyanobacteria (blue-green algae) over invasive green algae (Van der Westhuizen and Eloff 1983). When specifically used to combat the impact of invasive *Chlorella* on *Arthrospira* monocultures, water column pH values greater than 10 were shown to favor *Arthrospira* (Vonshak et al. 1983).

Toxic chemicals and biocides also are commonly added to mitigate the impacts of invading pests. A key limitation with this strategy is the negative collateral impact that these chemicals may have upon growth of the target algal strain. Chemicals can be screened to identify those that inhibit contaminants without impacting the growth of the target strain; however, this is not simple in most cases, particularly when the contaminant to be controlled is similar to the target algal strain. Nonetheless, repeated pulses of ammonia followed by dilution of the culture have been reported to be a good strategy to mitigate the impact of *Chlorella* on *Arthrospira* cultures (Richmond 2007). Ammonia spikes have also been used to arrest the development of invading grazers (*Stylonychia* sp.) in a culture of *Chlorella* (Grobbelaar 1981), as well as to control the growth of herbivorous rotifers (Lincoln et al. 1983). Chytrids have been shown to be controllable by the addition of Mg^{2+} or K^+ at concentrations of 10^{-2} moles or higher (Abeliovich and Dikbuck 1977).

Additions of hypochlorite also have been used to control contaminants in open ponds, especially in *Nannochloropsis* cultures to control protozoa (Richmond 2013). Quinine has been shown to effectively kill ciliates while not killing *Dunaliella* strains (Moreno-Garrido and Canavate 2001). Glyphosate has been added to *Nannochloropsis* cultures resistant to glyphosate to maintain selectivity (Vick 2010). Disinfectant and ozone shock have been used in *Nannochloropsis* culture to clear contaminants (Weissman et al. 2010; Weissman and Radaelli 2010). Metronidazole has been successfully used in disinfecting *Scenedesmus* sp. cultures invaded by the zooflagellate *Amphelidium* sp. (Heussler et al. 1978). Pesticides such as trichlorphon, decamethrin, tralocythrin, and buprofezin have also been used to eradicate zooplankton in microalgal suspensions (Wang et al. 2013). Fungicides such as Triton-N and Funginex have been used as control measures for some fungal contaminants of *Scenedesmus*, such as *P. scenedesmi* (Benderliev et al. 1993; Ravikumar 2014), and chytrid-like organisms have also been successfully controlled in the field using the pesticide Headline (McBride et al. 2014). The fungicide benomyl has been used to treat pathogenic fungi in cultures of *Euglena gracilis* (Mokrosnop and Zolotareva 2013). DCMU (3-(3,4-dichlorophenyl)-1,1-dimethylurea) has been proposed as an effective herbicide to manage contaminant algae growth in cultures of *Nannochloropsis* (Gonen-Zurgil et al. 1996). Formaldehyde has been used to control some ciliates in algae cultures (Moreno-Garrido and Canavate 2001).

As an alternative to deploying chemicals in the face of a crop protection challenge, biological agents can be utilized. For example, it is possible to add size-selective herbivores that consume algal contaminants that are smaller in size than the target strain. This approach has most successfully been demonstrated in cultures of *Arthrospira*, where the deployment of rotifers was effectively used to control the growth of nontarget unicellular algae (Mitchell and Richmond 1987). Similarly, strategic additions of predators to the cultivation system can potentially be used to control the population growth of undesirable zooplankton contaminants that would otherwise contribute to grazing losses from herbivorous cladocerans and rotifers. Microalgae may also have natural antibacterial abilities (Kellam and Walker 1989). As will be discussed in greater detail in the next section, we suggest that biological control methods should be a key component of integrated algal pest management in the future.

7.2.2 REACTIVE STRATEGIES: THE FUTURE

Section 7.2.1 focused on reviewing strategies that have already been used while reacting to pests in open algae ponds. Research into new methods for the detection and resolution of pest challenges will be needed to push algal biomass cultivation technology toward commercially viable scales. Two components are important to an effective and successful response to pest detection: first, one must carefully choose the agent(s) that are to be used (chemical, biological, or both). Second, one must also consider how the agent(s) are to be deployed (e.g., time of day, means of deployment). We present several approaches in the following texts.

The crop protection strategies that have historically been demonstrated to be effective both in agricultural and in some open pond settings primarily involve the deployment of chemicals that impact the target pest versus the cultivated algal species differently. The research needed here involves finding and optimizing chemicals for deployment in aquatic systems that minimize harm to the targeted cultivated algal species or strain(s). Some crop protection compounds are already applicable in aquatic systems, but most are not. Thus, our chemical armamentarium should be expanded by creating and formulating chemicals that are highly effective in open pond environments and have a broad effectiveness against key classes of biological contaminants. This task could be fairly straightforward for crop protection against pests that are evolutionary very different from the target algal strains, but more challenging for invading algal strains that are taxonomically similar to the cultivated target algal species. If chemicals are similarly lethal against both pests and target algal strains, engineering target strains to resist the chemical would be a good strategy to improve the therapeutic window of the deployed chemical. This is an area that could use both evolutionary and gene modification tools to develop new and useful biology.

Chemicals are not the only strategy that can be effectively deployed to mitigate pest challenges, however. Biological control is widely used in terrestrial agriculture and on public lands to manage unwanted pests ranging from invasive plants to mammals. Biological control of plant pathogens has been increasingly applied and, in some cases, is preferred over chemical means of control: where chemical residues present a problem; where regulation of use is too restrictive; or to avoid pest resistance to pesticides (Fravel 2005). Biological solutions in open algae ponds are an effective strategy whose potential has been demonstrated both conceptually and at the pilot scale, but their real-world potential and commercial viability at production-level scales have not yet been fully explored (Smith et al. 2010; Kazamia et al. 2012a). These biological solutions, as with their chemical brethren, also need to be discriminatory. For example, although the use of herbivorous rotifers is a logical tool for targeting invading algal pests that differ from the target algal strain in cell or colony size, the impact of rotifers on other kinds of pests has not yet been fully explored. For example, it is thought that chytrid zoospores in lake ecosystems are an important food source for grazing zooplankton (Kagami et al. 2014). Thus, using rotifers to blunt chytrid blooms may be feasible but needs further investigation. Amoebae, ciliates, and pest-specific viruses also could be similarly deployed as biological control agents.

Another form of biological control that holds promise is hyperparasitism. Organisms that prey on and parasitize microalgae have their own set of predatory parasites, termed hyperparasites. Metagenomic evidence suggests that hyperparasites are common in marine communities (James and Berbee 2012). The host range of hyperparasites has been determined to be narrow, with each species of hyperparasite infecting only closely related species. Hyperparasites that can infect important parasites of algae thus have potential as a novel biological control mechanism in algal production ponds. The examples of such relationships are common in freshwater environments and one marine example has been described (reviewed by Gleason et al. 2012). It is feasible that a dried hyperparasite stock could be used to inoculate fungal-infected algal production ponds in order to control the fungal parasite (Fravel 2005).

To deploy mitigation strategies, important questions first need to be answered: is indiscriminate dosing the most cost-effective strategy or is targeted dosing more effective and economically viable in the long term? For example, the indiscriminate deployment of insecticides has resulted in the list of known insecticide-resistant insects growing to 553 species by 2008 (Whalon et al. 2008; Smith and Crews 2014). One strategy that has been implemented in terrestrial crops to mitigate chemical resistance is the deployment of insect resistant management (IRM) strategies. The goal of IRM is to reduce the potential for pest organisms to become resistant to applied biocides. These strategies encourage rotating or mixing pesticides and the development of technologies that give more resolution on pest populations to inform discriminate deployment of pesticides (Andow et al. 1990; Denholm and Rowland 1992). There are also many other questions in the "how to deploy a strategy" category that are important to understand more thoroughly. For example, does dosing efficacy vary with pond variables such as temperature, light, and nutrient levels? In addition, practical questions also need to be addressed, such as how to effectively apply strategies over large areas. Will current agricultural methods (i.e., aerial spraying) translate effectively, or will new strategies need to be developed?

7.2.3 PREVENTATIVE MANAGEMENT: CURRENT SOLUTIONS

In contrast to the reactive crop protection solutions described earlier, preventative management seeks to prevent crop problems from arising. Here, we briefly review successful examples of these types of strategies.

The most successful documented examples of preventative management strategies involve the protection of high-value biomass through the use of extreme environments that help to select for dominance by the target algal strain. This method has been very successfully employed in the mass production of *Arthrospira*, *Dunaliella*, and *Chlorella* (Lee 2001; Borowitzka and Moheimani 2013). The key to these strategies is creating a pond environment in which the target strain has a higher fitness than any contaminant. In these situations, invading contaminants that enter the open pond will ultimately fail to thrive because the cultivation system's environmental conditions are unfavorable for their growth. For example, in the case of *Arthrospira* cultivation, the use of high alkalinity water supply typically selects for the target algal strain because it is naturally adapted to highly alkaline environments. High alkalinity, thus, limits the range of biotic contaminants that can flourish and possibly dominate over the open ponds. In the case of *Dunaliella*, high salinity is used to discourage the growth of potential contaminants. For *Chlorella*, the addition of acetic acid as a carbon source, coupled with limited exposure to open environments, typically ensures that biological contamination remains a small issue.

7.2.4 PREVENTATIVE MANAGEMENT: THE FUTURE

The preventative category of solutions for algal crop protection in open ponds has not been the focus of much research. More work is, therefore, needed to explore this very promising area, and these efforts will require both an understanding and implementation of evolutionary, ecological, and genetic modification approaches. Several preventive approaches that have great promise are discussed further in this section.

Biologically engineered solutions to crop protection are premised upon the idea that open pond monocultures of algae will inevitably become contaminated and unstable, and that polycultures in contrast may be less susceptible to contamination, as well as potentially be more productive and more stable over time than monocultures. If assembled correctly, mixed-species populations can perform functions that are difficult or even impossible for individual strains or species (Brenner et al. 2008). These attractive advantages rely in large part upon two organizing features: (1) Members of the consortium communicate with one another, whether by trading metabolites or by exchanging dedicated molecular signals. (2) This communication enables the second important feature, which is division of labor: the overall output of the consortium rests upon a combination of tasks performed by constituent individuals or subpopulations (Brenner et al. 2008). We, thus, note that multispecies systems have historically been employed to increase yields and efficiency in many industrial microbial-based processes, including anaerobic digestion, fermentation, and bioremediation (Sabra et al. 2010).

For the algal biomass production arena, we suggest that it may be best to seed ponds intentionally with a carefully selected suite of organisms so that the production system is comprised of a stable and highly productive consortium rather than a random biological assemblage that is far less stable and functional (Kazamia et al. 2014). This approach would seek to ensure that key niches are occupied, thereby preventing unwanted biological contaminants from colonizing the pond. The engineered consortium approach also could potentially reduce grazing losses through interference (see Figure 7.3), an idea whose viability has recently been demonstrated at laboratory scale (Shurin et al. 2013). In addition, maintaining high algal species richness in the pond cultivation system also may help reduce crop losses due to host-specific microbial pathogens (Figure 7.4). Moreover, it may even be possible to add zooplankton grazers that selectively remove life stages of microbial pathogens and yet do not significantly consume cells of the target algae. An example is shown in Figure 7.5 in which carefully selected rotifer populations are added that remove motile zoospore life stages of parasitic chytrid fungi, thereby limiting the reproduction and proliferation of this undesirable algal pathogen.

Other ideas for future crop protection involve the deployment of consortia of algae, some of which chemically modify the cultivation environment, along with

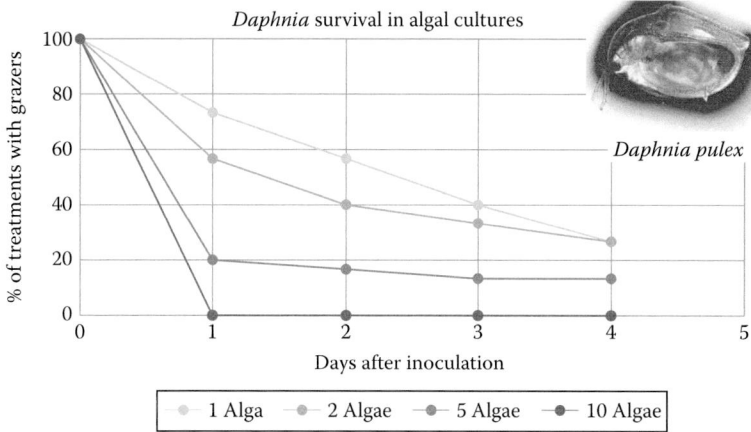

FIGURE 7.3 An illustration displaying the impact of algae diversity on predation of algae by water fleas (*Daphnia pulex*). In this experiment, predators were added to cultures of algae composed of differing amounts of diversity, and the number of cultures in which the water fleas survived was measured daily. The photograph is an image of *Daphnia* taken under the light microscope at 10× magnification. (From Kazamia, E. et al., *Ind. Biotechnol.*, 10, 184, 2014; drawing on data from Shurin, J. et al., *Ecol. Lett.*, 16, 1393, 2013.)

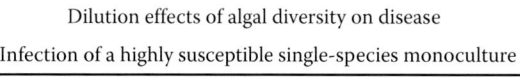

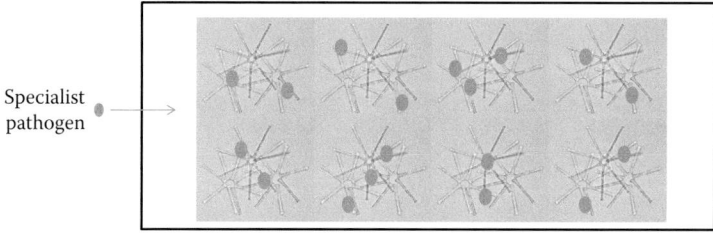

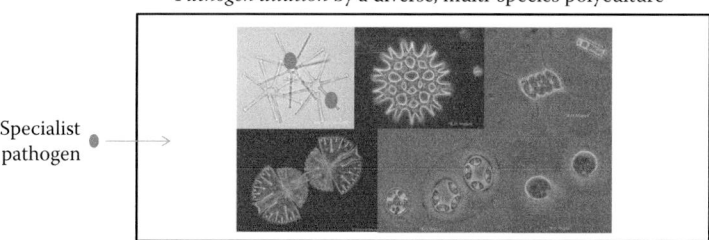

FIGURE 7.4 Generalized example of the potential effects of increasing algal diversity on disease prevalence, as determined by the fraction of total algal cells that are infected in the culture system. A specialist pathogen can heavily infest a monoculture of the susceptible diatom *Asterionella formosa*, but in a polyculture containing resistant algal species that dilute disease risk, infections should be restricted to the diatom.

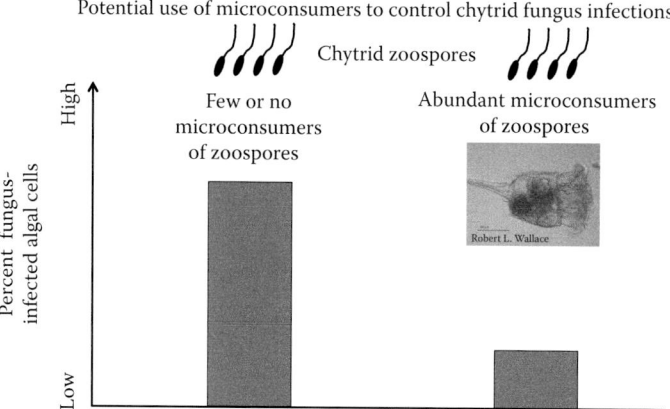

FIGURE 7.5 Hypothetical example of the potential for biological control of chytrid fungus pests. Many chytrids have a life cycle that includes a small, motile zoospore stage that is susceptible to predation by microconsumers. Careful introductions of selected species of protozoa and rotifers could be used to reduce chytrid reproduction and proliferation. (Modified from Schmeller, D. S. et al., *Curr. Biol.*, 24, 176, 2014.)

other strains or species that are resistant to those chemical changes (Kazamia et al. 2014). For example, *Pleurochrysis carterae* elevates the pH of the culture in which it is growing to high levels (~11), which are known to limit the growth of unwanted contaminants (Weisse and Stadler 2006). Some algae may also secrete chemicals such as dimethyl sulfide (DMS) into their environment that will actively kill competitors (Houdan et al. 2004). Polycultures should not solely focus on eukaryotes; algae–bacterial interactions can be positive, and having beneficial bacteria may actually prevent harmful bacteria from colonizing (Kazamia et al. 2012b). Also see the recent work of Nesme and Simonet (2014), on the soil resistome.

Another ecological engineering approach is to create an environment that impacts bacterial quorum sensing. Quorum sensing is the means by which bacteria regulate their own gene expression through intraspecies chemical signaling, and this process has been linked to disease induction in aquatic populations (reviewed by Natrah et al. 2011). In nature, some organisms, including microalgae, appear to have the ability to disrupt quorum sensing and thus disrupt the ability of the pathogen to regulate certain phenotypes that promote virulence factors. Preliminary work reports the disruption of quorum sensing by *Chlorella* sp. and *Chlamydomonas reinhardtii* (see Natrah et al. 2011). By disrupting quorum sensing of pathogenic bacteria, a microalgae host could regulate pathogen-induced disease. For example, when algal cultures attain certain densities, the pH of the culture medium is increased to levels where quorum sensing molecules are inhibited for certain species of bacteria such as *Pseudomonas* spp., which are then prohibited from efficiently communicating with each other (Yates et al. 2002).

For the long term, it is critical to create or find an environment not only that maximizes production in the short term but also that maximizes production stability.

Evolutionary biology suggests that in the scenario of continuous cultivation, selection will result in contaminating strains that grow faster at the expense of target phenotypes assuming a trade-off between target phenotypes and growth rate (Bull and Collins 2012). This is an example of an evolutionary unstable situation. It would be far preferable if the scenario of continuous cultivation selected for the target phenotype instead of against it. This can be achieved by structuring the environment of the algal pond cultivation system in a way that positively selects for the desired phenotype. If a pond does not do this, even though short-term productivities may be maximized, long-term stability will be compromised. An example of this sort of strategy can be found if you are considering a target phenotype of lipid storage. In this scenario, if ponds are only fed with nitrogen at night, a selective environment is created that advantages strains that store energy well (Bull and Collins 2012; Mooij et al. 2013). Thus, the selective pressure is against fast-growing strains with low lipid and for strains that store lipid.

An additional strategy involves genetic solutions. There is strong precedent for genetic solutions that are targeted to specific pests, but given the vagaries of open ponds these solutions may not be effective preventative strategies. If there are genetic solutions that are generic in their resistance to pests, such as engineering algal strains to grow better at high salt or alkalinity, these would be good approaches to creating an environment that is less susceptible to invasion and disruption.

We believe that the key to the future of crop protection in open algae ponds for the production of low-cost biomass is to create a cultivation environment in which a target strain grows faster than the pest organisms that can negatively impact its productivity. This could be achieved using a combination of both reactive and preventative management strategies. To effectively manage open algae ponds at the industrial scale, it will be imperative to consider the ponds not only from an industrial microbiology perspective but also from an ecological and evolutionary perspective.

Equally important, but frequently overlooked in developing solutions for contamination challenges, it is crucial both to quantify the economic consequences of the proposed solutions and to critically evaluate the compatibility of these proposed crop protection solutions with downstream processes such as harvesting, dewatering, chemical processing, and product profile. Considered solutions should be constrained to those that will support low-cost biomass production and also must be compatible with postproduction processes. This requires understanding the biology and physiology of the target algae as well as the technoeconomics and production process constraints of open pond algae production.

Open algae ponds for the production of low-cost biomass represent a transformative platform for multiple industries including for fuel, food, and possibly medicine. Having the ability to cheaply and consistently make large quantities of these products could revolutionize the way we think about energy, protein, and new drug production. For this potential to be usefully realized, the challenge of crop protection needs to be tackled and solved. The success of solving this challenge depends on effective and transparent collaborations between industry, academic, and government researchers in this area. Success will be driven by these collaborations and will benefit all involved.

ACKNOWLEDGMENTS

VS thanks Bob Honea and the University of Kansas Transportation Research Institute for the financial support of his algal biofuels research as well as his colleagues in KU's Feedstock to Tailpipe program. Many of the embedded photos shown in Figures 7.1, 7.4, and 7.5 were obtained from the free image library of the Association for the Sciences of Limnology and Oceanography and from Robert O. Megard (University of Minnesota, Twin Cities). Publication permissions for the remaining images were graciously provided by Wim van Egmond (professional photographer and curator of the Micropolitan Museum), Jed Fuhrman (University of Southern California), and Michelle A. Maier and Tawnya D. Peterson (Oregon Health & Science University).

REFERENCES

Abeliovich, A. and S. Dikbuck. 1977. Factors affecting infection of *Scenedesmus obliquus* by a *Chytridium* sp. in sewage oxidation ponds. *Appl. Environ. Microbiol.* 34:832–836.

Aguirre, A. M. and A. Bassi. 2013. Investigation of biomass concentration, lipid production, and cellulose content in *Chlorella vulgaris* cultures using response surface methodology. *Biotechnol. Bioeng.* 110:2114–2122.

Anderson, S., A. Stewart, and G. T. Allen. 1995. *Pseudosphaerita euglenae*, a fungal parasite of *Euglena* spp. in the Mangere Oxidation Ponds, Auckland, New Zealand. *NZ J. Mar. Freshw. Res.* 29:371–379.

Andow, D. A., P. M. Rosset, C. Caroll, and J. Vandermeer. 1990. Integrated pest management. In *Agroecology*, ed. S. R. Gliessman, pp. 413–439. New York: Springer.

Becker, E. W. 1994. *Microalgae: Biotechnology and Microbiology*. Cambridge, UK: Cambridge University Press.

Belay, A. 1997. Mass culture of *Spirulina* outdoors: The Earthrise Farms experience. In *Spirulina Platensis (Arthrospira): Physiology, Cell-Biology and Biotechnology*, ed. A. Vonshak, pp. 131–158. London, UK: Taylor & Francis.

Belay, A., Y. Ota, K. Miyakawa, and H. Shimamatsu. 1994. Production of high quality *Spirulina* at Earthrise farms. In *Algal Biotechnology in the Asia-Pacific Region*, ed. S. M. Phang, pp. 92–102. Kuala Lumpur, Malaysia: University of Malaya Press.

Benderliev, K., I. Pouneva, and N. Ivanova. 1993. Fungicide effect of Triton-N on *Phlyctidium*. *Biotechnol. Tech.* 7:335–338.

Benemann, J. R. 2008. Open ponds and closed photobioreactors—Comparative economics. In *Fifth Annual World Congress on Industrial Biotechnology and Bioprocessing*, Chicago, IL, p. 30.

Borowitzka, M. 2005. Culturing microalgae in outdoor ponds. In *Algal Culturing Techniques*, ed. R. A. Andersen, pp. 205–218. Burlington, MA: Elsevier.

Borowitzka, M. and N. Moheimani. 2013. Open pond culture systems. In *Algae for Biofuels and Energy*, eds. M. A. Borowitzka and N. R. Moheimani, pp. 133–152. Berlin, Germany: Springer.

Brenner, K., L. You, and F. H. Arnold. 2008. Engineering microbial consortia: A new frontier in synthetic biology. *Trends Biotechnol.* 26:483–489.

Bull, J. J. and S. Collins. 2012. Algae for biofuel: Will the evolution of weeds limit the enterprise? *Evolution* 66:2983–2987.

Burge, H. A. and C. A. Rogers. 2000. Outdoor allergens. *Environ. Health Perspect.* 108:653–659.

Carney, L. T. and T. W. Lane. 2014. Parasites in algae mass culture. *Aquat. Microb.* 5:278.

Charalambidou, I. and L. Santamaría. 2002. Waterbirds as endozoochorous dispersers of aquatic organisms: A review of experimental evidence. *Acta Oecol.* 23:165–176.

Chaumont, D. 1993. Biotechnology of algal biomass production: A review of systems for outdoor mass culture. *J. Appl. Phycol.* 5:593–604.

Cook, P. 1950. Large-scale culture of *Chlorella*. In *The Culture of Algae*, eds. J. Brunel, G. W. Prescott, and L. H. Tiffany, pp. 53–77. Yellow Springs, OH: Charles F. Kettering Foundation.

Daft, M. and W. Stewart. 1973. Light and electron microscope observations on algal lysis by bacterium CP-I. *New Phytol.* 72:799–808.

Day, J. G. and M. S. Stanley. 2012. Biological constraints on the production of microalgal-based biofuels. In *The Science of Algal Fuels*, eds. R. Gordon and J. Seckbach, pp. 101–129. Berlin, Germany: Springer.

Denholm, I. and M. Rowland. 1992. Tactics for managing pesticide resistance in arthropods: Theory and practice. *Annu. Rev. Ent.* 37:91–112.

Després, V. R., J. A. Huffman, S. M. Burrows et al. 2012. Primary biological aerosol particles in the atmosphere: A review. *Tellus B* 64:1–58.

Ehresmann, D. W. and M. T. Hatch. 1975. Effect of relative humidity on the survival of airborne unicellular algae. *Appl. Microbiol.* 29:352–357.

Fott, B. 1967. *Phlyctidium scenedesmi* spec. nova, a new chytrid destroying mass cultures of algae. *Zeitschrift für allgemeine Mikrobiologie* 7:97–102.

Fravel, D. 2005. Commercialization and implementation of biocontrol. *Annu. Rev. Phytopathol.* 43:337–359.

Gachon, C. M., T. Sime-Ngando, M. Strittmatter, A. Chambouvet, and G. H. Kim. 2010. Algal diseases: Spotlight on a black box. *Trends Plant Sci.* 15:633–640.

Gleason, F. H., L. T. Carney, O. Lilje et al. 2012. Ecological potentials of species of *Rozella* (Cryptomycota). *Fungal Ecol.* 5:651–656.

Gonen-Zurgil, Y., Y. Carmeli-Schwartz, and A. Sukenik. 1996. Selective effect of the herbicide DCMU on unicellular algae—A potential tool to maintain monoalgal mass culture of *Nannochloropsis*. *J. Appl. Phycol.* 8:415–419.

Green, A. J. and J. Figuerola. 2005. Recent advances in the study of long-distance dispersal of aquatic invertebrates via birds. *Divers. Distrib.* 11:149–156.

Grobbelaar, J. 1981. Infections: Experiences in mini-ponds. In *Wastewater for Aquaculture*, eds. J. U. Grobbelaar, C. J. Soeder, and D. F. Toerien, pp. 116–123. Bloemfontein, South Africa: University of the Orange Free State Publication Series.

Gummert, F., M. Meffert, and H. Stratmann. 1953. Nonsterile large-scale culture of *Chlorella* in greenhouse and open air. In *Algal Culture: From Laboratory to Pilot Plant*, ed. J. S. Burlew, pp. 166–176. Washington, DC: Carnegie Institution.

Heussler, P., S. Castillo, and F. Merino. 1978. Parasite problems in the outdoor cultivation of *Scenedesmus*. *Arch. Hydrobiol. Beih.* 11:223–227.

Hoffman, Y., A. Zarka, and S. Boussiba. 2002. Isolation and characterization of a parasitic chytrid from a culture of the chlorophyte *Haematococcus pluvialis*. In *The First Congress of the International Society for Applied Phycology*, p. 48. Book of Abstracts—*Algal Biotechnology*. Almería, Spain.

Houdan, A., A. Bonnard, J. Fresnel, S. Fouchard, C. Billard, and I. Probert. 2004. Toxicity of coastal coccolithophores (Prymnesiophyceae, Haptophyta). *J. Plankton Res.* 26:875–883.

Ibelings, B. W., A. De Bruin, M. Kagami, M. Rijkeboer, M. Brehm, and E. V. Donk. 2004. Host parasite interactions between freshwater phytoplankton and chytrid fungi (Chytridiomycota). *J. Phycol.* 40:437–453.

Ilkov, G. 1975. Population dynamic relationships during *Phlyctidium scenedesmi* development in *Scenedesmus acutus* cultures. *Appl. Microbiol.* 6:104–110.

Jacquet, S., X. Zhong, A. Parvathi, and A. S. P. Ram. 2013. First description of a cyano-phage infecting the cyanobacterium *Arthrospira platensis* (*Spirulina*). *J. Appl. Phycol.* 25:195–203.

James, T. Y. and M. L. Berbee. 2012. No jacket required—New fungal lineage defies dress code. *Bioessays* 34:94–102.

Kagami, M., T. Miki, and G. Takimoto. 2014. Mycoloop: Chytrids in aquatic food webs. *Front. Microbiol.* 5:1–9.

Kazamia, E., D. C. Aldridge, and A. G. Smith. 2012a. Synthetic ecology—A way forward for sustainable algal biofuel production? *J. Biotechnol.* 162:163–169.

Kazamia, E., H. Czesnick, T. T. V. Nguyen et al. 2012b. Mutualistic interactions between vitamin B12-dependent algae and heterotrophic bacteria exhibit regulation. *Environ. Microbiol.* 14:1466–1476.

Kazamia, E., A. S. Riseley, C. J. Howe, and A. G. Smith. 2014. An engineered community approach for industrial cultivation of microalgae. *Ind. Biotechnol.* 10:184–190.

Kellam, S. J. and J. M. Walker. 1989. Antibacterial activity from marine microalgae in labora-tory culture. *Br. Phycol. J.* 24:191–194.

Kim, W., J. M. Park, G. H. Gim et al. 2012. Optimization of culture conditions and compari-son of biomass productivity of three green algae. *Bioproc. Biosyst. Eng.* 35:19–27.

Kristiansen, J. 1996. Dispersal of freshwater algae—A review. *Hydrobiologia* 336:151–157.

Lee, Y. K. 2001. Microalgal mass culture systems and methods: Their limitation and poten-tial. *J. Appl. Phycol.* 13:307–315.

Leeuwen, C. H., G. Velde, J. M. Groenendael, and M. Klaassen. 2012. Gut travellers: Internal dispersal of aquatic organisms by waterfowl. *J. Biogeogr.* 39:2031–2040.

Letcher, P. M., S. Lopez, R. Schmieder et al. 2013. Characterization of *Amoeboaphelidium protococcarum*, an algal parasite new to the cryptomycota isolated from an outdoor algal pond used for the production of biofuel. *PLoS One* 8:e56232.

Lincoln, E. P., T. W. Hall, and B. Koopman. 1983. Zooplankton control in mass algal cultures. *Aquaculture* 32:331–337.

Liu, H., Y. Zhou, W. Xiao, L. Ji, X. Cao, and C. Song. 2012. Shifting nutrient-mediated inter-actions between algae and bacteria in a microcosm: Evidence from alkaline phospha-tase assay. *Microbiol. Res.* 167:292–298.

Loosanoff, V. L., J. E. Hanks, and A. E. Ganaros. 1957. Control of certain forms of zooplank-ton in mass algal cultures. *Science* 125:1092–1093.

Lukavsky, J. 1971. *Phlyctidium scenedesmi*, a chytrid destroying an outdoor mass culture of *Scenedesmus obliquus*. *Nova Hedwigia* 19:775–777.

Maynard, N. G. 1968. Aquatic foams as an ecological habitat. *Zeitschrift für allgemeine Mikrobiologie* 8:119–126.

McBride, R. C., S. Lopez, C. Meenach et al. 2014. Contamination management in low cost open algae ponds for biofuels production. *Ind. Biotechnol.* 10:221–227.

Mendes, L. B. B. and A. B. Vermelho. 2013. Allelopathy as a potential strategy to improve microalgae cultivation. *Biotechnol. Biofuels* 6:152.

Méndez, C. and E. Uribe. 2012. Control de *Branchionus* sp. y Amoeba sp. en cultivos de *Arthrospira* sp. *Latin American J. Aquat. Res.* 40:553–561.

Milliger, L. E. and H. E. Schlichting Jr. 1968. The passive dispersal of viable algae and pro-tozoa by an aquatic beetle. *Trans. Am. Microscop. Soc.* 87:443–448.

Mitchell, S. A. and A. Richmond. 1987. The use of rotifers for the maintenance of monoalgal mass cultures of *Spirulina*. *Biotechnol. Bioeng.* 30:164–168.

Mituya, A., T. Nyunoya, and H. Tamiya. 1953. Pre-pilot-plant experiments on algal mass cul-ture. In *Algal Culture: From Laboratory to Pilot Plant*, ed. J. S. Burlew, pp. 273–281. Washington, DC: Carnegie Institution.

Mokrosnop, V. and E. Zolotareva. 2013. Influence of fungicides on the growth of the microal-gal culture *Euglena gracilis* Klebs (Euglenophyta). *Int. J. Algae* 15:180–187.

Mooij, P. R., G. R. Stouten, J. Tamis, M. C. Van Loosdrecht, and R. Kleerebezem. 2013. Survival of the fittest. *Energy Environ. Sci.* 6:3404–3406.

Moreno-Garrido, I. and J. Canavate. 2001. Assessing chemical compounds for controlling predator ciliates in outdoor mass cultures of the green algae *Dunaliella salina*. *Aquacult. Eng.* 24:107–114.

Natrah, F., M. M. Kenmegne, W. Wiyoto et al. 2011. Effects of micro-algae commonly used in aquaculture on acyl-homoserine lactone quorum sensing. *Aquaculture* 317:53–57.

Nesme, J. and P. Simonet. 2014. The soil resistome: A critical review on antibiotic resistance origins, ecology and dissemination potential in telluric bacteria. *Environ. Microbiol.* 17:913–930.

Parsons, W. M., H. E. Schlichting, and K. W. Stewart. 1966. In-flight transport of algae and protozoa by selected *Odonata*. *Trans. Am. Microscop. Soc.* 85:520–527.

Polson, S. W., S. W. Wilhelm, and K. E. Wommack. 2011. Unraveling the viral tapestry (from inside the capsid out). *ISME J.* 5:165.

Post, F., L. Borowitzka, M. Borowitzka, B. Mackay, and T. Moulton. 1983. The protozoa of a Western Australian hypersaline lagoon. *Hydrobiologia* 105:95–113.

Ravikumar, R. 2014. Micro algae in open raceways. In *Algal Biorefineries*, eds. R. Bajpai, A. Prokop, and M. Zappi, pp. 127–146. Berlin, Germany: Springer.

Rego, D., L. Redondo, V. Geraldes, L. Costa, J. Navalho, and M. Pereira. 2014. Control of predators in industrial scale microalgae cultures with pulsed electric fields. *Bioelectrochemistry* 103:60–64.

Richardson, J. W., M. D. Johnson, X. Zhang, P. Zemke, W. Chen, and Q. Hu. 2014. A financial assessment of two alternative cultivation systems and their contributions to algae biofuel economic viability. *Algal Res.* 4:96–104.

Richmond, A. 2007. Biological principles of mass cultivation. In *Handbook of Microalgal Culture*, ed. A. Richmond, pp. 125–177. Oxford, UK: Blackwell Publishing.

Richmond, A. 2013. Biological principles of mass cultivation of photoautotrophic microalgae. In *Handbook of Microalgal Culture*, ed. A. Richmond, pp. 169–204. Oxford, UK: Blackwell Publishing.

Sabra, W., D. Dietz, D. Tjahjasari, and A. P. Zeng. 2010. Biosystems analysis and engineering of microbial consortia for industrial biotechnology. *Eng. Life Sci.* 10:407–421.

Schenk, P. M., S. R. Thomas-Hall, E. Stephens et al. 2008. Second generation biofuels: High-efficiency microalgae for biodiesel production. *Bioenerg. Res.* 1:20–43.

Schlichting Jr., H. E. 1974. Periodicity and seasonality of airborne algae and protozoa. *Phenol. Seasonal. Model.* 8:407–413.

Schmeller, D. S., M. Blooi, A. Martel et al. 2014. Microscopic aquatic predators strongly affect infection dynamics of a globally emerged pathogen. *Curr. Biol.* 24:176–180.

Sharma, N. K., A. K. Rai, S. Singh, and R. M. Brown. 2007. Airborne algae: Their present status and relevance. *J. Phycol.* 43:615–627.

Sheehan, J., T. Dunahay, J. Benemann, and P. Roessler. 1998. A look back at the U.S. Department of Energy's Aquatic Species Program: Biodiesel from algae. NREL/TP-580-24190. Golden, CO: National Renewable Energy Laboratory.

Shimamatsu, H. 2004. Mass production of *Spirulina*, an edible microalga. *Hydrobiologia* 512:39–44.

Shurin, J., R. Abbott, M. Deal et al. 2013. Industrial-strength ecology: Trade-offs and opportunities in algal biofuel production. *Ecol. Lett.* 16:1393–1404.

Sigee, D., R. Glenn, M. Andrews et al. 1999. Biological control of cyanobacteria: Principles and possibilities. *Hydrobiologia* 395:161–172.

Smith, V. H. and T. Crews. 2014. Applying ecological principles of crop cultivation in large-scale algal biomass production. *Algal Res.* 4:23–34.

Smith, V. H., B. S. Sturm, F. J. Denoyelles, and S. A. Billings. 2010. The ecology of algal biodiesel production. *Trends Ecol. Evol.* 25:301–309.

Tredici, M. 2004. Mass production of microalgae: Photobioreactors. In *Handbook of Microalgal Culture: Biotechnology and Applied Phycology*, ed. A. Richmond, pp. 178–214. Oxford, UK: Blackwell Publishing.

Van der Westhuizen, A. and J. Eloff. 1983. Effect of culture age and pH of culture medium on the growth and toxicity of the blue-green alga *Microcystis aeruginosa. Zeitschrift für Pflanzenphysiologie* 1102:157–163.

Van Etten, J. L., L. C. Lane, and R. H. Meints. 1991. Viruses and virus-like particles of eukaryotic algae. *Microbiol. Rev.* 55:586.

Vick, B. 2010. Glyphosate applications in aquaculture. US 2009004296.

Vonshak, A. 1997. *Spirulina*: Growth, physiology and biochemistry. In *Spirulina Platensis (Arthrospira)*, ed. A. Vonshak, pp. 43–65. London, UK: Taylor & Francis.

Vonshak, A., S. Boussiba, A. Abeliovich, and A. Richmond. 1983. Production of *Spirulina* biomass: Maintenance of monoalgal culture outdoors. *Biotechnol. Bioeng.* 25:341–349.

Vonshak, A. and A. Richmond. 1988. Mass production of the blue-green alga *Spirulina*: An overview. *Biomass* 15:233–247.

Waltz, E. 2009. Biotech's green gold? *Nat. Biotechnol.* 27:15–18.

Wang, H., W. Zhang, L. Chen, J. Wang, and T. Liu. 2013. The contamination and control of biological pollutants in mass cultivation of microalgae. *Bioresour. Technol.* 128:745–750.

Weisse, T. and P. Stadler. 2006. Effect of pH on growth, cell volume, and production of freshwater ciliates, and implications for their distribution. *Limnol. Oceanogr.* 51:1708–1715.

Weissman, J. and G. Radaelli. 2010. Systems and methods for maintaining the dominance of *Nannochloropsis* in an algae cultivation system. US 20100183744.

Weissman, J., G. Radaelli, and D. Rice. 2010. Systems and methods for maintaining the dominance and increasing the biomass production of *Nannochloropsis* in an algae cultivation system. US 20100196995.

Wen-Li, Z., Q. Xiuting, S. Jingfeng, X. Kezhi, and T. Xue-Xi. 2011. Ecological effect of Z-QS01 strain on *Chlorella vulgaris* and its response to UV-B radiation stress. *Proc. Environ. Sci.* 11:741–748.

Whalon, M., D. Mota-Sanchez, and R. Hollingworth. 2008. Analysis of global pesticide resistance in arthropods. In *Global Pesticide Resistance in Arthropods*, eds. M. E. Whalon, D. Monta-Sanchez, and R. M. Hollingworth, pp. 5–31. Cambridge, MA: CABI.

Yates, E. A., B. Philipp, C. Buckley et al. 2002. N-acylhomoserine lactones undergo lactonolysis in a pH-, temperature-, and acyl chain length-dependent manner during growth of *Yersinia pseudotuberculosis* and *Pseudomonas aeruginosa. Infect. Immun.* 70:5635–5646.

Zmora, O. and A. Richmond. 2008. Microalgae for aquaculture: Microalgae production for aquaculture. In *Handbook of Microalgal Culture*, ed. A. Richmond, pp. 365–379. Oxford, UK: Blackwell Publishing.

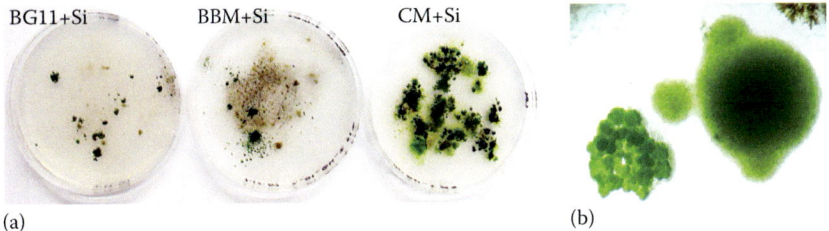

FIGURE 3.4 Algal colonies grown on agar-solidified media. (a) Aliquots of the same environmental freshwater sample were placed on agar-solidified media in regular standard 90 mm Petri dishes. Three different media as indicated on the top left of each plate were used, supplemented with silicate to allow for growth of diatoms. (b) An example of algal colonies grown on agar-solidified medium shown with a magnification of 60×.

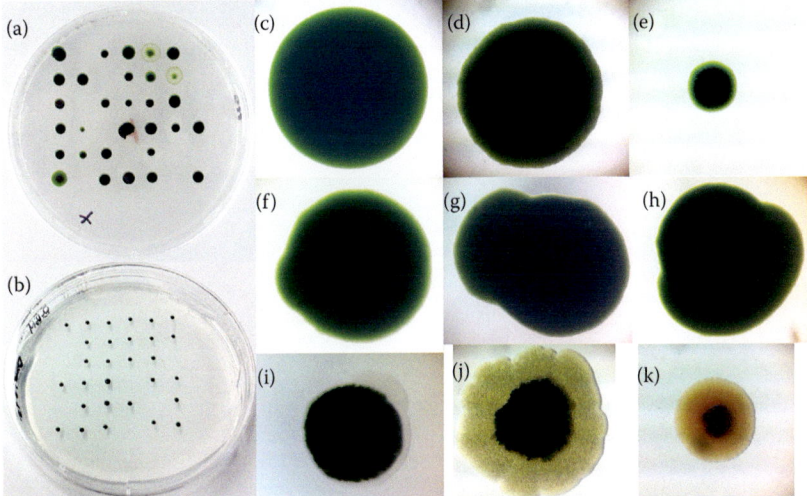

FIGURE 3.8 Examples of microalgal colonies growing on agar-solidified media in Petri dishes after FACS cell sorting. (a) Plate from a non-unialgal liquid culture. Although one strain seemed to dominate, several different colony colors and types indicate the presence of at least four different species. (b) One unialgal strain only, with most colonies being axenic. (c–k) 60× magnification images of algal colonies sorted from a single event that are (c–e) uni-algal and axenic, (f–h) non-unialgal, and (i–k) containing bacteria as indicated by the white, beige, or orange rings around the colonies.

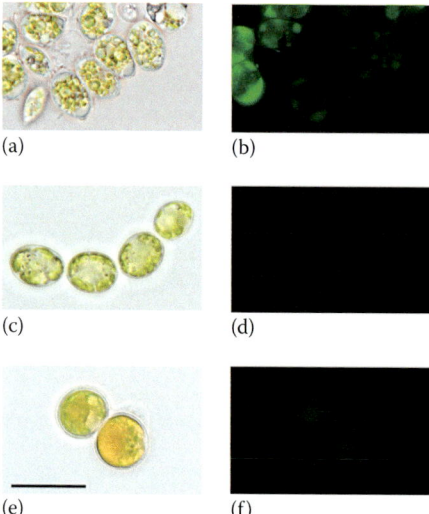

FIGURE 3.11 Cells from two of the strains deemed to be potential high producers, showing bright field (a, c, e) and Nile Red fluorescence images from the same pairwise field of view at 530 nm (b, d, f). (a, b) *Scenedesmus obliquus* strain EN0004 with oil bodies visible with fluorescence microscopy. (c, d) Strain EN1234 (a putative *Coelastrum*-related Desmid) from a young culture with few significant oil bodies visible. (e, f) Stressed cells of strain EN1234 taken from an older culture. Note the orange coloration under the bright field indicative of carotenoid accumulation (e) and apparent Nile Red fluorescence (f). The bar indicates 10 μm.

FIGURE 3.12 Identification of strains that overaccumulate carotenoids. (a) Sometimes, colonies of microalgae grown on plates would display a yellowish, orange, or reddish color indicating that the cells overaccumulated carotenoids. (b) More often, strains of liquid cultures from the first-level screen would display a yellowish, orange, or reddish color upon aging. (c) A culture displaying an apricot color indicating accumulation of carotenoids. (d) Microscopy of cells from culture shown in (c). The cells have an apricot color originating from very large oil bodies that sequester the carotenoids (data not shown). The bar represents 10 μm.

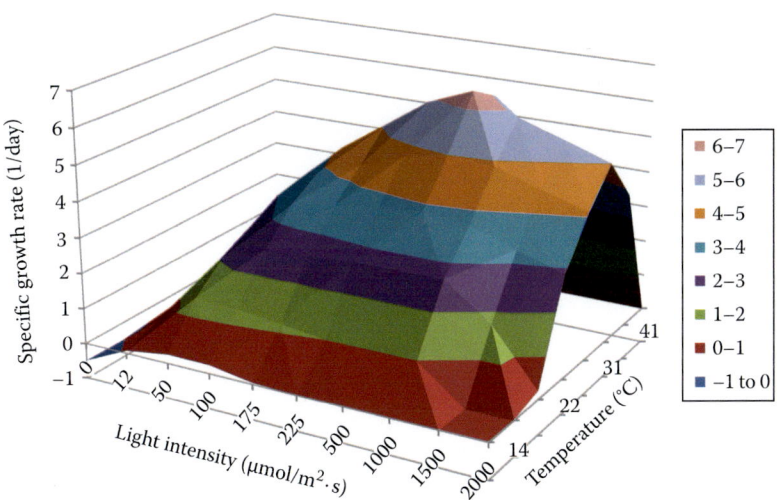

FIGURE 5.2 Specific growth rate (μ) of *Chlorella sorokiniana* (DOE 1412) as a function of light intensity (0–1950 μmol/m²·s) and temperature (14°C–41°C). Each value is the average of three to six measurements taken in replicate growth rate experiments.

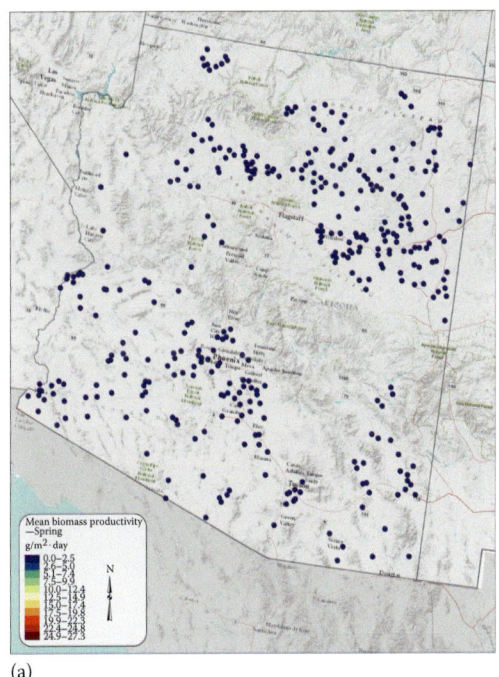

(a)

FIGURE 5.7 Long-term average seasonal biomass productivity of *Chlorella sorokiniana* (DOE 1412) using the BAT pond temperature and biomass growth models for potential pond sites in Arizona for (a) spring. Winter values, as for spring, were in the range 0.0–2.5 g/m²·day at all sites (dark blue), and therefore are not shown. *(Continued)*

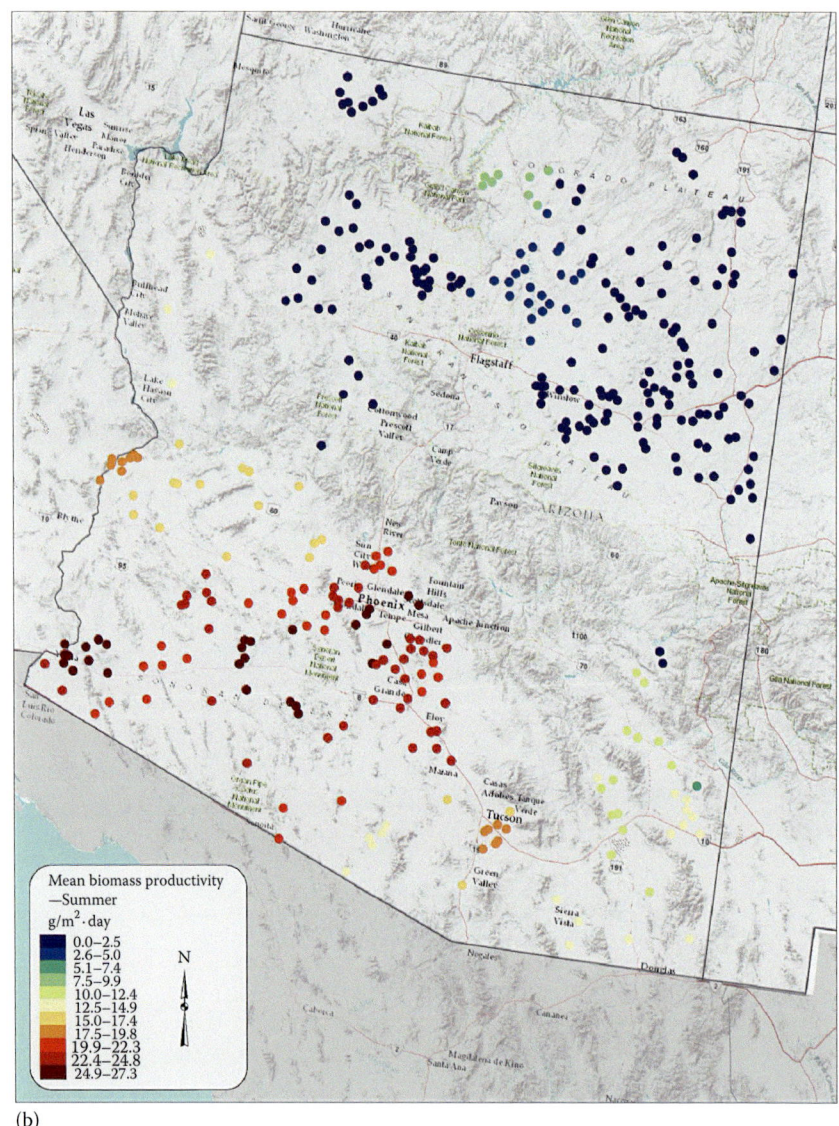

(b)

FIGURE 5.7 (*Continued*) Long-term average seasonal biomass productivity of *Chlorella sorokiniana* (DOE 1412) using the BAT pond temperature and biomass growth models for potential pond sites in Arizona for (b) summer. (*Continued*)

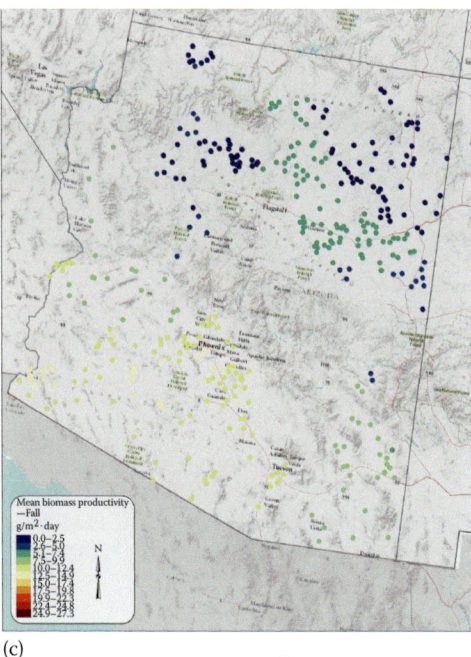

(c)

FIGURE 5.7 (Continued) Long-term average seasonal biomass productivity of *Chlorella sorokiniana* (DOE 1412) using the BAT pond temperature and biomass growth models for potential pond sites in Arizona for (c) fall.

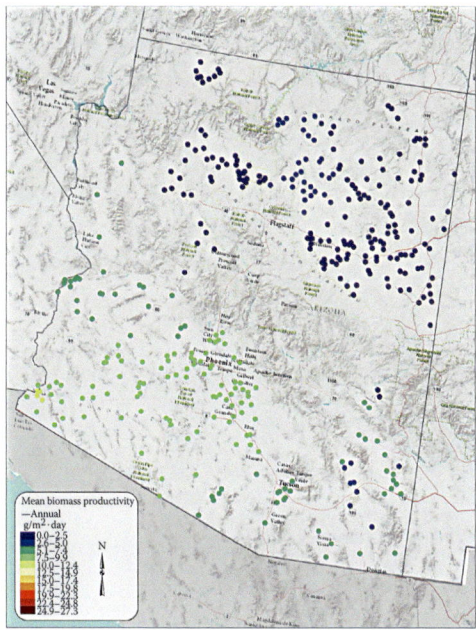

FIGURE 5.8 Average annual biomass productivity for *Chlorella sorokiniana* (DOE 1412) using the BAT pond temperature and biomass growth models for potential pond sites in Arizona.

FIGURE 5.10 LED-lighted and temperature-controlled raceway pond for culturing of microalgae under climate-simulated conditions.

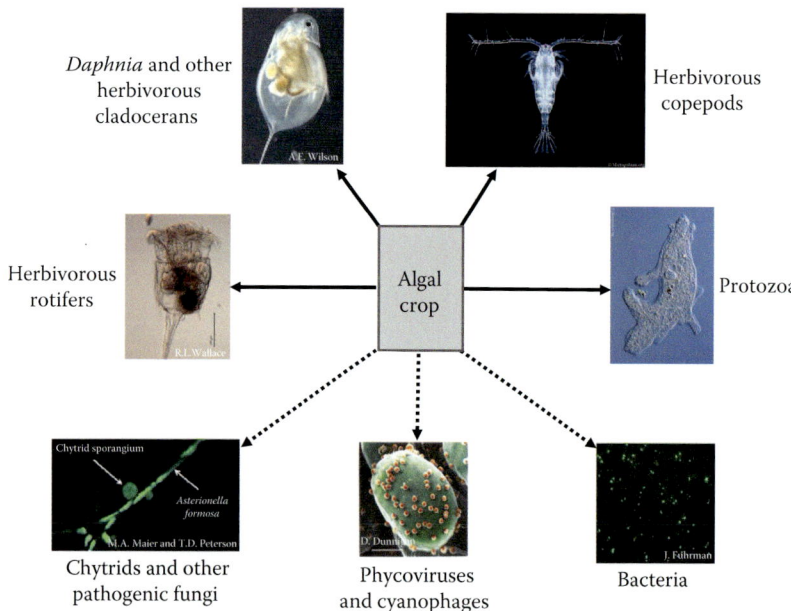

FIGURE 7.1 Examples of contaminating pest organisms that can impact the productivity of algae grown in open ponds. Herbivorous consumers are denoted by the solid arrows, and potential pathogens are denoted by the dotted lines.

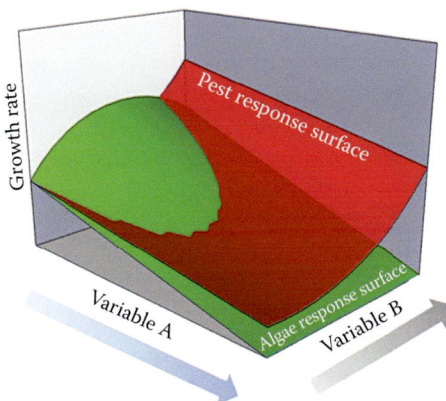

FIGURE 7.2 Hypothetical growth rate landscapes for algae and pests of algae across environmental gradients. This figure illustrates the concept that modifying the cultivation environment can differentially impact the relative growth rates of algae and their pests. Careful manipulations can be designed that drive the pond to a solution space in which the algal growth rate is higher than that of the pest, resulting in productive algal biomass cultivation systems.

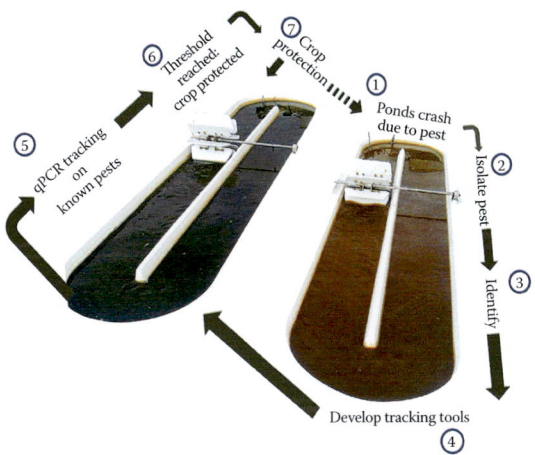

FIGURE 8.4 An example of a reactive pest response strategy aided by the use of quantitative PCR (qPCR). This strategy is triggered by qualitative microscope observations of pests in field samples and/or through bringing a pond sample into the lab and exposing it to conditions that would accelerate pest growth and precipitate a culture crash. If a new or unknown pest is observed, various microbiological techniques (e.g., plaque plating, baiting, selective media) are first used to isolate this pest organism. ITS or SSU rRNA sequencing is then used to identify the pest organism and to develop qPCR primers to track and quantify the genomic DNA of the targeted pest in established ponds. If isolation and culturing is not possible, metagenomic sequencing of the pond community may reveal enough of a pattern to identify a specific pest organism and provide sequence data for qPCR assay design. Specific thresholds can be set for the qPCR-determined abundance of each pest that is being monitored. Once this critical abundance threshold is reached in any given pond, a carefully chosen crop protection strategy (e.g., a selected pesticide application or the addition of a targeted biocontrol agent) is then implemented and the pest's abundance is consistently monitored to determine whether the control strategy is successful (see Chapter 7). (Reproduced from McBride, R. C. et al., *Ind. Biotechnol.*, 10, 221, 2014.)

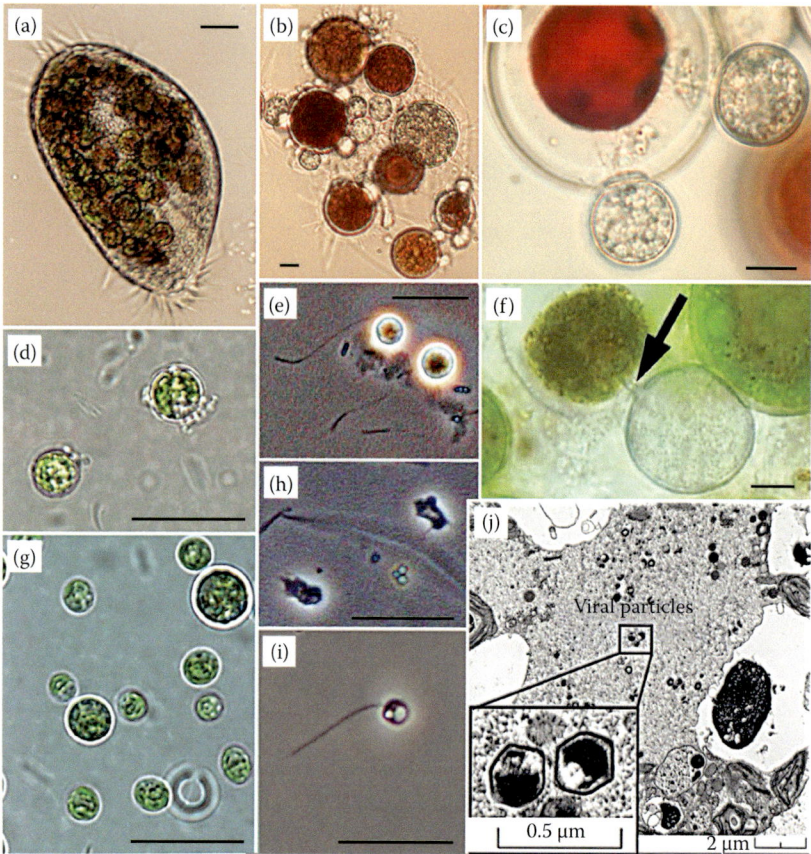

FIGURE 8.5 Biocontaminants observed in a mass algal culture: (a) Grazing ciliate, *Gastrostyla steinii*, with ingested whole *Haematococcus* cysts. (Photo courtesy of K. Sorensen.) (b) Sporangia of the parasitic blastocladian fungus, probably *Paraphysoderma sedebokerensis*, attached to *Haematococcus* growing in open ponds. (Photo courtesy of L. T. Carney.) (c) Higher magnification of *P. sedebokerensis* (isolate JEL821); note the fungal rhizoids extending into the *Haematococcus* cyst, showing signs of cell content depletion. (Photo courtesy of J. E. Longcore.) (d) Bacteria attached to lysing algal cells in a crashing culture correlated with increasing detection of the pathogenic bacterium, *Vampirovibrio chlorellavorus*, on multiple occasions using SSU rRNA sequencing and qPCR. (Photo courtesy of S. Qin.) (e) Motile *V. chlorellavorus*-like bacteria adjacent to algae cells in a culture that was confirmed by qPCR to contain *V. chlorellavorus*. (Photo courtesy of J. Wilkenfeld.) (f) Sporangium of isolate JEL812; the arrow is pointing to the rhizoidal connection to the *Haematococcus* cell. (Photo courtesy of J. E. Longcore.) (g) Algae cells prior to the appearance of the attachment and detection of *V. chlorellavorus*. (Photo courtesy of S. Qin.) (h) Amoeboid swarmer stages of *P. sedebokerensis* observed in open ponds growing *Haematococcus*. (Photo courtesy of J. Wilkenfeld.) (i) Motile zoospore stage of the parasitic chytrid isolate JEL812 (Rhizophydiales; Aquamycetaceae). (Photo courtesy of K. Sorensen.) (j) TEM photograph of virus particles inside an algal cell. (Photo courtesy of R. Roberson.) *Note:* Scale bars = 10 μm where not otherwise indicated.

(b)

FIGURE 9.5 Integrated algal and aquaculture research units at Clemson University (South Carolina).

(b)

FIGURE 9.8 Green algal enhancement with tilapia filtration (left tank) versus cyanobacterial enhancement with freshwater mussel filtration (right tank) (b).

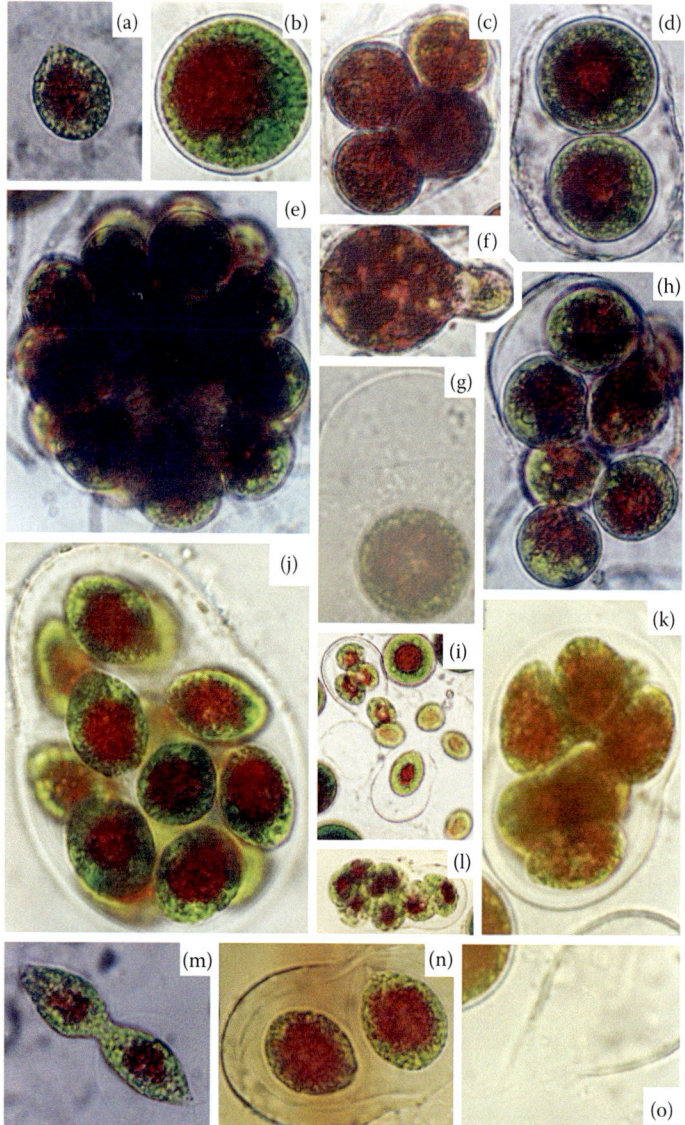

FIGURE 12.1 The cell forms of *Haematococcus pluvialis*: (a) Motile cell. (b) Nonmotile cell. (c, d) Sporangia with 2–4 aplanospores. (e) Sporangia with >20 aplanospores. (f) Vegetative reproduction by cell budding. (g, h) The moment of aplanospore release. (i) The moment of zoospore release. (j–l) The early process of zoospore release. (m) Vegetative reproduction by direct cell division. (n) Sporangia with two zoospores. (o) Theca after spore release.

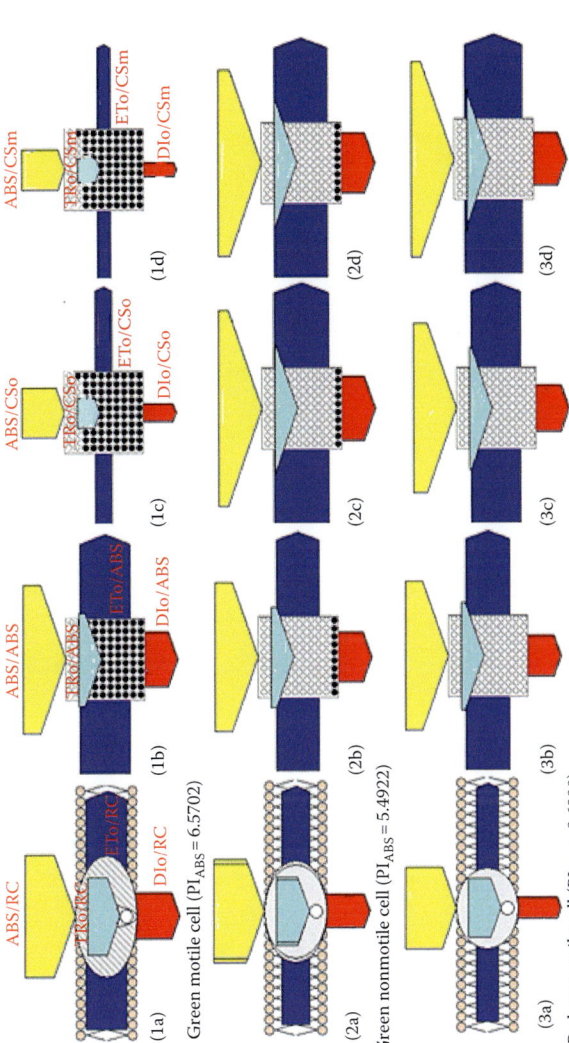

FIGURE 12.4 The photochemical parameters and energy pipeline models of different cell forms of *Haematococcus pluvialis*: (a) specific membrane model (per active reaction center, RCs); (b) phenomenological yield model; (c) phenomenological model (per excited cross section) at time zero; and (d) phenomenological model (per excited cross section, CSm) at the time of reaching the maximal fluorescence. The photochemical parameters were measured with a Handy PEA fluorometer (Hansatech Instruments). All the measurements were done with 10 minute dark-adapted samples at room temperature. The chlorophyll contents in green motile cells, green nonmotile cells, and red nonmotile cells were 0.26, 0.31, and 0.61 μg/10⁴ cell, respectively. And the astaxanthin contents in green motile cells, green nonmotile cells, and red nonmotile cells were 0.03, 0.12, and 0.48 μg/10⁴ cells, respectively. ABS, TRo, ETo, and DIo represent the energy fluxes for absorption (yellow), trapping (light blue), electron transport (dark blue), and dissipation (red). The relative value of each of the parameters (ABS/CSm, TRo/CSm, ETo/CSm, and DIo/CSm) can be seen from the width of its arrow. Active and inactive RCs are shown as open and solid circles, respectively.

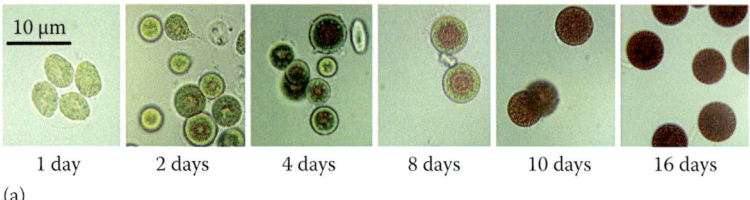

(a)

FIGURE 12.6 The variation of cell morphology, the maximal and minimal Fv/Fm and PI$_{ABS}$ in *Haematococcus pluvialis* during a culture period from astaxanthin-free motile cells to astaxanthin-enriched nonmotile cells in a tubular photobioreactor: (a) the cell morphological changes.

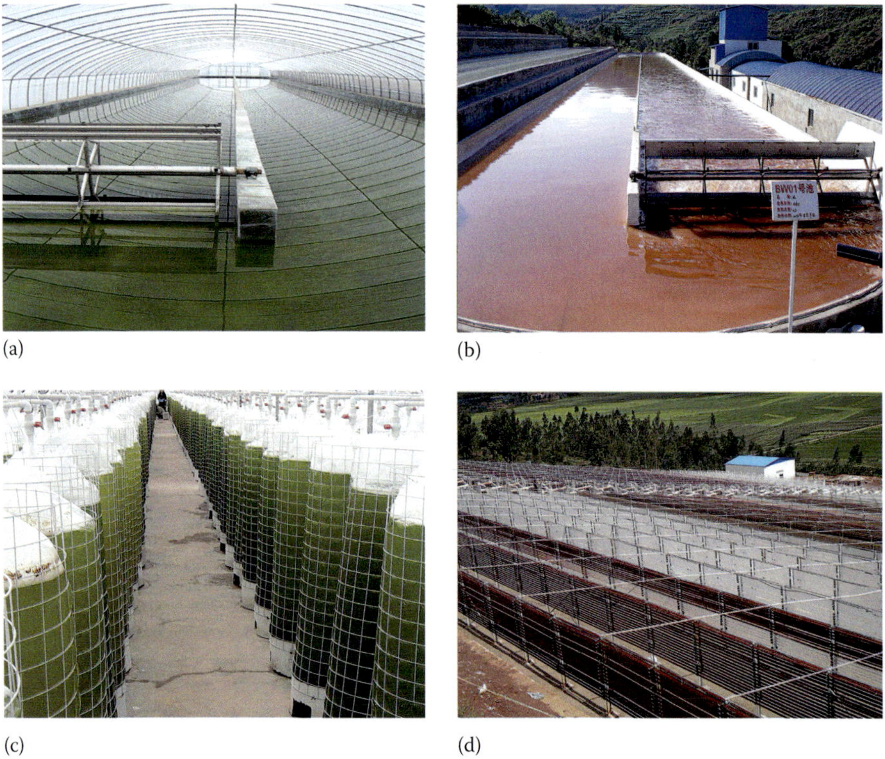

(a)   (b)

(c)   (d)

FIGURE 12.7 Photobioreactors used for industrial-scale production of *Haematococcus pluvialis*: (a) open raceway pond inside a greenhouse; (b) open raceway pond outdoors; (c) closed column photobioreactors inside a greenhouse; and (d) closed tubular photobioreactors outdoors.

8 Molecular Diagnostic Solutions in Algal Cultivation Systems

Laura T. Carney, Robert C. McBride,
Val H. Smith, and Todd W. Lane

CONTENTS

8.1 INTRODUCTION

One of the major challenges to achieving high rates of long-term production in microalgal mass cultures is the elimination or reduction of the impact of bio-contamination and culture losses (i.e., crashes) in production systems. Although there are both biotic and abiotic root causes of mass culture crashes, infection by deleterious organisms is the most important and least understood. In general, the diversity of pathogens, parasites, predators, and competing algal species (or *weed* species) has not been well characterized. Lost production days due to pond crashes can significantly lower annual production yields. In addition, depending on the scale and type of system, days to weeks of production can be lost while the system is disinfected and new inoculum and the growth medium is prepared. Depending on the design and operation of the production facility, there is a risk of spread or persistence of contamination and successive crashes. Despite a paucity of publically available data on the economic impact of biocontaminants on the nascent algae biomass industry, the consensus is that they constitute an economic barrier to commercialization (Davis et al. 2012; Gao et al. 2012). Some insight into the potential magnitude of the financial impact may be gained from the aquaculture-for-food industry, which loses several billion U.S. dollars annually (Subasinghe et al. 2001; FAO 2010) due to bacterial and fungal infections (Defoirdt et al. 2004; Ding and Ma 2005; Ramaiah 2006).

There are a wide variety of deleterious species including bacteria (Cole 1982; Fukami et al. 1997), viruses (Dunigan et al. 2006), parasites (e.g., Letcher et al. 2013), fungi (Fott 1967; Hoffman et al. 2008; Li et al. 2010), and herbivorous zooplankton (e.g., Park et al. 2011). In addition, there are algal weed species (Pienkos and Darzins 2009; Bull and Collins 2012) that can reduce the value of the biomass by supplanting the desired species. Early detection of deleterious species is the key to informed pond management strategies. Thus, successful large-scale algae cultivation will require routine, detailed, fast, and cost-effective identification of potentially deleterious species before they become established in the culture and knowledge on how their populations may change with time and environmental conditions.

8.2   DIAGNOSTIC METHODS

Identification of algal pests is still largely dependent on microscopy (e.g., Rasconi et al. 2009). Although optical methods can produce near-real-time data on the presence of contaminants and cannot be avoided when describing novel biocontaminants, microscopy is a method that is labor intensive, has low throughput, and requires a certain level of expert knowledge to recognize contaminant species. Thus, a routine monitoring program relying solely on traditional microscopy is not desirable. As a consequence of these limitations, a variety of advanced optical methods have recently been developed such as flow cytometry coupled with digital imaging and image recognition (reviewed by Álvarez et al. 2011; Day et al. 2012) and hyperspectral confocal florescence microscopy (Collins et al. 2014). While such methods are advances over standard microscopic methods, these newer methods, as well as many molecular techniques, are still dependent on *a priori* knowledge of deleterious species.

Molecular diagnostic techniques can, in general, be divided into two separate types of processes. First is the initial identification of known and novel etiological agents: viruses, grazers, pathogens, and parasites. Once identified and characterized, this information is used to develop assays against these agents that enable the second process, namely, the detection of such agents in algal cultures. Initial identification of etiological agents is still, fundamentally, based on the postulates first formulated by Robert Koch in the late nineteenth century. Classical methods of fulfilling Koch's postulates have depended upon the isolation and culturing of the etiological agent and the demonstration that the disease state is unambiguously correlated with the presence of the pathogen via reinfection and reisolation.

These methods remain the "gold standard" for pathogen identification; however, in the last two decades, culture-free molecular methods of identification and detection have become more commonplace in clinical diagnostics. There have been a number of instances in which the etiological agent was not culturable and was only recognized by its molecular signature (for a review, see Fredricks and Relman 1996; Miller et al. 2013). Although, in most cases, identifications that are based solely on molecular signatures are considered to be presumptive until confirmed by other evidence, the rise of molecular diagnostics has led to a rethinking of Koch's postulates to reflect the use of these methods (Fredricks and Relman 1996). It is now reasonable to reformulate Koch's postulates for molecular diagnosis of algal production system infections:

1. A nucleic acid sequence belonging to the putative pathogen, predator, or parasite should be present in an infected algal mass culture.
2. In healthy mass cultures, copies of the contaminant-associated nucleic acid sequences should occur below a management threshold value specified for that pathogen/pest.
3. When sequence detection predates mass culture infection, increase in sequence copy number should correlate with increased contaminant density, loss of algal biomass, decrease in productivity, or aberrant pond performance.
4. The organism, identified by sequence-based analysis, should have properties that are consistent with, and capable of, generating the phenomena observed in the infected pond.

Several factors can confound the identification of pest species by molecular analysis of algal mass culture infections: (1) Other sequences may become prevalent in crashed or infected ponds including those related to species that feed upon detritus or dead algal cells. In these cases, it is critical to establish temporal relationships and thus the frequency of pond sampling is an important consideration. A reasonable strategy is to collect and archive samples on a frequent and routine basis to enable the analysis of a time course of infection should a deleterious event occur. (2) Environmental conditions and abiotic stressors may play a significant role in mediating pond crashes and can be in themselves the root cause of pond crashes. (3) Pond infections by multiple microbial agents are possible, although many crashes appear to be caused by a single agent (Hu, pers. comm. 2014). (4) The correlation of an increase of a biocontaminant and a decrease of the target strain does not confirm the biocontaminant as the causative agent.

There are three main types of molecular techniques that are employed for culture diagnostics: probe hybridization, target amplification (quantitative polymerase chain reaction [qPCR]), and DNA sequencing. Each of these techniques, which are described in greater detail in the following text, take advantage of the information contained in specific, well-characterized genetic regions for identification and/or detection of the target organisms. The information contained in these regions can be used to identify known and novel agents and to develop taxon-specific hybridization probes or PCR primers for the detection of these agents.

8.2.1 MOLECULAR TARGETS FOR DIAGNOSTIC METHODS

There are several molecular markers that have been employed for molecular analysis of clinical and environmental samples (Hoef-Emden 2012) including the large and small subunit ribosomal RNA (LSU and SSU rRNA, respectively) genes, the internal transcribed spacer (ITS) of the ribosomal RNA genes, and the mitochondrial cytochrome oxidase (cox) gene. The best known of these targets is the SSU rRNA gene (see Figure 8.1). The prokaryotic SSU rRNA gene has nine hypervariable regions dispersed along its length (Van De Peer et al. 1996). Of these, the sequence information contained in the individual hypervariable regions 3 or 6 (Huse et al. 2008) or a fragment covering hypervariable regions 1 through 3 or 4 (Kim et al. 2011) are the most useful for phylogenetic determination. Sequence data from individual hypervariable regions are often sufficient for genus-level distinction, whereas data from multiple hypervariable regions or the entire SSU rRNA gene can result in species-level distinction.

The eukaryotic SSU rRNA gene also contains nine variable regions, of which V4 and V9 are the most informative and are often used in combination (Amaral-Zettler et al. 2009; Stoeck et al. 2010; Pawlowski et al. 2011; Orsi et al. 2013). The V4 region is the longest with the highest degree of length variation and sequence heterogeneity (Nickrent and Sargent 1991). The information contained in either region is generally sufficient for the genus-level identification of an organism. The shorter V9 region is less useful partially because it is sometimes not present in truncated versions of eukaryotic SSU rRNA genes found in sequence databases. Although phylogenetic analysis based on SSU rRNA gene sequence is more common, similar analysis based on the LSU rRNA gene is also carried out (Ludwig and Schleifer 1994; Hunt et al. 2006; Steven et al. 2012). The LSU rRNA gene is longer and contains two variable regions, D1 and D2 (see Figure 8.2), which can be used for phylogenetic analysis (Sonnenberg et al. 2007; Putignani 2008).

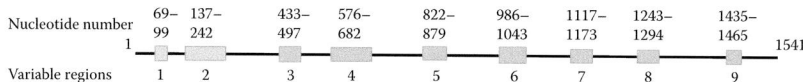

FIGURE 8.1 Map of prokaryotic SSU rRNA gene showing variable regions (nucleotide numbering corresponds to the *Escherichia coli* SSU rRNA gene).

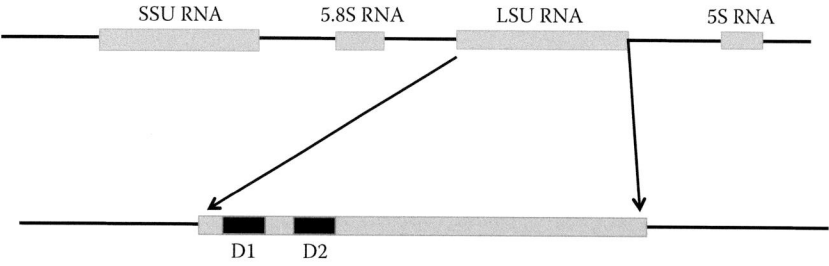

FIGURE 8.2 Map of the variable regions in the eukaryotic LSU gene.

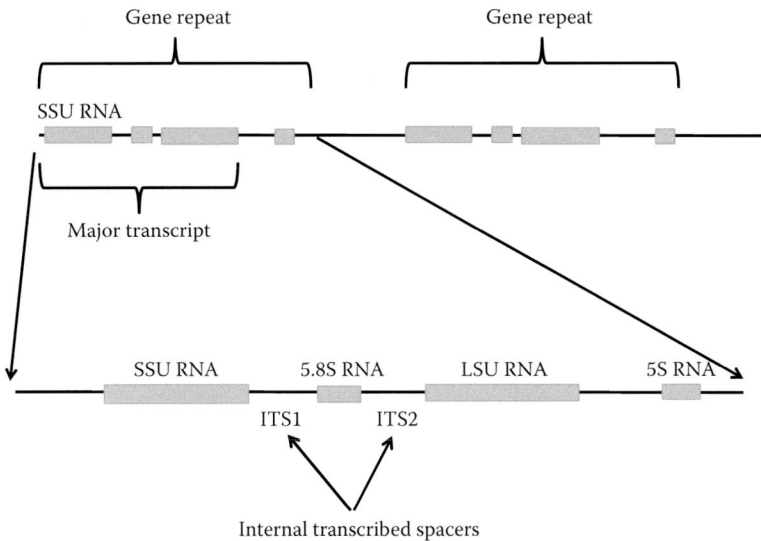

FIGURE 8.3   Map of the entire eukaryotic rRNA transcript structure showing ITS regions.

The ITS region, located between the rRNA genes encoding the SSU and the LSU (see Figure 8.3), is commonly used for genus- or species-level discrimination in fungal (Lindahl et al. 2013) and microalgal (Leliaert et al. 2014) phylogenetics. There is a very large database of fungal ITS sequences (see Section 8.2.7). In eukaryotes, the ITS region consists of two hypervariable spacers, ITS1and ITS2, which flank the gene encoding the 5.8S ribosomal subunit. A full-length amplicon including both ITS1 and ITS2 regions and the 5.8S subunit is approximately 650 bp in length, which is beyond the current read length limits of many of the next-generation sequencers (see Table 8.1). Because of this inability to cover the entire ITS region, individual

TABLE 8.1

Current Next-Generation Sequencer Capabilities

Sequencer	Read Lengths (nt)	Read Number[a]	Run Time
Illumina HiSeq	300	6×10^9	2–11 days
Illumina MiSeq	600	$1.2–1.5 \times 10^7$	39 hours
454 GS	400	7×10^4	10 hours
GS FLX	700	1×10^6	23 hours
Ion Torrent PGM	Up to 400	Up to 6×10^6	2–7 hours
Ion Torrent PI	Up to 200	Up to 3.3×10^8	2–4 hours
Pacific biosciences	4300–5000	$2.5–5.0 \times 10^3$	~2 hours
Applied biosystems solid 5500	50–100	1×10^8	1–7 days

nt, nucleotides.

[a]   Read number: number of individual sequences obtained from a run.

ITS regions have been analyzed by next-generation sequencing (Lindner et al. 2013). Reports indicate that community analyses based on ITS1 versus ITS2 yield different taxonomic compositions from each other as well as from those based on the full-length ITS region (Bazzicalupo et al. 2013; Blaalid et al. 2013). In the complex communities that may be present in algal mass culture systems, the potential for incorrect identification or detection can be mitigated by the use of multiple assays targeting different regions.

The 5′ terminus of the mitochondrial cytochrome oxidase gene *cox1* has also been used as a molecular barcode region primarily for metazoans (Bucklin et al. 2011). There are caveats to the use of *cox1*: the amplification of nuclear encoded pseudogenes can lead to overestimation of species diversity (Song et al. 2008) and it has proven difficult to develop universal primers (Saunders and McDevit 2012).

8.2.2 Sample Preparation for Molecular Diagnostics

Most molecular diagnostic methods such as sequencing, hybridization, or PCR require cell lysis and extraction of nucleic acids from the biomass sample. Differential extraction of nucleic acids, or failure thereof, from one group of organisms can skew apparent relative abundances. A variety of protocols and commercial kits for the extraction of nucleic acids from recalcitrant organisms have been developed and some comparisons of these methods are reported in the literature (Purdy 2005; Koid et al. 2012). However, new methods and commercial kits are routinely developed and it is advisable to test the effectiveness of a variety of protocols on the sample types of interest using the intended molecular diagnostic method (quantitative polymerase chain reaction [qPCR], sequencing, etc.). Many lysis and extraction protocols feature a combination of both chemical and mechanical lysis methods. There are several forms of mechanical lysis including sonication, the French pressure cell, the nitrogen bomb, cryopulverization, and the bead beater. For field applications, bead beating is arguably the most convenient form of mechanical lysis. This is largely due to the potential low cost of instrumentation and ability to handle small sample volumes. For particularly recalcitrant samples, an alternative method is cryopulverization in which the sample is first flash frozen in liquid nitrogen and then pulverized with a cold mortar and pestle. There are a variety of commercial devices that automate this process. Because of the potential for a high degree of complexity in samples and, therefore, the likelihood that they could contain species which are difficult to lyse, cryopulverization may be a particularly attractive option. However, more work is needed to mitigate the bias involved in nucleic acid extraction during microbiome analysis in order to transform these efforts from relative to quantitative analyses.

8.2.3 Identification of Contaminants by DNA Sequencing

Next-generation sequencing (NGS) has been employed extensively to identify unknown pathogens in a diversity of environments ranging from humans, to vineyards, to beehives. NGS-based approaches have recently been applied to the characterization of the predator, pathogen, and parasite loads of underperforming and crashed pilot-scale ponds and photobioreactors (Carney et al. 2014; Poorey et al.

In prep.). The sequencing strategies employed in such analyses are dependent on the read length and number that are in turn dependent on the technical specifications of the sequencing system employed. There are a number of sequencing systems in use today, and the most common are summarized in Table 8.1.

Generally, sequencers fall into two broad categories, high-volume machines that are found in core facilities and commercial providers, and less expensive, lower volume "personal" sequencers. Personal sequencers—including the Illumina MiSeq, the 454 GS, and the Ion Torrent PGM—have the advantage of relatively rapid run times and lower cost per run than the larger machines. Pacific Biosciences sequencers are targeted for specific applications requiring a relatively small number of long reads. The costs associated with the actual sequencing can be reduced by the creation of multiplexed sequencing libraries allowing for the sequencing of more than one sample per lane. Various systems that take advantage of each type of sequencer have been developed for the creation of multiplexed libraries (McKenna et al. 2008; Caporaso et al. 2011; Whiteley et al. 2012).

Next-generation sequencing strategies have evolved as the read lengths have progressively improved. At the time of writing, it is possible to obtain reads that span regions of up to 600 nt (Illumina MiSeq, using paired-end kits). Taking prokaryotic SSU rRNA gene analysis as an example, the MiSeq is now capable of completely sequencing amplicons covering hypervariable regions 1–3. An alternative strategy for obtaining full-length rRNA sequence coverage, with next-generation sequencers, entails amplifying the full-length gene and then shotgun sequencing and assembling the amplicon. This strategy is more complex because it requires an assembly step that is not necessary for the analysis of shorter amplicons. The advantage is that more information is available upon which to base phylogenetic assignments.

The handling of the raw sequence data from the sequencer is device dependent, and instruments generally come with data process and analysis software packages. Raw sequencing reads must pass through a series of quality control steps to remove low-quality and primer sequences and to mask low-complexity regions. Once this initial processing is complete, there are a number of software packages for microbiome analysis including the QIIME (http://qiime.org/index.html) (Caporaso et al. 2010), mothur (Schloss et al. 2009), and RDP pyrosequencing pipeline designed for the analysis of 454 sequencing datasets.

In samples that contain high algae biomass densities, prokaryotic rRNA gene sequencing libraries may be dominated by amplicons derived from the algal chloroplast. In these cases, primers devised to exclude amplification of plastid SSU rRNA sequences can be used in single step or in nested PCR amplifications (Chelius and Triplett 2001; Rastogi et al. 2010). It should be noted that these primers also exclude the amplification of cyanobacterial sequences. Other strategies for reducing the burden of noninformative sequences include the use of the so-called blocking primers that bind to and prevent amplification of unwanted sequences. Such primers contain modified nucleotides that prevent primer extension (Vestheim and Jarman 2008). In theory, blocking primer strategies may be more generally applicable and could be applied to additional amplification targets aside from rRNA genes.

The advantages of next-generation sequencing–based methods of identification are that they are culture independent and do not rely on isolation of the organism.

There are three major disadvantages to sequencing-based systems for routine pond surveillance. The first is the cost of the equipments and reagents. The second is the technical sophistication required to carry out the library preparation and the data analysis. Finally, the time required to go from sample to answer is too long for routine surveillance. There are a number of commercial entities that provide contract library preparation and sequencing services mitigating some drawbacks. Because sequencing library preparation protocols require amplification, next-generation sequencing suffers from the same potential sources of bias as PCR-based detection methods. In addition, each next-generation sequencing system displays a different level of bias against templates with high and low GC ratios (Quail et al. 2012).

In cases where a deleterious organism or weed species is of sufficient abundance in the contaminated culture or has been isolated or enriched in culture, it is possible to clone and carry out dideoxynucleotide terminator sequencing (Sanger et al. 1977) of the desired region. With Sanger sequencing, it is generally possible to achieve 700 nt of sequence data per primer extension reaction. Thus, paired-end reactions should be sufficient to sequence both DNA strands of the ITS region and single strands of most, if not all, of the full-length SSU rRNA genes. By providing full sequencing coverage of the region of interest, Sanger sequencing can result in a high degree of taxonomic distinction of the target organism and maximal information for the design of PCR primer or oligonucleotide probes for the future detection and quantification of the deleterious species in algal mass culture.

8.2.4 Detection of Contaminants by PCR-Based Methods

Quantitative PCR (qPCR) is one of the more common molecular diagnostic techniques for the detection of known deleterious species (for review, see Botes et al. 2013). qPCR can be an essential tool for conducting reactive management strategies (see Section 7.2.1; McBride et al. 2014), where a biocontaminant is isolated, identified, and targeted for routine monitoring (Figure 8.4). Two different reporter systems for qPCR are in general use: fluorogenic dyes or fluorescent oligonucleotide probes. The first system (Ponchel et al. 2003) utilizes dyes, such as SYBR green, which fluoresce when bound to the double-stranded product of PCR reactions allowing quantification of the product. The limitation of this system is that the dye binds nonspecifically to any double-stranded DNA including primer dimers. The major advantage of the fluorogenic dye–based system is that it is less expensive than the probe-based system, both to design and optimize the probes and to run the reactions. Probe-based systems (often referred to as TaqMan) utilize a fluorescently labeled oligonucleotide that binds to the desired target product (Holland et al. 1991). This allows for greater specificity and enables quantification of the target even in the presence of nontarget amplicons. In addition, probe-based systems can be multiplexed for the detection of multiple species. The disadvantage of such systems is that the reactions are more expensive because of the requirement for the labeled oligonucleotide probe and are technically more challenging, as optimizing multiplexed probes is substantively more complex and time-consuming than single primers and more sensitive to interference from unpredictable environmental samples.

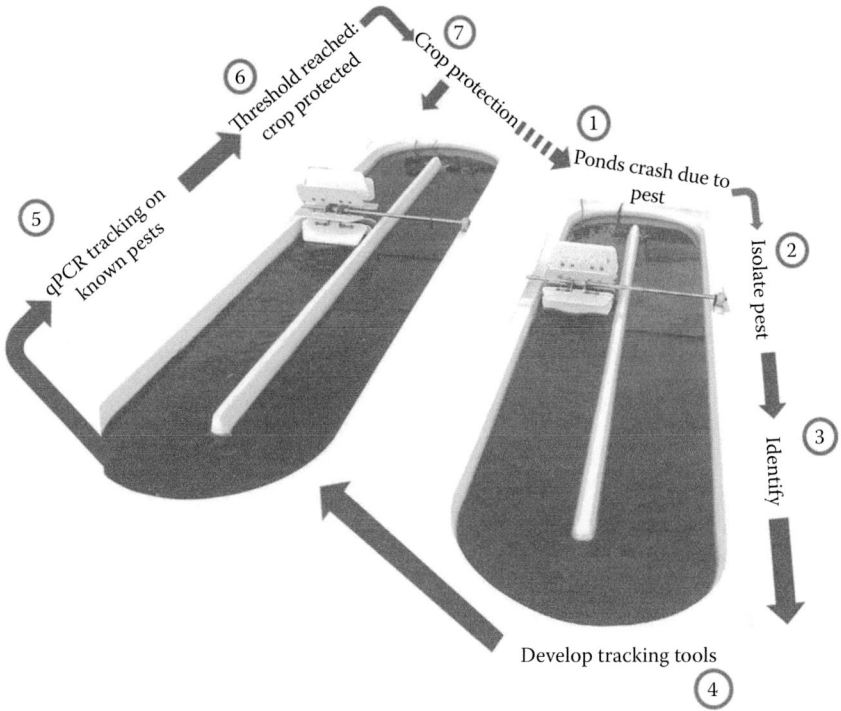

FIGURE 8.4 **(See color insert.)** An example of a reactive pest response strategy aided by the use of quantitative PCR (qPCR). This strategy is triggered by qualitative microscope observations of pests in field samples and/or through bringing a pond sample into the lab and exposing it to conditions that would accelerate pest growth and precipitate a culture crash. If a new or unknown pest is observed, various microbiological techniques (e.g., plaque plating, baiting, selective media) are first used to isolate this pest organism. ITS or SSU rRNA sequencing is then used to identify the pest organism and to develop qPCR primers to track and quantify the genomic DNA of the targeted pest in established ponds. If isolation and culturing is not possible, metagenomic sequencing of the pond community may reveal enough of a pattern to identify a specific pest organism and provide sequence data for qPCR assay design. Specific thresholds can be set for the qPCR-determined abundance of each pest that is being monitored. Once this critical abundance threshold is reached in any given pond, a carefully chosen crop protection strategy (e.g., a selected pesticide application or the addition of a targeted biocontrol agent) is then implemented and the pest's abundance is consistently monitored to determine whether the control strategy is successful (see Chapter 7). (Reproduced from McBride, R. C. et al., *Ind. Biotechnol.*, 10, 221, 2014.)

qPCR has several advantages, the most important of which is the potential for high sensitivity. Under the appropriate conditions, qPCR can approach single-cell detection levels. In practice, this level of sensitivity can be difficult to obtain and care must be taken in the preparation of samples to avoid cross-contamination. qPCR reactions can be multiplexed for the detection of multiple species within a single reaction, and given the appropriate equipment, thousands of such reactions could be run on a daily basis. In addition to bias introduced during sample preparation, there are two major sources of PCR bias: the choice of primers for amplification and the reaction conditions used for amplification. Specific primers for detection are designed for either universal or taxa-specific amplification. It has, however, proven difficult to create primer sets that amplify all targeted sequences with equal efficiency. This bias can be characterized and potentially limited by choice and testing of a variety of primers against a diversity of target species or near neighbors. It may be desirable to compare libraries created with alternative primer sets targeting either the same region or different molecular barcodes. PCR amplification bias at high and low GC ratios is also a potentially significant source of error introduced during the generation of sequencing libraries (Aird et al. 2011). This can be controlled or eliminated by adjustment of the PCR cycle parameters and choice of polymerase. An additional limitation to qPCR as a detection technology includes the initial costs of the equipment.

Fulbright et al. (2014) have demonstrated the application of cleaved amplified polymorphic sequences (CAPS) to the detection of weed species in algal mass culture systems. CAPS is based on restriction fragment length polymorphisms in PCR-amplified products. In this procedure, a target region, such as the SSU rRNA gene, is amplified from DNA extracted from the mass culture (test sample). The product of the PCR reaction is then cleaved with one or more restriction enzymes and the lengths of the resulting fragments are determined by electrophoretic analysis and compared to those derived from the digestion of the same target amplified from the desired algal strain (control sample). The presence of restriction fragments in the test sample that are not present in the control is indicative of biocontamination. The major advantages of this method are that it can detect contamination without the need for specific probes and that it does not require the more expensive equipment and reagents needed for qPCR. The disadvantages are that the method does not actually identify the contaminant species, and it is not as sensitive as qPCR.

8.2.5 Detection of Contaminants by Microarray-Based Methods

Microarrays generally consist of glass slides with oligonucleotide probes arrayed in spots on the surface. Each spot contains a specific probe that is designed to hybridize to a particular target sequence. Samples are prepared by coupling a fluorescent probe to extracted nucleic acids, that are then applied to the array. DNA fragments that are complementary to probes on the array will bind to that specific spot and then are detected by fluorescence. Microarray-based systems have been developed using arrayed oligonucleotide probes targeted to the SSU rRNA.

These so-called phylochips were originally developed for prokaryotic community structure analysis, and there is an extensive literature on their use (Loy et al. 2002). Custom phylochips can be developed for the detection of organisms of interest and have been adapted to eukaryotic SSU rRNA analysis and applied to the detection of marine eukaryotes, including microalgae and pathogenic protozoa (Metfies et al. 2007). Probe design is a particular challenge in the development of microarrays. All probes on a particular microarray must behave in the same manner under a given set of hybridization conditions. There are a number of databases and tools to assist in the design of probes (see Section 8.2.7). The microarray technique, by itself, is not as sensitive as qPCR methods, but this can be improved by amplification of the target by PCR or by multiple displacement amplification (Binga et al. 2008). A potential drawback to phylochip-based pond diagnostics is the expense of microarray development and production, particularly in situations when there are several species to be monitored. Macroarray systems, based on dot-blot technology, can be utilized for rapid optimization and validation of probes or for applications where only a modest number of species are to be detected.

8.2.6 DETECTION OF CONTAMINANTS BY FLUORESCENT *IN SITU* HYBRIDIZATION METHODS

Fluorescent *in situ* hybridization (FISH) has been applied to the identification and enumeration of morphologically indistinguishable species such as eukaryotic picoplankton (Not et al. 2002), Bacteria (Amann et al. 1991), and Archaea (DeLong et al. 1999). The method entails hybridization of permeabilized samples with fluorescently labeled oligonucleotide probes followed by fluorescent imaging or flow cytometry. The analysis can be multiplexed through the utilization of mixtures of probes with differently colored fluorescent tags. Signal-amplification methods such as catalyzed reporter deposition (CARD), also referred to as tyramide signal amplification (TSA), have been developed (Lebaron et al. 1997). In CARD–FISH, horseradish peroxidase (HRP) is conjugated to the probe. In the presence of hydrogen peroxide, HRP converts tyramide (conjugated to a fluorophore) to a reactive intermediate that interacts with aromatic compounds in the cell (chiefly tyrosine and tryptophan) forming an insoluble complex that is deposited around the probe, greatly enhancing the signal. This method has been used in conjunction with rRNA-targeted probes to eukaryotic algae (Simon et al. 1995), zoosporic fungi (Jobard et al. 2010), and marine bacteria (Pernthaler et al. 2002). The major strengths of FISH techniques are the single-cell resolution of the method and the flexibility to utilize either fluorescence microscopy or flow cytometry as the output device. FISH is a technically sophisticated method that may limit its application to small-scale or research operations or for the production of high-value products. It does not have the same potential for high-throughput, multiplexed analysis as other methods such as PCR and, thus, seems unlikely to be applied to routine mass culture diagnostics or surveillance.

8.2.7 BIOINFORMATICS FOR MOLECULAR DIAGNOSIS

There are a number of sequence data repositories for rRNA and mitochondrial cytochrome oxidase genes. The SILVA database (http://www.arb-silva.de/) is a curated repository of SSU and LSU rRNA gene sequences from Archaea, Bacteria, and Eukarya (Quast et al. 2013). The greengenes database contains LSU rRNA sequences from Bacteria and Archaea (http://greengenes.secondgenome.com/downloads) (DeSantis et al. 2006). Both databases can be downloaded from their websites in a flat file format for use with the various microbiome analysis software packages. The Ribosomal Database Project (RDP) (http://rdp.cme.msu.edu/)

(Cole et al. 2009) maintains a database of bacterial and archaeal SSU sequences. UNITE (Abarenkov et al. 2010) is a searchable database of primarily fungal ITS regions that can be downloaded in flat file format from its home website (http://unite.ut.ee/) or from the QIIME website. As the name suggests, ITS2 database (Koetschan et al. 2012) is a searchable database of specifically ITS2 sequences (http://its2.bioapps.biozentrum.uni-wuerzburg.de/), and ITSoneDB (http://itsonedb.ba.itb.cnr.it/) focuses on the ITS1 region. The BOLD database (Ratnasingham and Hebert 2007) (http://www.barcodinglife.org) contains both sequence data and a registry of primers for *cox1*. GenBank is the NIH depository of publically available annotated sequences (Benson et al. 2012) (http://www.ncbi.nlm.nih.gov/genbank/) and can be queried to identify sequences that are not found in other databases.

Once the deleterious agent or weed species of concern has been identified, and sufficient sequence information has become available, bioinformatics analysis is employed in the development of oligonucleotide primers and probes for the detection and quantification of the agent in algal mass culture. ProbeBase (http://probebase.csb.univie.ac.at/) is an online database of probes and primers targeting rRNA (Loy et al. 2007). The rRNA gene databases SILVA, greengenes, and RDP host web-based tools to assist in the development of probes by checking them for homology to sequences in their respective databases. The web tool probeCheck (http://131.130.66.200/cgi-bin/probecheck/content.pl?id=home) (Loy et al. 2008) provides a single interface for the comparison of rRNA subunit–targeted probe sequences against multiple databases.

8.2.8 Detection Using Remote Sensing

There is much work to be done in the detection and recognition of pest challenges, and the most exciting advances will be those technologies that can be implemented in a low-cost manner and that provide simultaneous information on multiple problems. Coupled with recent advances in computing and geopositioning technologies, remote sensing data obtained from ground-, air-, and space-based platforms are now capable of providing detailed spatial and temporal information on plant responses to their local environment that is needed for site-specific precision agriculture (Pinter et al. 2003). We suggest here that reflectance monitoring of the open pond surface holds similar promise for the development of rapid strategies for the early detection of grazing- and disease-induced stress in open algal cultures (Reichardt et al. 2014). Multispectral image analysis can be used to measure algal biomass concentrations, detect invasive species, and monitor culture health in real time (Murphy et al. 2014). As algal biomass production moves to commercial scales, we suggest that the use of remote sensing tools will be essential components of real-time crop management systems. In the future, these optical tools could potentially be deployed using small, mobile devices that could quickly sample the entire cultivation system landscape. Following the initial detection of a potential crop growth issue, these devices could immediately be returned to perceived trouble spots for problem confirmation. Mobile devices also could be used for monitoring the behavior of algae in localized subsections of the cultivation facility in which targeted crop protection measures have been implemented.

8.3   CASE STUDIES OF PROBLEMATIC CONTAMINANTS

Molecular techniques are very useful in characterizing the community within an algae cultivation system and detecting biocontaminants before they become an irreversible problem. In cases where a natural community of algal strains is allowed to colonize a cultivation system (e.g., *polyculture* processes), molecular tools can be quite useful in characterizing the algal populations. In the following sections, we describe some potentially deleterious biocontaminants and provide examples where molecular techniques were used to detect them.

8.3.1   MESOGRAZERS

Growing algae to scale will undoubtedly lead to some infestations by algivorous predators (i.e., grazers). Metazoan rotifers and herbivorous ciliates are perhaps the most commonly reported biocontaminants in large-scale cultivation systems. For example, once established, rotifers cause extremely rapid biomass loss as they are able to double their density in 24 hours (Yúfera and Navarro 1995; Sarma et al. 2001) while ingesting 200 algal cells minute^{-1} rotifer^{-1} (Hirayama and Ogawa 1972). Rotifers and some ciliates also form thick-walled resting stages (i.e., cysts) that are resistant to desiccation and chemical treatment such as chlorine. However, other lesser known grazers are also problematic, including amoebae, heliozoans, and vampyrellids. Further characterization of large-scale cultures will be needed, to establish the specific threats that these and other potential algal grazers pose, to commercial algae productivity.

A grazing ciliate has been frequently observed in open raceways in Arizona. A combined strategy of microscopy, isolation, culture, and sequencing identified this organism as *Gastrostyla steinii* (Figure 8.5a), informing appropriate measures to reduce its impact (Carney, unpublished data). Replicate raceway ponds in Texas growing *Nannochloropsis salina* were reported to simultaneously crash, presumably from different organisms. Microbiome analysis using NGS confirmed that one crash was due to the outbreak of the rotifer, *Brachionus plicatilis*, while the other crashed raceway had become dominated by a mix of the grazing gastrotrich, *Chaetonotus*, and the *Cytophagia* bacterium, *Aureibacter* (Carney et al. in review, Algal Res.). That simultaneous crashes occurred in adjacent ponds—but due to very different types of biocontaminants—suggests the complexity of the problem.

8.3.2   FUNGAL PARASITES

Fungal parasites are particularly insidious in algal culture (reviewed by Carney and Lane 2014). While acute fungal infections cause obvious rapid pond crashes, chronic infections cause reduced productivity over longer time periods (Carney et al. 2014). In natural freshwater environments, chytrids are well known to parasitize microalgae at infection rates that exceed 90%, causing severe biomass losses (Kagami et al. 2007). Chytrid infections of green algae have been described within cultivation systems as well (both open and closed) and can severely and rapidly reduce algae productivity (Fott 1967; Hoffman et al. 2008). For example, only 3 days may be required for a chytrid parasite to infect 100% of algae cells within a *Haematococcus* culture (Hoffman et al. 2008).

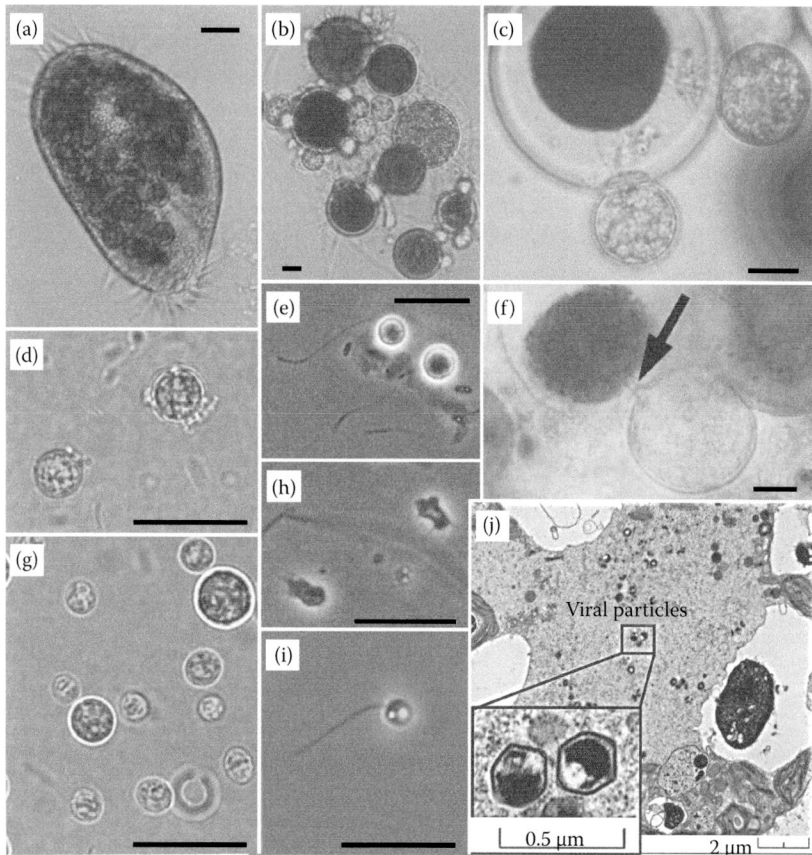

FIGURE 8.5 (See color insert.) Biocontaminants observed in a mass algal culture:
(a) Grazing ciliate, *Gastrostyla steinii*, with ingested whole *Haematococcus* cysts. (Photo
courtesy of K. Sorensen.) (b) Sporangia of the parasitic blastocladian fungus, probably
Paraphysoderma sedebokerensis, attached to *Haematococcus* growing in open ponds. (Photo
courtesy of L. T. Carney.) (c) Higher magnification of *P. sedebokerensis* (isolate JEL821);
note the fungal rhizoids extending into the *Haematococcus* cyst, showing signs of cell
content depletion. (Photo courtesy of J. E. Longcore.) (d) Bacteria attached to lysing algal
cells in a crashing culture correlated with increasing detection of the pathogenic bacterium,
Vampirovibrio chlorellavorus, on multiple occasions using SSU rRNA sequencing and qPCR.
(Photo courtesy of S. Qin.) (e) Motile *V. chlorellavorus*-like bacteria adjacent to algae cells
in a culture that was confirmed by qPCR to contain *V. chlorellavorus*. (Photo courtesy of
J. Wilkenfeld.) (f) Sporangium of isolate JEL812; the arrow is pointing to the rhizoidal connection to the *Haematococcus* cell. (Photo courtesy of J. E. Longcore.) (g) Algae cells prior
to the appearance of the attachment and detection of *V. chlorellavorus*. (Photo courtesy of
S. Qin.) (h) Amoeboid swarmer stages of *P. sedebokerensis* observed in open ponds growing
Haematococcus. (Photo courtesy of J. Wilkenfeld.) (i) Motile zoospore stage of the parasitic
chytrid isolate JEL812 (Rhizophydiales; Aquamycetaceae). (Photo courtesy of K. Sorensen.)
(j) TEM photograph of virus particles inside an algal cell. (Photo courtesy of R. Roberson.)
Note: Scale bars = 10 μm where not otherwise indicated.

In a prototype-enclosed photobioreactor at a wastewater treatment plant in California, the microbial community was analyzed during a potential crash of a freshwater green alga growing in wastewater effluent that had been chlorinated and dechlorinated (Carney et al. 2014). Microbiome analysis revealed a small spike in the proportion of *Rhizophydium*, a known algal parasite that correlated with a decrease in algae biomass. After a few days, this chytrid was replaced by another chytrid that was saprobic (i.e., feeding on dead algae biomass). This change in chytrid community was consistent with the end of an infection period when dead algal biomass accumulates.

Recently, a new member of Cryptomycota, *Amoeboaphelidium protococcarum*, was discovered infecting the green alga *Scenedesmus* in commercial ponds in New Mexico (Letcher et al. 2013). The abundance of this aphelid was inversely correlated with that of the algae, a potential disaster considering the parasite was previously unknown. It was isolated and cultured, and finally a multigene analysis was used to place the parasite within the Cryptomycota. Sister taxa included other parasites, which were known to have infected algae ponds.

Two aquatic fungi parasitized *Haematococcus* growing in open raceways in Arizona, isolate JEL821 identified as *Paraphysoderma sedebokerensis* (Blastocladiomycota; Figure 8.5b,c,h) and isolate JEL812 (Figure 8.5f,i) identified as belonging to the family Aquamycetaceae in the Rhizophydiales (Chytridiomycota). *P. sedebokerensis* was described by Hoffman et al. (2008) and Gutman et al. (2009) as infecting *Haematococcus pluvialis*. The isolate JEL812 may be cosmopolitan and has been observed infecting mass algal cultures in the past (JEL317; Joyce Longcore, pers. comm.). Both fungi were isolated and examined via microscopy for identifying characteristics. Identification was verified and qPCR assays were designed based on 18S SSU rRNA sequence data, which are used routinely to monitor raceways to inform crop protection strategies.

8.3.3 Pathogenic Bacteria

The role of bacteria in algae-dominated communities is not straightforward. Bacteria are reported as both beneficial (Cole 1982; Kazamia et al. 2012) and pathogenic (Cole 1982; Fukami et al. 1997). Deciphering which strains under which conditions are pathogenic will be important for algae cultivation management. *Bdellovibrio*-like bacteria have been reported to lyse the green alga, *Chlorella* (Coder and Starr 1978; Cole 1982). We detected this genus in an enclosed photobioreactor growing freshwater green algae in California (Carney et al. 2014). Genera from the group *Cytophagia* may also be able to infect algae and are known to attach to cyanobacteria, secrete lytic substances to then dissolve the cell wall, causing lysis (reviewed by Rashidan and Bird 2001). These strains are quite common during natural cyanobacterial bloom formation (Rashidan and Bird 2001). We have repeatedly detected cytophage sequences in declining green algae cultures (Carney et al. 2014; Carney et al. in review, *Algal Res.*), and it is possible that they have lytic effects on more than just cyanobacteria. The pathogenic bacterium *Vampirovibrio chlorellavorus* was detected by metagenomic analysis in open raceways in Arizona and correlated with high levels of bacterial attachment to algal cells and lysis observed using microscopy (Figure 8.5d,e,g). Because *V. chlorellavorus* is an obligate parasite (Coder and Goff 1986), isolation and culture techniques proved difficult. Once sequence data were obtained, a qPCR assay was designed and used to monitor the presence of *V. chlorellavorus* and validate crop protection strategies to avoid future crashes (Carney, unpublished data).

8.3.4 ALGAL VIRUSES

Algal viruses may be the greatest source of uncharacterized genetic diversity in the world (Dunigan et al. 2006). Those which were able to be cultured include 40 representatives, infecting 11 microalgae species, and all were characterized as large dsDNA viruses (family Phycodnaviridae; Nagasaki and Bratbak 2010). To date, pond crashes due to viral infection have not been frequently reported; however, virus particles have been detected inside algae cells growing in mass culture systems using TEM (Figure 8.5j; Roberson and Carney, unpublished data). Algal viruses play a significant role in the collapse of algal blooms in natural ecosystems, and it is inevitable that viral pathogens will be discovered as impacting algal mass culture once viral presence becomes more routinely monitored for. Although PCR amplification of specific sequences related to the Phycodnaviridae has been demonstrated in marine samples (e.g., Chen and Suttle 1995), there have been no universal molecular barcode regions developed for viruses in general. Thus, the identification of novel viruses is dependent on shotgun sequencing of purified viral nucleic acid fractions followed by genomic assembly. Methods have been developed in both freshwater and marine systems to concentrate viral particles and prepare the nucleic acids for sequencing (Lawrence and Steward 2010).

8.4 CONCLUSIONS

As algal mass culture systems grow in size, the financial impact of contamination will, concomitantly, become more severe. The development of rapid, sensitive, high-throughput methods of contaminant identification and detection is of paramount importance for the further economic development of this industry. Many of the molecular methods, such as next-generation sequencing, PCR, and microarray analysis described here, have the characteristics that can fill this unmet need. The choice of methods will largely depend on the cost of such methods, the scale of production, and the value of the product being produced. Operations that are producing low-value products, such as fuel, must amortize these costs over larger scales, while production of more valuable products, such as nutraceuticals, would have greater choices in surveillance methods at smaller scales. Clearly, there is significant need for research and development of low-cost molecular detection methods.

Two key innovations would facilitate the development of molecular approaches in the management of algae biomass production: (1) The uniform extraction of DNA from populations in environmental samples would transform molecular diagnosis of pond communities from qualitative to quantitative. Qualitative analyses or relative analyses are useful, but developing more advanced extraction technology is sure to speed this field. (2) The development of molecular targets that monitor genetic responses of the algae target to distress would be key in moving forward.

Instead of monitoring unique pest signatures, tracking algae stress response signatures may circumvent the need to have *a priori* knowledge of the challenging organisms.

Once biocontaminants are detected by classical or more advanced monitoring methods, a rapid response is usually required in order to reduce or eliminate them. More detailed information on algae crop protection, including common responses in use today and novel approaches that are currently in development, is presented in Chapter 7.

ACKNOWLEDGMENTS

This chapter was improved greatly by the comments received from M. Strittmatter, S. Slocombe, J. Benemann, J. Longcore, and J. Wilkenfeld. Figure 8.5, specifically, was much improved by edits received from J. Longcore. Sandia National Laboratories is a multiprogram laboratory managed and operated by Sandia Corporation, a wholly owned subsidiary of Lockheed Martin Corporation, for the U.S. Department of Energy's National Nuclear Security Administration under Contract #DE-AC04-94AL85000. Research at Sandia National Laboratories was supported by the BioEnergy Technology Office of the Office of Energy Efficiency and Renewable Energy, U.S. Department of Energy under Award #NL0022897.

REFERENCES

Abarenkov, K., R. H. Nilsson, K. H. Larsson et al. 2010. The UNITE database for molecular identification of fungi—Recent updates and future perspectives. *New Phytol.* 186:281–285.

Aird, D., M. G. Ross, W. S. Chen et al. 2011. Analyzing and minimizing PCR amplification bias in Illumina sequencing libraries. *Genome Biol.* 12:R18.

Álvarez, E., A. Lopez-Urrutia, E. Nogueira, and S. Fraga. 2011. How to effectively sample the plankton size spectrum? A case study using FlowCAM. *J. Plankton Res.* 33:1119–1133.

Amann, R., N. Springer, W. Ludwig, H. D. Görtz, and K. H. Schleifer. 1991. Identification *in situ* and phylogeny of uncultured bacterial endosymbionts. *Nature* 351:161–164.

Amaral-Zettler, L. A., E. A. McCliment, H. W. Ducklow, and S. M. Huse. 2009. A method for studying protistan diversity using massively parallel sequencing of V9 hypervariable regions of small-subunit ribosomal RNA genes. *PLoS One* 4:e6372.

Bazzicalupo, A. L., M. S. Balint, and I. Schmitt. 2013. Comparison of ITS1 and ITS2 rDNA in 454 sequencing of hyperdiverse fungal communities. *Fungal Ecol.* 6:102–109.

Benson, D. A., M. Cavanaugh, K. Clark et al. 2012. GenBank. *Nucleic Acids Res.* 41(Database issue):D36–D42.

Binga, E. K., R. S. Lasken, and J. D. Neufeld. 2008. Something from (almost) nothing: The impact of multiple displacement amplification on microbial ecology. *ISME J.* 2:233.

Blaalid, R., S. Kumar, R. H. Nilsson, K. Abarenkov, P. M. Kirk, and H. Kauserud. 2013. ITS1 versus ITS2 as DNA metabarcodes for fungi. *Mol. Ecol. Resour.* 13:218–224.

Botes, M., M. de Kwaadsteniet, and T. E. Cloete. 2013. Application of quantitative PCR for the detection of microorganisms in water. *Anal. Bioanal. Chem.* 405:91–108.

Bucklin, A., D. Steinke, and L. Blanco-Bercial. 2011. DNA barcoding of marine metazoa. *Annu. Rev. Mar. Sci.* 3:471–508.

Bull, J. J. and S. Collins. 2012. Algae for biofuel: Will the evolution of weeds limit the enterprise. *Evolution* 66:2983–2987.

Caporaso, J. G., J. Kuczynski, J. Stombaugh et al. 2010. QIIME allows analysis of high-throughput community sequencing data. *Nat. Methods* 7:335–336.

Caporaso, J. G., C. L. Lauber, W. A. Walters et al. 2011. Global patterns of 16S rRNA diversity at a depth of millions of sequences per sample. *Proc. Natl. Acad. Sci. USA* 108:4516–4522.

Carney, L. T. and T. W. Lane. 2014. Parasites in algae mass culture. Special issue of *Front. Microbiol.* 5:278.

Carney, L. T., S. S. Reinsch, P. D. Lane et al. 2014. Microbiome analysis of a microalgal mass culture growing in municipal wastewater in a prototype OMEGA photobioreactor. *Algal Res.* 4:52–61.

Carney, L. T., J. Wilkenfield, P. Lane et al. In review. Pond Crash Forensics: Presumptive identification of pond crash agents by next generation sequencing in replicate raceway mass cultures of *Nannochloropsis salina* Algal Res.

Chelius, M. K. and E. W. Triplett. 2001. The diversity of Archaea and Bacteria in association with the roots of *Zea mays* L. *Microbial Ecol.* 41:252–263.

Chen, F. and C. A. Suttle. 1995. Amplification of DNA polymerase gene fragments from viruses infecting microalgae. *Appl. Environ. Microbiol.* 61:1274–1278.

Coder, D. M. and L. J. Goff. 1986. The host range of the chlorellavorous bacterium ("*Vampirovibrio chlorellavorus*"). *J. Phycol.* 22:543–546.

Coder, D. M. and M. P. Starr. 1978. Antagonistic association of the chlorellavorous bacterium ("*Bdellovibrio*" *chlorellavorus*) with *Chlorella vulgaris*. *Curr. Microbiol.* 1:59–64.

Cole, J. J. 1982. Interactions between bacteria and algae in aquatic ecosystems. *Annu. Rev. Ecol. Syst.* 13:291–314.

Cole, J. R., Q. E. Wang, E. Cardenas et al. 2009. The Ribosomal Database Project: Improved alignments and new tools for rRNA analysis. *Nucleic Acids Res.* 37:D141–D145.

Collins, A. M., H. D. T. Jones, R. C. McBride, C. Behnke, and J. A. Timlin. 2014. Host cell pigmentation in *Scenedesmus dimorphus* as a beacon for nascent parasite infection. *Biotechnol. Bioeng.* 111:1748–1757.

Davis, R., D. Fishman, E. D. Frank et al. 2012. *Renewable Diesel from Algal Lipids: An Integrated Baseline for Cost, Emissions and Resource Potential from a Harmonized Model*. Argonne, IL: Argonne National Laboratory.

Day, J. G., N. J. Thomas, U. E. M. Achilles-Day, and R. J. G. Leakey. 2012. Early detection of protozoan grazers in algal biofuel cultures. *Bioresour. Technol.* 114:715–719.

Defoirdt, T., N. Boon, P. Bossier, and W. Verstraete. 2004. Disruption of bacterial quorum sensing: An unexplored strategy to fight infections in aquaculture. *Aquaculture* 240:69–88.

DeLong, E. F., L. T. Taylor, T. L. Marsh, and C. M. Preston. 1999. Visualization and enumeration of marine planktonic archaea and bacteria by using polyribonucleotide probes and fluorescent *in situ* hybridization. *Appl. Environ. Microbiol.* 65:5554–5563.

DeSantis, T. Z., P. Hugenholtz, N. Larsen et al. 2006. Greengenes, a chimera-checked 16S rRNA gene database and workbench compatible with ARB. *Appl. Environ. Microbiol.* 72:5069–5072.

Ding, H. Y. and J. H. Ma. 2005. Simultaneous infection by red rot and chytrid diseases in *Porphyra yezoensis* Ueda. *J. Appl. Phycol.* 17:51–56.

Dunigan, D. D., L. A. Fitzgerald, and J. L. Van Etten. 2006. Phycodnaviruses: A peek at genetic diversity. *Virus Res.* 117:119–132.

FAO. 2010. The state of world fisheries and aquaculture 2010. Rome, Italy: FAO Fisheries and Aquaculture Department, Food and Agriculture Organization of United Nations. Retrieved October 2, 2012 from the World Wide Web: http://www.fao.org/docrep/013/i1820e/i1820e00.htm.

Fott, B. 1967. *Phlyctidium scenedesmi* spec. nova, a new chytrid destroying mass cultures of algae. *Z. Allg. Mikrobiol.* 7:97–102.

Fredricks, D. N. and D. A. Relman. 1996. Sequence-based identification of microbial pathogens: A reconsideration of Koch's postulates. *Clin. Microbiol. Rev.* 9:18–33.

Fukami, K., T. Nishijima, and Y. Ishida. 1997. Stimulative and inhibitory effects of bacteria on the growth of microalgae. *Hydrobiology* 358:185–191.

Fulbright, S. P., M. K. Dean, G. Wadle, P. J. Lammers, and S. Chisolm. 2014. Molecular diagnostics for monitoring contaminants in algal cultivation. *Algal Res.* 4:41–51.

Gao, Y., C. Gregor, Y. Liang, D. Tang, and C. Tweed. 2012. Algae biodiesel—A feasibility report. *Chem. Central J.* 6:S1.

Gutman, J., A. Zarka, and S. Boussiba. 2009. The host-range of *Paraphysoderma sedebokerensis*, a chytrid that infects *Haematococcus pluvialis*. *Eur. J. Phycol.* 44:509–514.

Hirayama, K. and S. Ogawa. 1972. Fundamental studies on the physiology of the rotifer for its mass culture. I. Filter feeding of rotifer. *Bull. Jpn. Soc. Sci. Fish.* 38:1207–1214.

Hoef-Emden, K. 2012. Pitfalls of establishing DNA barcoding systems in protists: The Cryptophyceae as a test case. *PLoS One* 7:e43652.

Hoffman, Y., C. Aflalo, A. Zarka, J. Gutman, T. Y. James, and S. Boussiba. 2008. Isolation and characterization of a novel chytrid species (phylum Blastocladiomycota), parasitic on the green alga *Haematococcus*. *Mycol. Res.* 112:70–81.

Holland, P. M., R. D. Abramson, R. Watson, and D. H. Gelfand. 1991. Detection of specific polymerase chain reaction product by utilizing the 5′–3′ exonuclease activity of *Thermus aquaticus* DNA polymerase. *Proc. Natl. Acad. Sci. USA* 88:7276–7280.

Hunt, D. E., V. Klepac-Ceraj, S. G. Acinas, C. Gautier, S. Bertilsson, and M. F. Polz. 2006. Evaluation of 23S rRNA PCR primers for use in phylogenetic studies of bacterial diversity. *Appl. Environ. Microbiol.* 72:2221–2225.

Huse, S. M., L. Dethlefsen, J. A. Huber et al. 2008. Exploring microbial diversity and taxonomy using SSU rRNA hypervariable tag sequencing. *PLoS Genet.* 4:e1000255.

Jobard, M., S. Rasconi, and T. Sime-Ngando. 2010. Fluorescence *in situ* hybridization of uncultured zoosporic fungi: Testing with clone-FISH and application to freshwater samples using CARD-FISH. *J. Microbiol. Methods* 83:236–243.

Kagami, M., A. de Bruin, M. Rijkeboer, B. W. Ibelings, and E. Van Donk. 2007. Parasitic chytrids: Their effect on phytoplankton communities and food-web dynamics. *Hydrobiology* 578:113–129.

Kazamia, E., H. Czesnick, T. T. Van Nguyen et al. 2012. Mutualistic interactions between vitamin B12-dependent algae and heterotrophic bacteria exhibit regulation. *Environ. Microbiol.* 14:1466–1476.

Kim, M., M. Morrison, and Z. Yu. 2011. Evaluation of different partial 16S rRNA gene sequence regions for phylogenetic analysis of microbiomes. *J. Microbiol. Methods* 84:81–87.

Koetschan, C., T. Hackl, T. Müller, M. Wolf, F. Förster, and J. Schultz. 2012. ITS2 database IV: Interactive taxon sampling for internal transcribed spacer 2 based phylogenies. *Mol. Phylogenet. Evol.* 63:585–588.

Koid, A., C. William, W. C. Nelson, A. Mraz, and K. B. Heidelberg. 2012. Comparative analysis of eukaryotic marine microbial assemblages from 18S rRNA gene and gene transcript clone libraries by using different methods of extraction. *Appl. Environ. Microbiol.* 78:3958.

Lawrence, J. E. and G. F. Steward. 2010. Purification of viruses by centrifugation. In *Manual of Aquatic Viral Ecology*, eds. S. W. Wilhelm, M. G. Weinbauer, and C. A. Suttle, pp. 166–181. Waco, TX: ASLO.

Lebaron, P., P. Catala, C. Fajon, F. Joux, J. Baudart, and L. Bernard. 1997. A new sensitive, whole-cell hybridization technique for detection of bacteria involving a biotinylated oligonucleotide probe targeting rRNA and tyramide signal amplification. *Appl. Environ. Microbiol.* 63:3274–3278.

Leliaert, F., H. Verbruggen, P. Vanormelingen et al. 2014. DNA-based species delimitation in algae. *Eur. J. Phycol.* 49:179–196.

Letcher, P. M., S. Lopez, R. Schmieder et al. 2013. Characterization of *Amoeboaphelidium protococcarum*, an algal parasite new to the Cryptomycota isolated from an outdoor algal pond used for the production of biofuel. *PLoS One* 8(2):e56232.

Li, W., T. Zhang, X. Tang, and B. Wang. 2010. Oomycetes and fungi: Important parasites on marine algae. *Acta Oceanol. Sin.* 5:74–81.

Lindahl, B. D., R. H. Nilsson, L. Tedersoo et al. 2013. Fungal community analysis by high-throughput sequencing of amplified markers—A user's guide. *New Phytol.* 199:288–299.

Lindner, D. L., T. Carlsen, R. H. Nilsson, M. Davey, T. Schumacher, and H. Kauserud. 2013. Employing 454 amplicon pyrosequencing to reveal intragenomic divergence in the internal transcribed spacer rDNA region in fungi. *Ecol. Evol.* 3:1751–1764.

Loy, A., R. Arnold, P. Tischler, T. Rattei, M. Wagner, and M. Horn. 2008. probeCheck—A central resource for evaluating oligonucleotide probe coverage and specificity. *Environ. Microbiol.* 10:2894–2898.

Loy, A., A. Lehner, N. Lee et al. 2002. Oligonucleotide microarray for 16S rRNA gene-based detection of all recognized lineages of sulfate-reducing prokaryotes in the environment. *Appl. Environ. Microbiol.* 68:5064–5081.

Loy, A., F. Maixner, M. Wagner, M. Horn. 2007. probeBase—An online resource for rRNA-targeted oligonucleotide probes: New features 2007. *Nucleic Acids Res.* 35:D800–D804.

Ludwig, W. and K. H. Schleifer. 1994. Bacterial phylogeny based on 16S and 23S rRNA sequence analysis. *FEMS Microbiol. Rev.* 15:155–173.

McBride, R. C., S. Lopez, C. Meenach et al. 2014. Contamination management in low cost open algae ponds for biofuels production. *Ind. Biotechnol.* 10:221–227.

McKenna, P., C. Hoffmann, N. Minkah et al. 2008. The macaque gut microbiome in health, lentiviral infection, and chronic enterocolitis. *PLoS Pathog.* 4:e20.

Metfies, K., M. Berzano, C. Mayer et al. 2007. An optimized protocol for the identification of diatoms, flagellated algae and pathogenic protozoa with phylochips. *Mol. Ecol. Notes* 7:925–936.

Miller, R. R., V. Montoya, J. L. Gardy, D. M. Patrick, and P. Tang. 2013. Metagenomics for pathogen detection in public health. *Genome Med.* 5:81.

Murphy, T. E., K. Macon, and H. Berberoglu. 2014. Rapid algal culture diagnostics for open ponds using multispectral image analysis. *Biotechnol. Prog.* 30:233–240.

Nagasaki, K. and G. Bratbak. 2010. Isolation of viruses infecting photosynthetic and nonphotosynthetic protists. In *Manual of Aquatic Viral Ecology*, eds. S. W. Wilhelm, M. G. Weinbauer, and C. A. Suttle, pp. 82–91. Waco, TX: American Society of Limnology and Oceanography.

Nickrent, D. L. and M. L. Sargent. 1991. An overview of the secondary structure of the V4 region of eukaryotic small-subunit ribosomal RNA. *Nucleic Acid Res.* 19:227–235.

Not, F., N. Simon, I. C. Biegala, and D. Vaulot. 2002. Application of fluorescent *in situ* hybridization coupled with tyramide signal amplification (FISH-TSA) to assess eukaryotic picoplankton composition. *Aquat. Microb. Ecol.* 28:157–166.

Orsi, W., J. F. Biddle, and V. Edgcomb. 2013. Deep sequencing of subseafloor eukaryotic rRNA reveals active fungi across marine subsurface provinces. *PLoS One* 8:e56335.

Park, J. B. K., R. J. Craggs, and A. N. Shilton. 2011. Wastewater treatment high rate algal ponds for biofuel production. *Bioresour. Technol.* 102:35–42.

Pawlowski, J., R. Christen, B. Lecroq et al. 2011. Eukaryotic richness in the abyss: Insights from pyrotag sequencing. *PLoS One* 6:e18169.

Pernthaler, A., J. Pernthaler, and R. Amann. 2002. Fluorescence *in situ* hybridization and catalyzed reporter deposition for the identification of marine bacteria. *Appl. Environ. Microbiol.* 68:3094–3101.

Pienkos, P. T. and A. Darzins. 2009. The promise and challenges of microalgal-derived bio-fuels. *Biofuels Bioprod. Biorefin.* 3:431–440.

Pinter, P. J., J. L. Hatfield, J. S. Schepers et al. 2003. Remote sensing for crop management. *Prog. Eng. Rem. Sens.* 69:647–664.

Ponchel, F., C. Toomes, K. Bransfield et al. 2003. Real-time PCR based on SYBR-Green I fluorescence: An alternative to the TaqMan assay for a relative quantification of gene rearrangements, gene amplifications and micro gene deletions. *BMC Biotechnol.* 3:18.

Poorey, K., L. T. Carney, P. D. Lane et al. Microbial community structure dynamics of long term open algae cultivation systems using amplicon and 16S shotgun sequencing. *Microbial Biol.* In preparation.

Purdy, K. J. 2005. Nucleic acid recovery from complex environmental samples. *Methods Enzymol.* 397:271–292.

Putignani, L., M. G. Paglia, E. Bordi, E. Nebuloso, L. P. Pucillo, and P. Visca. 2008. Identification of clinically relevant yeast species by DNA sequence analysis of the D2 variable region of the 25–28S rRNA gene. *Mycoses* 51:209–227.

Quail, M. A., M. Smith, P. Coupland et al. 2012. A tale of three next generation sequencing platforms: Comparison of Ion Torrent, Pacific Biosciences and Illumina MiSeq sequencers. *BMC Genom.* 13:341.

Quast, C., E. Pruesse, P. Yilmaz et al. 2013. The SILVA Ribosomal RNA Gene Database Project: Improved data processing and web-based tools. *Nucleic Acids Res.* 41:D590–D596.

Ramaiah, N. 2006. A review on fungal diseases of algae, marine fishes, shrimps and corals. *Indian J. Mar. Sci.* 35:380–387.

Rasconi, S., M. Jobard, L. Jouve, and T. Sime-Ngando. 2009. Use of Calcofluor white for detection, identification and quantification of phytoplanktonic fungal parasites. *Appl. Environ. Microbiol.* 75:2545–2553.

Rashidan, K. K. and D. F. Bird. 2001. Role of predatory bacteria in the termination of a cyanobacterial bloom. *Microbiol. Ecol.* 41:97–105.

Rastogi, G., J. J. Tech, G. L. Coaker, and J. H. J. Leveau. 2010. A PCR-based toolbox for the culture independent quantification of total bacterial abundances in plant environments. *J. Microbiol. Methods* 83:127–132.

Ratnasingham, S. and P. D. N. Hebert. 2007. BOLD: The barcode of life data system. *Mol. Ecol. Notes* 7:355–364.

Reichardt, T. A., A. M. Collins, R. C. McBride, C. A. Behnke, and J. A. Timlin. 2014. Spectroradiometric monitoring for open outdoor culturing of algae and cyanobacteria. *Appl. Optics* 53:F31–F45.

Sanger, F., S. Nicklen, and A. R. Coulson. 1977. DNA sequencing with chain-terminating inhibitors. *Proc. Natl. Acad. Sci. USA* 74:5463–5467.

Sarma, S. S. S., P. S. L. Jurado, and S. Nandin. 2001. Effect of three food types on the population growth of *Brachionus calyciflorus* and *Brachionus patulus* (Rotifera: Brachionidae). *Rev. Biol. Trop.* 49:77–84.

Saunders, G. W. and D. C. McDevit. 2012. Methods for DNA barcoding photosynthetic protists emphasizing the macroalgae and diatoms. In *DNA Barcodes: Methods and Protocols,* Methods in Molecular Biology, Vol. 858, eds. W. J. Kress and D. L. Erickson, pp. 207–222. Berlin, Germany: Springer.

Schloss, P. D., S. L. Westcott, T. Ryabin et al. 2009. Introducing mothur: Open-source, platform-independent, community-supported software for describing and comparing microbial communities. *Appl. Environ. Microbiol.* 75:7537–7541.

Simon, N., N. LeBot, D. Marie, F. Partensky, and D. Vaulot. 1995. Fluorescent *in situ* hybridization with rRNA-targeted oligonucleotide probes to identify small phytoplankton by flow cytometry. *Appl. Environ. Microbiol.* 61:2506–2513.

Song, H., J. E. Buhay, M. F. Whiting, and K. A. Crandall. 2008. Many species in one: DNA barcoding overestimates the number of species when nuclear mitochondrial pseudo-genes are coamplified. *Proc. Natl. Acad. Sci. USA* 105:13486–13491.

Sonnenberg, R., A. W. Nolte, and D. Tautz. 2007. An evaluation of LSU rDNA D1-D2 sequences for their use in species identification. *Front. Zool.* 4:6.

Steven, B., S. McCann, and N. L. Ward. 2012. Pyrosequencing of plastid 23S rRNA genes reveals diverse and dynamic cyanobacterial and algal populations in two eutrophic lakes. *FEMS Microbiol. Ecol.* 82:607–615.

Stoeck, T., D. Bass, M. Nebel et al. 2010. Multiple marker parallel tag environmental DNA sequencing reveals a highly complex eukaryotic community in marine anoxic water. *Mol. Ecol.* 19:21–31.

Subasinghe, R. P., M. G. Bondad-Reantaso, and S. E. McGladdery. 2001. Aquaculture develop-ment, health and wealth. In *Aquaculture in the Third Millennium. Technical Proceedings of the Conference on Aquaculture in the Third Millennium*, eds. R. P. Subasinghe, P. Bueno, M. J. Phillips, C. Hough, S. E. McGladdery, and J. R. Arthur, pp. 167–191. Bangkok, Thailand: NACA and FAO.

Van de Peer, Y., S. Chapelle, and R. De Wachter. 1996. A quantitative map of nucleotide sub-stitution rates in bacterial rRNA. *Nucleic Acids Res.* 24:3381–3391.

Vestheim, H. and S. N. Jarman. 2008. Blocking primers to enhance PCR amplification of rare sequences in mixed samples—A case study on prey DNA in Antarctic krill stomachs. *Front. Zool.* 5:12.

Whiteley, A. S., S. Jenkins, I. Waite et al. 2012. Microbial 16S rRNA Ion Tag and community metagenome sequencing using the Ion Torrent (PGM) Platform. *J. Microbiol. Methods* 91:80–88.

Yúfera, M. and N. Navarro. 1995. Population growth dynamics of the rotifer *Brachionus plicatilis* cultured in non-limiting food conditions. *Hydrobiology* 313/314:399–405.

9 Terrestrial Agriculture and Aquaculture Waste Treatment

Gregory Schwartz and David E. Brune

CONTENTS

9.1 INTRODUCTION: AGRICULTURE AND AQUACULTURE WASTE

The United States has more than 330 million acres (134 million ha) of agricultural croplands, which produce an abundant supply of food, feed, fiber, fuels, and other products. In 2011, 13 million tons of nitrogen and 4 million tons of phosphate fertilizers were applied to agricultural land for crop production (USDA 2012a). From these, 1.3 million tons of nitrogen and 0.7 million tons of phosphates were converted into waste at animal feed operations (Lory et al. 2006). Anually, aquaculture in the United States generates wastewater containing about 40,000 tons of nitrogen fertilizer. Globally, 5 million tons per year of nitrogen is discharged into the environment as wastewater from aquaculture processes.

Agriculture intensification is a necessity moving into the future; as much as 80% of agriculture growth in developed countries will come from intensification and not from farming new land (Bruinsma 2009). Similarly, consolidation and intensification of concentrated animal feed operations (CAFOs) (EPA 1999), in part due to regional regulatory requirements, has led to increasing concerns regarding contamination of water resources, particularly from nutrient pollution (EPA 2005; Burkholder et al. 2007).

Water quality boards across the country are charged with protecting the nation's water resources and are adopting water quality compliance programs requiring water and waste dischargers to meet increasingly stringent local, regional, and national standards. For example, in northern California, multiple agencies have jointly developed "water quality plans" for the dairy industry to improve nutrient and manure management, to address air and water quality issues and reduce greenhouse gas emissions, and to produce value-added products (CDFA 2012). Implementation of such programs can be expensive and might require farmers to submit annual reports, develop and implement nutrient management budgets, and initiate daily, weekly, and/or monthly monitoring (Cady and Francesconi 2010). The contamination of natural bodies of surface waters, groundwater, and even coastal waters (dead zones) has led to increasingly stringent environmental regulations for farmers across the United States and, internationally, with the goal of achieving a sustainable agriculture (Neumeier and Mitloehner 2013).

Agricultural activities that cause nutrient pollution include improper, excessive, or poorly timed application of irrigation water and fertilizer, poorly located or managed animal feeding operations, and overgrazing (EPA 2005). Some of the pathways for nutrient loss include soil erosion, runoff, volatilization, denitrification, and leaching (Ribaudo et al. 2011).

9.2 ALGAE FOR AGRICULTURAL WASTE TREATMENT

The natural coupling of algae culture with agriculture and aquaculture provides for a holistic and sustainable nutrient management technology. Microalgae have the ability to assimilate nitrogen and phosphorus down to very low concentrations and package these into a recoverable nutrient-dense biomass (Woertz et al. 2009; Boelee et al. 2011; Abdel-Raouf et al. 2012). Recovery and recycling of these wasted nutrients would vastly improve the efficiency, energy footprint, and sustainability of agricultural production. Beyond the environmental benefits, the relationship of algae culture with agriculture and aquaculture leads to economic benefits from the production of saleable products and reduced waste treatment costs, compared to the energy intensive alternatives currently available.

Three major types of agriculture pollution discussed herein are agriculture runoff, discharges from CAFOs and the effluents from anaerobic digestion (AD) processes (see Table 9.1). Agriculture runoff has low levels of nutrients, though often higher than allowed by discharge limits mandated for environmental protection. CAFO and AD effluents are typically highly concentrated in nutrients as well as high in organic matter, measured as chemical oxygen demand (COD) and total suspended solids (TSS). In AD effluents, much of the initial organic matter in the waste is broken down by anaerobic microorganisms, and much of the remaining organic matter can be recalcitrant to further breakdown. Much of the organic nitrogen present in the original waste is converted to ammonia through AD.

TABLE 9.1

Typical Water Quality Characteristics of Agriculture Wastewater

	Total Nitrogen (mg N L^{-1})	Total Phosphorus (mg P L^{-1})	Total Suspended Solids (mg L^{-1})	Soluble COD (mg L^{-1})
Runoff[a]	11–22	0.7–2.0	100–350	—
CAFO[b]	1,200	300	53,200	19,300
Digester[b]	225–2,370	25–240	2,670–31,600	375–4,900

[a] Massingill et al. (2008).
[b] Wilkie and Mulbry (2002).

9.2.1 AGRICULTURE RUNOFF

The recovery efficiency of nutrients used in agriculture (e.g., recovered in the harvest divided by that applied to the land) averages, overall for all crops in the United States, about 75% for N and 55%–95% for applied P, depending on the crop (USDA 2012b). Runoff from farms is the leading source for impairments of surveyed rivers and lakes (EPA 2005), carrying not only nutrients but also pesticides, biological contaminants and particulates (Cai 2012).

In some irrigation practices, tail water (runoff from irrigated fields), tile water (from shallow underground drainage systems), or moderate runoff from storm events can be captured for treatment before being discharged into the environment. One treatment option for such agricultural runoff is through microalgae cultivation to recover the nutrients in an algal biomass. Carlberg et al. (2008) evaluated the use of algal pond raceway production systems to remediate agriculture runoff and estimated that to remove 90% of the phosphorus from these water sources, 650 L min^{-1} ha^{-1} of algae pond (70 gallons per minute per acre) would be remediated using this process.

For a specific example, most of the tail waters from 600,000 acres (243,000 ha) of irrigated farmland in the Imperial and Coachella valleys in southern California drain to one of three main tributaries to the Salton Sea (an inland sea without an outlet). A flow-through high-rate open pond algae growth reactor, the "controlled eutrophication process" (CEP), was used to reduce nutrient levels in the agriculture tail waters of the Whitewater River flowing into the Salton Sea. Specifically P, deemed the key factor causing eutrophication of the Salton Sea, was targeted for 80%–90% reduction. Table 9.2 shows seasonal nutrient uptake rates achieved for such a process along with hydraulic residence times (HRT) used to achieve these nutrient reductions (Carlberg et al. 2008).

A 2-week average N and P removal peaked at an uptake rate of about 1.4–1.5 g N m^{-2} d^{-1} and 0.17–0.23 g P m^{-2} d^{-1} during the summer and fall seasons and about 0.5 g N m^{-2} d^{-1} and 0.07 g P m^{-2} d^{-1} during winter. Algal productivity increased with loading rate, but the low concentrations of nutrients in this agricultural wastewater limited loading rates to 2.6 g N m^{-2} d^{-1} due to washout at higher rates. Actual reductions in N levels averaged 50%–80% during the

TABLE 9.2
Seasonal Nutrient Loading Rates, Algae Production Rates, and Nutrient Uptake Rates for Algae Growth Reactors Fed Agriculture Tail and Tile Water

		Winter	Spring	Summer	Fall
Input	TN (g m^{-2} d^{-1})	0.86	1.5–1.76	1.5–2.61	1.5–2.61
	TP (g m^{-2} d^{-1})	0.09	0.18	0.17–0.29	0.17–0.29
Output	Algal biomass (g m^{-2} d^{-1})	4.5–6.5	5.3–12.7	6.2–13.8	6.0–10.2
	Effluent N conc. (mg L^{-1})	4.9–9.2	1.9–8.6	0.9–11.9	4.3–9.5
	Effluent P conc. (mg L^{-1})	0.1–0.3	0.1–0.5	0.0–0.7	0.2–1.0
	% N reduction	49–60	51–64	51–82	55–81
	% P reduction	73–89	73–94	61–98	46–86
Uptake	g N m^{-2} d^{-1} (mean)	0.33–0.51	0.77–0.96	0.84–1.42	0.57–1.58
		(0.42)	(0.9)	(1.18)	(1.17)
	g P m^{-2} d^{-1} (mean)	0.06–0.07	0.11–0.13	0.11–0.23	0.09–0.17
		(0.066)	(0.12)	(0.14)	(0.12)

summer and fall and 50%–60% during the winter and spring. P reduction was typically between 70% and 90% but fell to 45%–50% when the loading rate was above 0.46 g P m^{-2} d^{-1} in the late fall, as seen in Figure 9.1 (Massingill et al. 2008; Schwartz et al. 2010).

9.2.2 CONCENTRATED ANIMAL FEED OPERATIONS

Concentrated animal feed operations (CAFOs) produce large amounts of manures with high nutrient contents, requiring disposal. Current practices call for the storage of liquid effluents (e.g., from flush operations) in a lagoon and irrigation of the surrounding land with lagoon water. This allows for natural processes to breakdown suspended solids, inorganic nutrients, and pathogens. However, due to intensification and urban sprawl, CAFOs typically have an increasing number of animals and produce more manure, while having less land, and as a result, they are no longer able to safely dispose of most of their waste by land application. Excess application rates lead to negative impacts on the surface waters, air, and ecosystem in general, even human health. The detrimental impacts of CAFOs on the environment have led to increasingly stringent laws and regulations, resulting in manure management practices. Manure is the single largest contributor of total nitrogen pollution in 113 watersheds in the United States (EPA 1999).

An alternative to land application of CAFO effluents is to grow algal biomass for nutrient capture and remediation, which would require less acreage, due to the higher nutrient assimilation capacity of microalgae (Kebede-Westhead et al. 2006; Mulbry et al. 2008b). Wilkie and Mulbry (2002) estimated that conventional land application of manure for about 20 dairy cows, for corn and rye silage production, would require

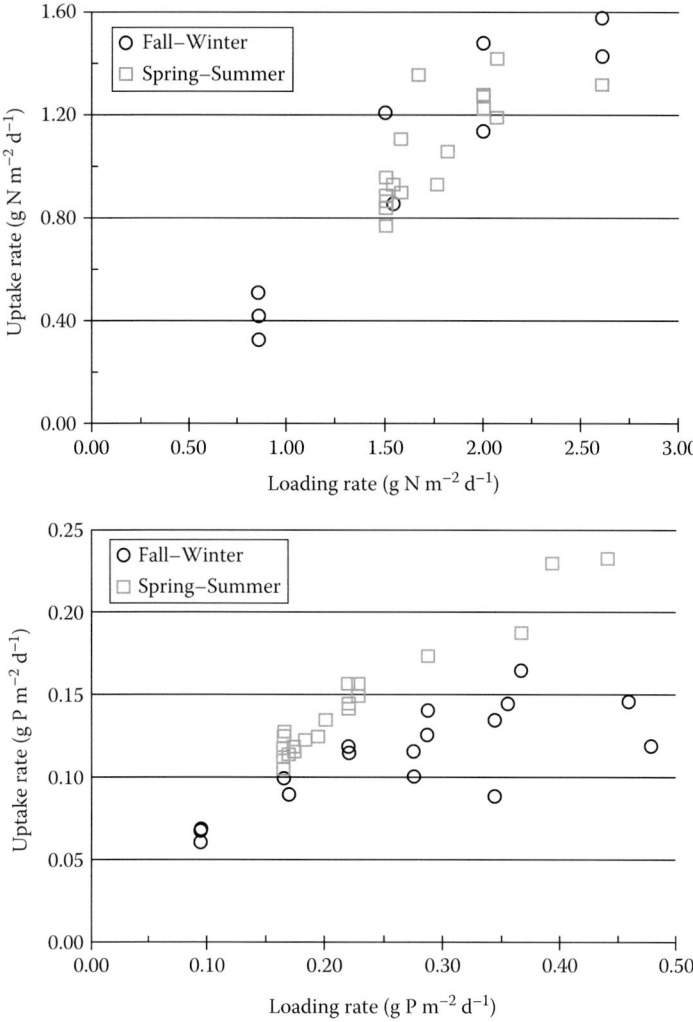

FIGURE 9.1   Algal nitrogen and phosphorus uptake rate from agricultural runoff as a function of nutrient loading rate.

0.135 ha of land per cow compared to 0.05 ha per cow for benthic algae production (over 60% reduction in land use). Even greater reductions would be possible with raceway pond systems.

   Investigating the use of benthic algae, Mulbry et al. (2008b) found that algal productivity increased with increased loading rate when fed dairy manure effluent, while Kebede-Westhead et al. (2006), with a similar system, reported that algal productivity increased only slightly with loading rate when fed piggery waste effluent (see Table 9.3). N and P removal rates of greater than 80% and 70%, respectively,

TABLE 9.3

Benthic Algae Production Using Dairy and Swine Manure Effluent

		Dairy Manure Effluent[a]				Swine Manure Effluent[b]			
Input	TN (g m^{-2} d^{-1})	0.51	0.83	1.73	2.5	0.27	0.48	0.7	1.39
	TP (g m^{-2} d^{-1})	0.08	0.12	0.24	0.4	0.08	0.14	0.21	0.42
Output	Algal biomass (g m^{-2} d^{-1})	8.3	11	18.2	25.1	7.1	9.4	8.7	9.6
	Algal N content (%)	4.8	5.4	6.9	6.3	3.7	4.3	5.4	5.7
	Algal P content (%)	0.69	0.8	0.9	0.86	0.85	1.1	1.3	1.8
	% N recovery	83	72	51	57	98	95	67	40
	% P recovery	91	79	80	62	76	77	55	41
Uptake	g N m^{-2} d^{-1} (mean)	0.4	0.59	1.26	1.58	0.26	0.4	0.47	0.55
	g P m^{-2} d^{-1} (mean)	0.06	0.09	0.16	0.22	0.06	0.1	0.11	0.17

[a] Mulbry et al. (2008b).
[b] Kebede-Westhead et al. (2006).

could be achieved at lower loading rates, 0.5 g N m^{-2} d^{-1} and 0.2 g P m^{-2} d^{-1}. As the N and P loading rates increased, above 1.39 g N m^{-2} d^{-1} or 0.4 g P m^{-2} d^{-1}, the removal efficiency decreased to between 40% and 60% for both dairy and swine manure effluent and both nutrients. The overall N and P uptake rates continually increased with increasing nutrient inputs (see Figure 9.2) up to 1.58 g N m^{-2} d^{-1} and 0.22 g P m^{-2} d^{-1} for dairy manure effluent and up to 0.55 g N m^{-2} d^{-1} and 0.17 g P m^{-2} d^{-1} for swine manure effluent.

Massingill et al. (2005) were unsuccessful in growing planktonic algae directly on dairy wash water, with an input of 2 g N m^{-2} d^{-1}, due to high biochemical oxygen demand (BOD) levels (590 mg L^{-1}, with a ratio of 6:1 BOD/TN; BOD measures the O$_2$ required for bacterial break down of the organic substances present, TN is total nitrogen). After an aerobic pretreatment step was used to lower the BOD/TN ratio to 1:1, and the same input of 2 g N m^{-2} d^{-1}, an algal productivity of 17 g algae m^{-2} d^{-1} was achieved. It was estimated that at high ratios of BOD/TN, the systems will shift toward algal/bacterial and then solely bacterial populations. Waste from manure tends to have high BOD levels, thus limiting the loading rate, which dictate the maximum uptake rate and thus determine system size.

The development of algal/bacterial consortia has been investigated due to some advantages they bring over strictly algal systems. CAFO waste, in general, is dark colored, inhibiting light penetration and has high organic matter loadings, conditions that promote the growth of bacteria over algae. Advantages by such consortia include a potential for nitrification and denitrification, breakdown of organic matter and the subsequent release of ammonia and carbon dioxide, bacterial utilization of O$_2$ from photosynthesis, and the potential for bioflocculation (Benemann et al. 2003; Gonzalez-Fernandez et al. 2011; Pires et al. 2013). Algal/bacterial systems potentially offer a cost-effective solution for systems requiring the treatment of nitrogen, phosphorus, and carbon (De Godos et al. 2009). With raceway ponds, nitrogen

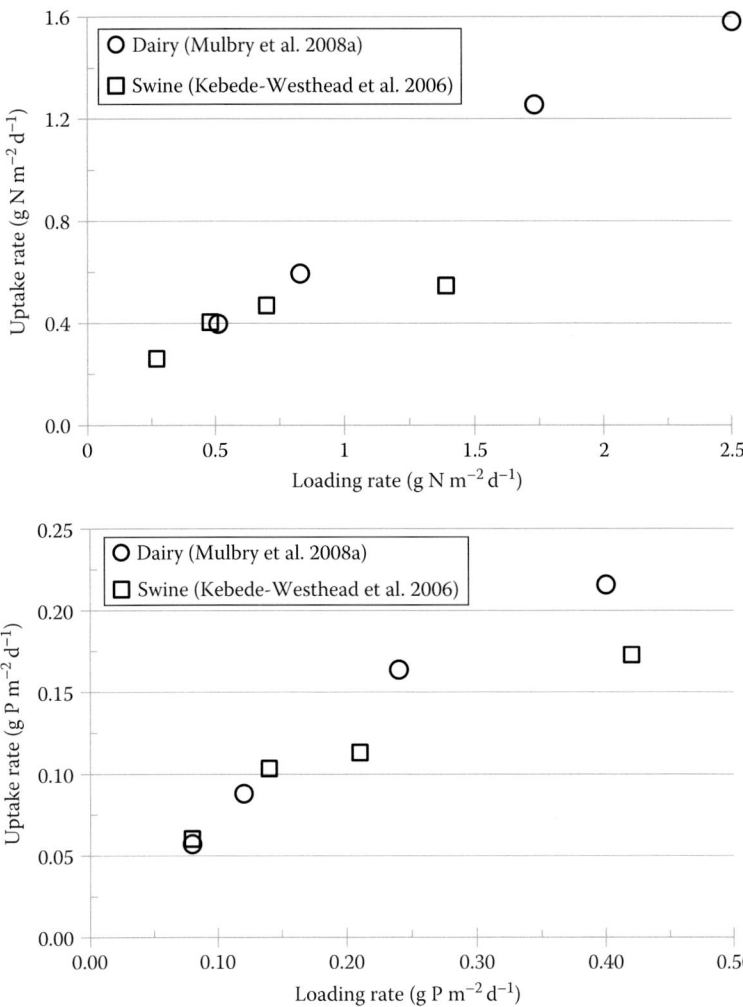

FIGURE 9.2 Algal nitrogen and phosphorus uptake rate from CAFO effluent as a function of nutrient loading rate.

removal from piggery wastewater in Spain ranged from 0.2–0.7 g N m^{-2} d^{-1} in winter to 2.9–5.7 g N m^{-2} d^{-1} in summer, algal biomass productivity averaged 5.8 g m^{-2} d^{-1} in winter and 24.5 g m^{-2} d^{-1} in summer, suggesting, based on a N content of 8%, a maximum algal removal rate of, on average, about 0.5 and 2 g N m^{-2} d^{-1}, winter and summer, respectively. Denitrification was suspected in summer, accounting for the high nitrogen removal rates. In this study, less than 10% reduction in phosphorus was observed, a major potential disadvantage (De Godos et al. 2009).

Travieso et al. (2006), using a mixture of sewage and piggery wastewater, modeled algal and bacterial biomass volatile suspended solids (VSS) as a function of dilution rate (d^{-1}). It was found that the highest algal biomass to bacterial biomass ratios were at the lowest dilution rate, or longest hydraulic residence time (HRT). Algal biomass productivity was as high as 38.2 g m^{-2} d^{-1} at a dilution rate of 0.125–0.25 d^{-1} (8–4 day HRT). Overall, high nutrient removal efficiencies were achieved with HRTs ranging from 8 to 14 days, depending on season.

9.2.3 ANAEROBIC DIGESTION EFFLUENT

A second option for manure management is the incorporation of anaerobic digestion (AD) for production of bioenergy. AD converts organic materials into CH_4, CO_2, and NH_3 along with bacterial biomass and residues. Similar to CAFO waste, excessive land application of anaerobic digestion effluent results in N and P contamination to groundwater or eutrophication of surface waters (Cai 2012). The effluent from AD is better suited for algal production than raw manure due to the biological decomposition from the digestion process. However, high ammonia levels and color require dilution.

Typical dairy manure digester effluent may contain 2200 mg L^{-1} of ammonia and 250 mg L^{-1} of total phosphorus and achieve a 40% reduction in COD after AD (Wang et al. 2010). However, effluents from other waste can range widely, between 125 and 3500 mg L^{-1} of total N and 20–380 mg L^{-1} of total P, depending on the animals, waste management (flush, scrape, etc.), and the digester operation itself.

Mulbry et al. (2008a) compared benthic algae productivity for raw swine manure, raw dairy manure, and anaerobic digester effluent and found no significant differences in productivity as a function of loading rate (see Figure 9.3).

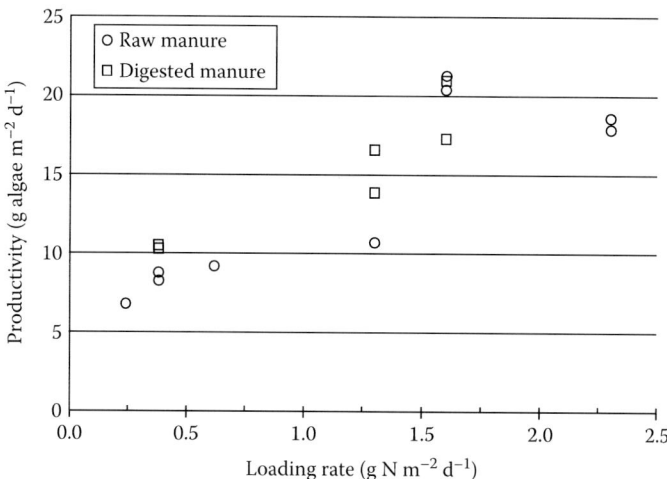

FIGURE 9.3 Algae growth rate as a function of the loading rate for two different agriculture wastewaters: raw manure effluent and anaerobically digested manure effluent. (From Mulbry, W. et al., *J. Appl. Phycol.*, 20, 1079, 2008a.)

9.3 ALGAE IN AQUACULTURE WASTEWATER TREATMENT

Global seafood supply is estimated at 16.4 kg per person per year or about 110 million tons per year. Aquaculture production of seafood has grown from 14% of total demand in 1986 to nearly 50% of global consumption today. Catfish, tilapia, shrimp, and salmon have all shifted from primarily wild caught to mostly aquaculture produced during this period. U.S. seafood consumption averages 15.8 lb (7.2 kg) per person (less than half the global average) or 2.5 million tons per year with a retail value of $80 billion per year. About a quarter of U.S. seafood consumption is marine shrimp with 86% of shrimp imported from Asian farms, while only 14% is wild caught. Increasing consumer demand of aquaculture products will continue to drive growth in this industry and with it increased pressure for control of water and waste discharges.

One ton of fish and shellfish production typically requires 2 tons of aquaculture feed. Aquaculture feeds average 35% protein (5.6% N) or 112 kg N fed per ton of wet weight fish produced. The vast majority of this N (85%–90%) is released into the aquaculture system water as a waste discharge (see Figure 9.4).

In the United States, approximately 50,000 tons per year of N is discharged into U.S. surface water and groundwater, based on a use of ~1 million tons per year of aquaculture feeds. As with conventional agriculture, nitrogen and phosphorus release represents a threat to local environmental quality, and water supplies, leading to increasing regulatory pressures to eliminate such discharges. As with conventional agriculture, N and P releases represent a threat to local environmental quality and water supplies, leading to increasing regulatory pressure to eliminate such discharges. U.S. aquaculture has been driven by state and federal agencies to move to zero-discharge practices, in practice totally eliminating waste discharge. In such systems, pond seepage, denitrification, volatilization, and/or land application has served to dispose of excess N.

With 54 million tons per year of global aquaculture production, about 6.0 million tons per year of N is being discharged from aquaculture, posing an increasing threat to freshwater and coastal aquatic environments. China's water discharge from shrimp culture ponds adds 10 times more waste to coastal zones than all other industrial

79%–88% Nitrogen discharged as a pollutant

Soy, corn, and fish meal nitrogen inputs

12%–21% Protein nitrogen converted to fish or shrimp

FIGURE 9.4 Aquaculture production capturing 10%–25% of feed nitrogen with 75%–90% potentially released as a pollutant.

waste combined (Stokstad 2010). As aquaculture continues to grow in importance, widespread adoption of more sustainable practices will gain increasing attention. Even with zero discharge, up to 90% of input nitrogen in the feed is wasted. Except where this waste can be returned to land farming, an infrequent opportunity, this is an unsustainable practice in which fossil fuels support grain and fish meal production and transport.

Algal culture integrated into aquaculture production offers the greatest potential for sustainable recovery and reuse of wasted aquaculture nutrients. Algae are capable of rapid growth assimilating N, P, and other nutrients at low levels, thereby ensuring high water quality for any recycled water or discharge, providing a healthy environment for continuous fish and shellfish culture. Algal culture is solar driven, potentially freeing aquaculture production from external energy inputs.

9.3.1 INTEGRATED ALGAL CULTURE AND AQUACULTURE IN THE PARTITIONED AQUACULTURE SYSTEM

Integrated algal/aquaculture systems can be designed to serve as high-rate, engineered aquatic production systems mimicking natural ecosystems, recycling nutrients and eliminating waste nutrient discharge, improving overall production efficiency. The partitioned aquaculture system (PAS), developed at Clemson University (Brune and Wang 1998; Brune et al. 2002, 2003, 2004a,b, 2012) is such a process, utilizing paddle wheel mixed raceway "high-rate" ponds, first developed by Oswald at UC Berkeley in the 1950s (Oswald 1988). The PAS is used to grow dense algal cultures on the nutrients released from aquaculture operations (see Figure 9.5). As with other algal wastewater treatment processes (Oswald 1988), the algae provide the oxygen used by bacteria to break down and consume organic waste, with inorganic nutrients thus released that are then assimilated by the algae.

One innovative aspect of the PAS is that the algal waters are diverted through fish or shrimp enclosures, providing oxygenated water when needed as well as serving as a potential food source for the cultured aquatic animals. By 1995, the Clemson team researchers had built and operated 0.13 ha (1/3 acre) prototypes of the PAS.

The principal advantage of the high-rate pond and PAS is a gain in algal productivity brought about by improved ammonia assimilation and oxygen production rates. The PAS incorporates other aquaculture intensification techniques into key aspects of the high-rate pond. In particular, manipulation of algal population, algal density control, and reductions in zooplankton numbers are all possible through tilapia coculture (Brune et al. 2012). By 1996, Clemson researchers were able to achieve catfish production levels of 11 t ha^{-1} yr^{-1} (about 10,000 lb per acre per year) in the 0.13 ha ponds, and this was further improved with an increase of 60% in 1998. Total annual yield from a subsequent and larger 0.8 ha (2-acre) PAS prototype ultimately reached 42,000 lb (23.8 t ha^{-1} yr^{-1}) of catfish along with coproduced tilapia at 4,500 lb (2.6 t ha^{-1} yr^{-1}) (Brune et al. 2012).

The key process controlling aquaculture pond ammonia concentrations in PAS is algal N and C assimilation. In a typical unmixed aquaculture pond, algal production

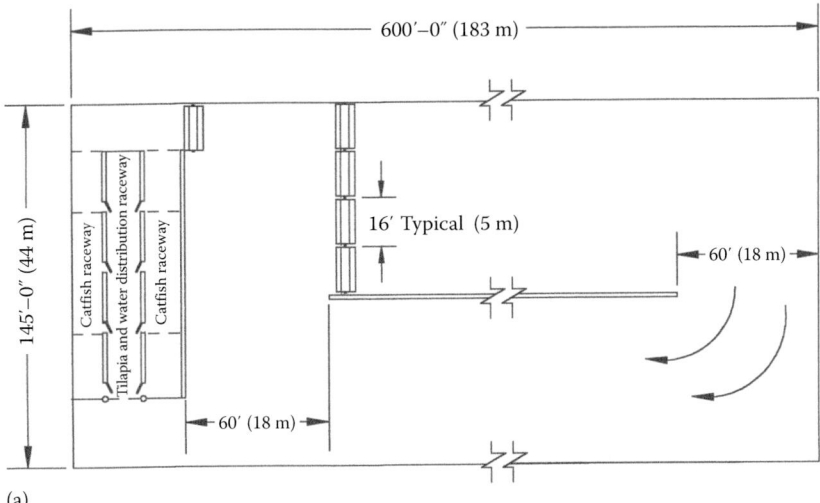

(a)

(b)

FIGURE 9.5   (See color insert.) Integrated algal and aquaculture research units at Clemson University (South Carolina).

is limited to 1–3 g C m^{-2} d^{-1}, which translates to 0.5 mg L^{-1} d^{-1} of N removal. This corresponds to a feed loading of 90–112 kg ha^{-1} d^{-1}, and this restricts pond fish production to 4–6 t ha^{-1} d^{-1} per season. This can be improved by harvesting the algal biomass either by sedimentation or utilizing filter feeders. Steps such as these will generate net N removal from the water column and also net oxygen addition, directly leading to enhanced fish carrying capacity of the PAS pond environment. A comparison of PAS units and conventional ponds fed at similar levels resulted in reduced ammonia levels in the former case of just under 5 mg L^{-1} c.f. 16 mg L^{-1} (Baumgarner et al. 2005; Figure 9.6).

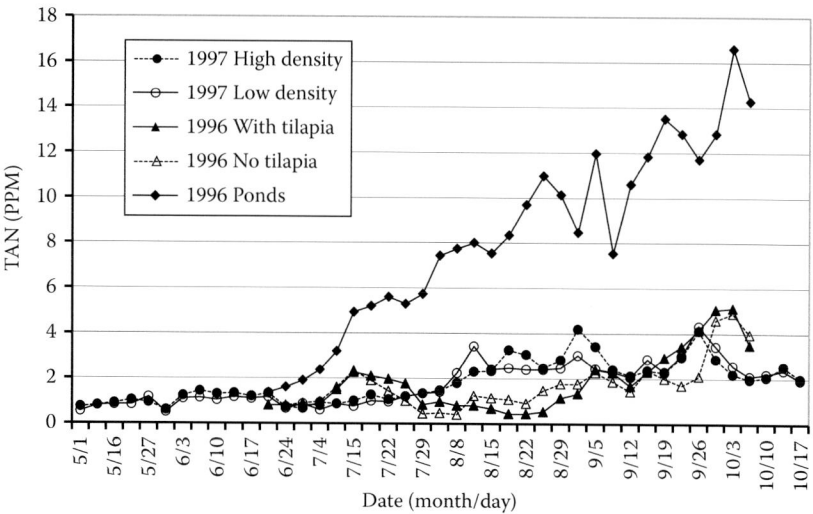

FIGURE 9.6   Ammonia nitrogen concentrations versus time in conventional ponds and PAS units at feed application rates ranging from 168 to 225 kg ha^{-1} (50–200 lb per acre per day).

9.3.2   Integrated Algal Culture in the Mississippi Split-Pond System

Wastewater treatment needs in conventional catfish ponds in southern United States presented another opportunity for the PAS-type technology. Researchers at Mississippi State University and the National Warmwater Aquaculture Center developed and demonstrated a lower-cost version of the PAS, called the "split pond" constructed by retrofitting existing earthen ponds (see Figure 9.7; Brune et al. 2012). The split pond has a relatively smaller algal production area (about 70%–80% of the total) and larger fish-holding area than the PAS with fish held at 5–10 times the density of traditional ponds. Aside from the advantage of confining the fish into a small area of the pond, which facilitates feeding, inventory, harvest, health management, and protection, the feed loading and fish production rates were significantly greater in the split pond than in traditional ponds. Through 7 years of study, net annual catfish production in the split pond ranged from 17 to almost 20 t ha^{-1} (based on the total water area for the system): two to four times that achieved in traditional ponds and only marginally less than in the typical PAS systems.

Feed conversion ratios (weight of feed fed divided by net fish weight gain) are about 1.8 and total ammonia-nitrogen concentrations seldom exceed 1 mg L^{-1}, except when phytoplankton communities crash or in late autumn, when cooler water temperatures slow nitrogen assimilation by the algae. Split-pond systems do not typically employ active algal harvest. The primary processes in maintaining water quality in the split pond is enhanced algal growth rate, coupled with algal sedimentation, degradation, and nitrification and denitrification in the deeper zone of the split ponds.

(a)

(b)

FIGURE 9.7   Paddle wheel at a Mississippi split-pond levee (a) and return-flow sluiceway (b).

9.3.3 Harvesting and Utilization of Algal Biomass from Freshwater Aquaculture

Algal biomass produced in the PAS and split-pond systems may be harvested and utilized, thus maximizing the algal production capacity of the high-rate pond, and its role in maintaining water quality for fish and shellfish production. Nile tilapia, *Oreochromis niloticus*, coculture was proved to be effective at stabilizing algal cultures by reducing zooplankton densities and grazing, by helping maintain optimal algal densities (based on Secchi disk visibilities of 15–18 cm) (Brune et al. 2012) and eliminating cyanobacteria (Turker et al. 2002, 2003a,b,c), a major issue as cyanobacteria are associated with catfish off-flavors. Alternatively, shellfish selectively remove green algae, driving the pond algal populations toward cyanobacteria dominance (Stuart et al. 2001, Figure 9.8). Tilapia grazing on algae in the PAS also helps maintain the algal cells in a rapid growth phase, with doubling times of 2–3 days for a tilapia density of 10%–25% of total fish biomass (the remainder being channel catfish) compared to 6–10 days in the control (no tilapia) (Stuart et al. 1999). Rapidly growing algal cells produce more net O_2 during the day and consume less at night.

In the initial PAS design, considerable quantities of algal biomass settled in the ponds, then decayed, thus recycling soluble N, P, and other nutrients back into the

(a)

(b)

FIGURE 9.8 (See color insert.) Algal cell age reduced with tilapia filtration (a) and green algal enhancement with tilapia filtration (left tank) versus cyanobacterial enhancement with freshwater mussel filtration (right tank) (b).

culture, while also consuming O_2 (Drapcho and Brune 2000). To avoid this, the excess algal biomass produced in the system must be harvested and removed. At C fixation rate of 10 g C m^{-2} d^{-1}, corresponding to 1.8 g N m^{-2} d^{-1}, a PAS feed assimilation capacity in excess of 450 kg ha^{-1} d^{-1} (400 lb per acre per day), suggests catfish yields of 33.6 t ha^{-1} (30,000 lb per acre) (Brune et al. 2012). However, harvesting proved to be a challenge. Chemical flocculation worked well but costs were prohibitive. However, it was discovered that tilapia populations could be used to concentrate algal cells into rapidly settling fecal pellets easily removed from the system. A moving belt under a cage of tilapia was effective in capturing and removing this biomass. This approach proved successful in both South Carolina and California as being the most cost-effective. Field trials showed that 90%–95% of algal biomass production can be harvested using tilapia. The harvested algal biomass can be subjected to anaerobic digestion (AD) to produce methane gas for energy, or even potentially recycled into feed, potentially providing aquaculture protein self-sufficiency.

9.3.4 HARVESTING AND UTILIZATION OF ALGAL BIOMASS IN MARINE AQUACULTURE

In 2011, a saltwater shrimp PAS was installed at the University of Missouri, demonstrating integration of PAS algal production for water quality control coupled with harvest and conversion of excess algal and bacterial biomass into brine shrimp (*Artemia*) and tilapia biomass. This system showed that tilapia and brine shrimp could be cocultured with the target high-value species, the Pacific white shrimp (*Litopenaeus vannamei*), providing human food and fish meal replacement and potential for biofuels and bioenergy coproduction (see Figure 9.9).

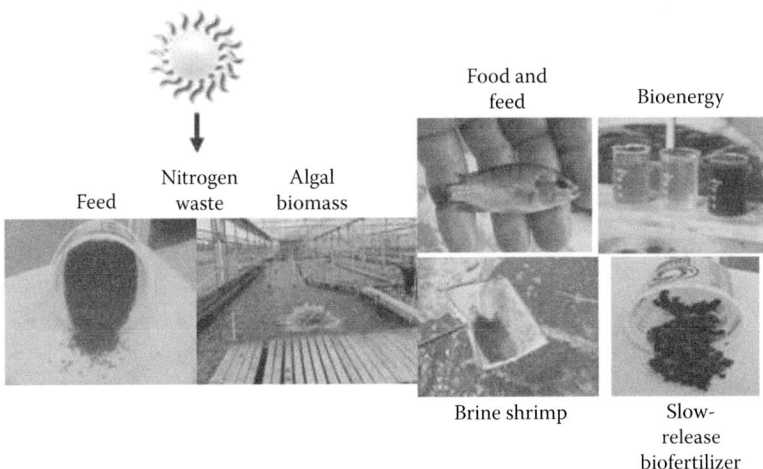

FIGURE 9.9 Nutrients released from aquaculture feeds converted into algal biomass in high-rate PAS, harvested by tilapia and brine shrimp for food, feed, fertilizer, and fuels.

The PAS high-rate algal ponds were shown to be capable of maintaining water quality in zero-discharge aquaculture system yielding in excess of 20,000 lb per acre (22 t ha^{-1}) over 120 days, along with algal biomass production of 10 tons per acre per year (25 t ha^{-1} yr^{-1}) yielding 5 tons per year fish meal replacement as *Artemia* biomass. Furthermore, the remaining algal biomass could potentially produce 4 kW per acre (10 kW ha^{-1}) of stationary power (as biogas) and 350–700 gallons of liquid fuel per acre per year (3000–7000 L ha^{-1}). The value of the shrimp production was estimated to range from $100,000 to $300,000 per acre per year ($247,000–$741,000 ha^{-1} yr^{-1}), with the *Artemia* value at $1,500 per acre per year ($3,700 ha^{-1} yr^{-1}) and biofuels at $1,800 per acre per year ($4,500 ha^{-1} yr^{-1}). Although the cash flow is projected at 95% from shrimp production, even a modest animal feed and bioenergy coproduction could provide an enhanced profit margin to such an operation and provide a bridge to continued improvement in algal feed and biofuels production.

9.4 CONCLUSIONS

Although much research has been conducted and is ongoing, there has been limited movement toward commercialization of algal waste treatment technologies in most agricultural sectors. This has been primarily due to lack of economic incentives to manage waste discharges, which thus makes any waste treatment unaffordable. Until farming is accountable for its impacts, farmers will likely continue to discharge untreated, or partially treated, water and waste into the environment.

Algae processes can offer economic and environmental benefits to farmers, including decreased impacts on the groundwater and surface water and a more sustainable and lower-cost farming practice through the recycling of nutrients. However, such systems must balance nutrient loading and water requirements. A CAFO operation that produces waste as in Table 9.4 could be loaded into an algal growth reactor at a rate of about 1 L m^{-2} d^{-1} using an annual average assimilation rate of just over 1 g N m^{-2} d^{-1}. Depending on the waste source, local evaporation rates, and water recycle efficiency, dilution water may be required. Mixed algal/bacterial consortiums may be useful in treating elevated organic levels from animal production.

TABLE 9.4

Estimated Freshwater Requirement for Algal Production Systems to Treat Agricultural Wastewater

	Estimated N Conc. (mg L^{-1})	Treatment Capacity[a] (L m^{-2} d^{-1})	Potential Evaporation[b] (L m^{-2} d^{-1})	Freshwater Required (L m^{-2} d^{-1})
Agricultural runoff	20	40	5.0	None
CAFO effluent	1200	0.7	5.0	4.3
Anaerobic digester effluent	2500	0.3	5.0	4.7

[a] Treatment capacity calculated using mean N uptake rates of 0.8 g N m^{-2} d^{-1}.
[b] Evaporation rates estimated from annual mean of 1.8 L m^{-2} d^{-1}.

High-rate algal pond nutrient assimilation rates are dependent on season and loading rate. Nutrient uptake rates peak in the summer with as much as 2 g N m^{-2} d^{-1} and 0.23 g P m^{-2} d^{-1}. Annual averages are much lower, at about half these levels, and winter time averages even lower, depending on location. Thus, effective nutrient reduction will require seasonally adjusted nutrient reduction objectives.

If lower nutrient concentrations are required for local surface water discharge permits, a two-stage algal system may be utilized, ensuring optimal algal productivity and nutrient uptake, providing reduced nutrient concentrations suitable for discharge. The two-stage algae treatment would consist of a mass removal stage followed by a polishing stage. Mass removal units would be operated with higher soluble nutrient levels to maintain high productivity and high mass removal rates for the total nitrogen and phosphorus. Polishing units would be operated to reduce N and P concentrations as low as possible at the expense of a lower overall productivity and lower total mass nutrient uptake. This two-step sequential process would provide the lowest final nitrogen or phosphorus effluent discharge concentrations.

The primary constraint to widespread commercialization of algal culture for treatment of agricultural and aquaculture wastewater is installation and operational costs relative to the low value of products and coproducts. The greatest potential for short-term utilization of economic high-rate algal culture is as applied to low-cost zero-discharge aquaculture such as the "split-pond" configuration applicable to catfish culture across the southern United States.

The "partitioned aquaculture system," an integration of high-rate algal culture, zero-discharge aquaculture, and aquatic-animal algal harvest and conversion, offers potential for economical yield of algal by-products as well as reduction in fossil-fuel requirements for fish and shellfish production. As feed and fuel costs increase, integrated processes of this type will likely become more attractive to commercial aquaculturists.

REFERENCES

Abdel-Raouf, N., A. Al-Homaidan, and I. Ibraheem. 2012. Microalgae and wastewater treatment. *Saudi J. Biol. Sci.* 19:257–275.

Baumgarner, B. L., T. E. Schwedler, A. G. Eversole, D. E. Brune, and J. A. Collier. 2005. Production characteristics of channel catfish stocked at two densities in the partitioned aquaculture system. *J. Appl. Aquacult.* 17:75–83.

Benemann, J. R., J. C. Van Olst, M. J. Massingill, J. A. Carlberg, J. C. Weissman, and D. E. Brune. 2003. The controlled eutrophication process: Using microalgae for CO$_2$ utilization and agricultural fertilizer recycling. *Proceedings of the Sixth International Conference on Greenhouse Gas Technologies.* Kyoto, Japan, October 4, 2002.

Boelee, N. C., H. Temmink, M. Janssen, C. J. N. Buisman, and R. H. Wijffels. 2011. Nitrogen and phosphorus removal from municipal wastewater effluent using microalgal biofilms. *Water Res.* 45:5925–5933.

Bruinsma, J. 2009. The resource outlook to 2050: By how much do land, water and crop yields need to increase by 2050? *Expert Meeting on How to Feed the World in 2050.* Rome, Italy: Food and Agriculture Organization of the United Nations. ftp://ftp.fao.org/agl/aglw/docs/ResourceOutlookto2050.pdf (accessed December 2013 and 2015).

Brune, D. E., J. A. Collier, and T. E. Schwedler. 2002. Partitioned aquaculture system. United States Patent No. 6,192,833.

Brune, D. E., G. Schwartz, J. A. Collier, A. G. Eversole, and T. E. Schwedler. 2004a. Partitioned aquaculture systems. In *Biology and Culture of Channel Catfish*, eds. C. S. Tuckers and J. A. Hargreaves, pp. 561–584. Amsterdam, the Netherlands: Elsevier.

Brune, D. E., G. Schwartz, A. G. Eversole, J. A. Collier, and T. E. Schwedler. 2003. Intensification of pond aquaculture and high rate photosynthetic systems. *Aquacult. Eng.* 28:65–86.

Brune, D. E., G. Schwartz, A. G. Eversole, J. A. Collier, and T. E. Schwedler. 2004b. *Partitioned Aquaculture Systems.* Stoneville, MS: Southern Regional Aquaculture Center Publication.

Brune, D. E., C. Tucker, M. Massingill, and J. Chappell. 2012. Partitioned aquaculture systems. In *Aquaculture Production Systems*, ed. J. H. Tidwell, pp. 308–342. Oxford, UK: Wiley-Blackwell.

Brune, D. E. and J. K. Wang. 1998. Recirculation in photosynthetic aquaculture systems. *Aquacult. Mag.* 24:63–71.

Burkholder, J., B. Libra, P. Weyer et al. 2007. Impacts of waste from concentrated animal feeding operations on water quality. *Environ. Health Perspect.* 115:308–312.

Cady, C. W. and M. Francesconi. 2010. Water quality regulations for dairy operators in California's Central Valley—Overview and compliance cost analysis. Sacramento, CA: California Department of Food and Agriculture.

Cai, T. 2012. Nutrient recovery from wastewater streams by microalgae: Status and prospects. *Renew. Sust. Energ. Rev.* 19:360–369.

Carlberg, J., J. VanOlst, M. Massingill, and G. Schwartz. 2008. Reduction of phosphorus and nitrogen in the whitewater river and Salton sea. Grant Agreement #04-394-557-0. CRWQCB Final Report, Los Angeles, CA, May 14, 2008.

CDFA. 2012. California cost of production annual. California Department of Food and Agriculture, Sacramento, CA. https://www.cdfa.ca.gov/dairy/pdf/Annual/2012/ProdCost Annual2012.pdf (accessed December 2013).

De Godos, I., S. Blanco, P. Encina, E. Becares, and R. Munoz. 2009. Long-term operation of high rate algal ponds for the bioremediation of piggery wastewaters at high loading rates. *Bioresour. Technol.* 100:4332–4339.

Drapcho, C. M. and D. E. Brune. 2000. Modeling of oxygen dynamics in the partitioned aquaculture system. *J. Aquacult. Eng.* 21:151–162.

EPA. 1999. Feedlots preliminary data summary: Feedlots point source category study. EPA-821-R-99-002. Washington, DC.

EPA. 2005. Protecting water quality from agricultural runoff. Agricultural nonpoint source fact sheet. EPA 841-F-05-001. Washington, DC. http://www.epa.gov/polluted-runoff-nonpoint-source-pollution/nonpoint-source-fact-sheets (accessed December 2013 and 2015).

Gonzalez-Fernandez, C., B. Riano-Irazabal, B. Salces, S. Blanco, and M. Gonzalez. 2011. Effects of operational condition on the degradation of organic matter and development of microalgae-bacteria consortia when treating swine slurry. *Appl. Microbiol. Biotechnol.* 90:1147–1153.

Kebede-Westhead, E., C. Pizarro, and W. Mulbry. 2006. Treatment of swine manure effluent using freshwater algae: Production, nutrient recovery, and elemental composition of algal biomass at four effluent loading rates. *J. Appl. Phycol.* 18:41–46.

Lory, J., R. Massey, and B. Joern. 2006. Using manure as a fertilizer for crop production. Washington, DC: USEPA. Water.epa.gov. http://water.epa.gov/type/watersheds/named/msbasin/upload/2006_8_25_msbasin_symposia_ia_session8.pdf (accessed December 2013).

Massingill, M., J. Carlberg, J. Van Olst et al. 2008. Large-scale microalgae cultivation in agricultural wastewaters for biofixation of CO_2 and greenhouse gas abatement. DOE Award # DE-FG02-04ER83988. DOE SBIR Phase II Final Report, Washington, DC, May 10, 2008.

Massingill, M. J., J. Van Olst, and G. Schwartz. 2005. Management of CAFO discharges utilizing Controlled Eutrophication Process (CEP) ponds for liquid waste storage and conversion to bioproducts and slow release biofertilizers. USEPA SBIR Phase I Final Report, Washington, DC. EP-D-05-011.

Mulbry, W., S. Kondrad, and J. Buyer. 2008a. Treatment of dairy and swine manure effluents using freshwater algae: Fatty acid content and composition of algal biomass at different manure loading rates. *J. Appl. Phycol.* 20:1079–1085.

Mulbry, W., S. Kondrad, C. Pizarro, and E. Kebede-Westhead. 2008b. Treatment of dairy manure effluent using freshwater algae: Algal productivity and recovery of manure nutrients using pilot-scale algal turf scrubbers. *Bioresour. Technol.* 99:8137–8142.

Neumeier, C. J. and F. M. Mitloehner. 2013. Cattle biotechnologies reduce environmental impact and help feed a growing planet. *Anim. Front.* 3:36–41.

Oswald, W.J. 1988. Microalgae and wastewater treatment. In *Microalgal Biotechnol.*, ed. Borowitzka and Borowitzka. Cambridge University Press, Cambridge, UK, pp. 305–328.

Pires, J., C. Alvim-Ferraz, F. Martins, and M. Simoes. 2013. Wastewater treatment to enhance the economic viability of microalgae culture. *Environ. Sci. Pollut. Res.* 20:5096–5105.

Ribaudo, M., L. Hansen, M. Livingston, R. Mosheim, J. Williamson, and J. Delgado. 2011. Nitrogen in agricultural systems: Implications for conservation policy. USDA-ERS Economic Research Report No. 127, Washington, DC. http://papers.ssrn.com/sol3/papers.cfm?abstract_id=2115532 (accessed December 2013 and 2015).

Schwartz, G., M. Massingill, J. C. Levin, J. M. Carlberg, and J. C. Van Olst. 2010. Microalgae for environmental remediation: Case study—Phosphorus removal at the Salton sea. *Proceedings of 83rd Annual Water Environment Federation,* New Orleans, LA, Session 32.

Stokstad, E. 2010. Down on the shrimp farm. *Science* 18:1504–1505.

Stuart, K. R., A. G. Eversole, and D. E. Brune. 1999. Effects of flow rate and temperature on the algal uptake by freshwater mussels. In *Proceedings of the First Mollusk Conservation Society Symposium*, pp. 219–224. Columbus, OH: Ohio Biological Survey.

Stuart, K. R., A. G. Eversole, and D. E. Brune. 2001. Filtration of green algae and cyanobacteria by freshwater mussels in the partitioned aquaculture system. *J. World Aquacult. Soc.* 32:105–111.

Travieso, L., F. Benitez, E. Sanchez, R. Borja, and M. Colmenarejo. 2006. Production of biomass (algae-bacteria) by using a mixture of settled swine and sewage as substrate. *J. Environ. Sci. Health* 41:415–429.

Turker, H., A. G. Eversole, and D. E. Brune. 2002. Partitioned aquaculture systems; tilapia filter green algae and cyanobacteria. *Advocate* 5:68–69.

Turker, H., A. G. Eversole, and D. E. Brune. 2003a. Effect of Nile tilapia, *Oreochromis niloticus* (L.), size on phytoplankton filtration rate. *Aquacult. Res.* 34:1087–1091.

Turker, H., A. G. Eversole, and D. E. Brune. 2003b. Effects of temperature and phytoplankton concentration on Nile tilapia, *Oreochromis niloticus*, filtration rates. *Aquacult. Res.* 34:453–459.

Turker, H., A. G. Eversole, and D. E. Brune. 2003c. Filtration of green and cyanobacteria by Nile tilapia, *Oreochromis niloticus*, in the partitioned aquaculture system. *Aquaculture* 215:93–101.

USDA. 2012a. Chemical inputs: Fertilizer use & markets. Washington, DC. http://www.ers.usda.gov/topics/farm-practices-management/chemical-inputs/fertilizer-use-markets (accessed December 2013 and 2015).

USDA. 2012b. Crop and livestock practices: Nutrient management. Washington, DC. http://www.ers.usda.gov/topics/farm-practices-management/chemical-inputs/fertilizer-use-markets (accessed December 2013 and 2015).

Wang, L., Y. Li, P. Chen et al. 2010. Anaerobic digested dairy manure as a nutrient supplement for cultivation of oil-rich green microalgae *Chlorella* sp. *Bioresour. Technol.* 101:2623–2628.

Wilkie, A. and W. Mulbry. 2002. Recovery of dairy manure nutrients by benthic freshwater algae. *Bioresour. Technol.* 84:81–91.

Woertz, I., A. Feffer, T. Lundquist, and Y. Nelson. 2009. Algae grown on dairy and municipal wastewater for simultaneous nutrient removal and lipid production for biofuel feedstock. *J. Environ. Eng.* 135:1115–1122.

10 Supply of CO_2 to Closed and Open Photobioreactors

F. Gabriel Acién, Jose Maria Fernández-Sevilla, and Emilio Molina Grima

CONTENTS

10.1 INTRODUCTION

Production of microalgae started in the 1950s at the Massachusetts Institute of Technology producing *Chlorella* for human consumption using a closed photobioreactor. At the same time, open raceway reactors were used at the University of California, Berkeley, to perform wastewater treatment using microalgae. Since this time, diverse processes have been developed using microalgae to produce valuable products and services. Great efforts were made in the 1980s and 1990s by the Aquatic Species Program in the United States and the RITE project in Japan to develop processes capable of producing biofuels from microalgae. Nevertheless, with the technology available and production costs at the time, it was not possible to compete with petrofuels. Since then, processes for the production of valuable products have been developed such as the production of feed for aquaculture, human nutraceuticals, specialty foods, and carotenoids, among others (Borowitzka 2013). Recently, great interest has emerged in the use of microalgae as CO_2 fixation systems for fossil fuel emissions, thus contributing to the mitigation of GHG emissions, while also producing biofuels. Utilization of microalgae cultures as a CO_2 biofixation method was proposed in the 1960s (Oswald and Golueke 1960); however, this does not amount to a sequestration strategy unless the biomass produced can be stored for any significant time. Thus, one way for microalgae to contribute to a reduction in CO_2 emissions is by producing biofuels to replace fossil fuels used today or, alternatively, by allowing the production of other commodities or by-products from flue gases, which provides a net CO_2 benefit in comparison with alternatives, allowing credits for carbon capture systems (Acién et al. 2012a). From several projects, values up to 50 $gCO_2/m^2 \cdot day$ have been reported, 10 times higher than the capacity of temperate forests (Benemann 1997; Otsuki 2001; Li et al. 2013; Singh and Ahluwalia 2013). However, potential commercial applications are scarce due to high microalgae biomass production costs, applications of microalgae remaining mainly in the markets of human foods and nutraceuticals, in addition to wastewater treatment (Acien et al. 2012b; Craggs et al. 2012; Benemann 2013).

Whatever the application of microalgae, the production of biomass requires sufficient supply of carbon. CO_2 is the preferred form in which to supply inorganic carbon because addition as carbonates is not cost-effective, and also, CO_2 supply allows control of the culture pH. With approximately 48% of biomass dry matter as carbon, to produce 1 kg of biomass, 1.8 kg of CO_2 must be supplied to the algal culture. Moreover, the CO_2 concentration in the culture should not fall below a critical value or the availability of a carbon source will limit photosynthesis. The limits are species, media, and process specific; for instance, a CO_2 concentration in bulk liquid of at least 0.065 mM and pH 8.5 resulted in optimal productivity of some marine and saline water diatoms in outdoor ponds (Weissman et al. 1998). This value is analogous to a minimum partial pressure of 0.2 kPa (0.076 mM) reported for the freshwater strain *Chlorella* sp. (Doucha et al. 2005). To avoid growth limitation by CO_2, it must be supplied in excess and it is generally accepted that supplementation of 1%–3% of CO_2 in air is sufficient. However, supplying CO_2 in excess is not sustainable both environmentally and economically; hence, loss of CO_2 in the exhaust gas needs to be minimized. Often, the system used to supply CO_2 to microalgae cultures is not adequately designed; more attention must be focused on the overall design of the reactor to accomplish this requirement, to avoid low CO_2 utilization efficiency.

Depending on the final application of the biomass and type of photobioreactor used, different CO_2 sources can be exploited from expensive pure CO_2 (0.1–0.3 €/kg) to cheap flue gases containing 4%–15% CO_2. CO_2 can be supplied to microalgal cultures by (1) continuous bubbling or (2) on-demand injection. With continuous bubbling of flue gases, the medium becomes acidic, and maximum CO_2 use efficiencies of 4.2% and 8.1% have been reported (Hu et al. 1998; Zhang et al. 2001). With on-demand injection of flue gases, the CO_2 use efficiency increases up to 32.8% in open shallow photobioreactors (Doucha et al. 2005) and up to 50% in closed photobioreactors (Camacho et al. 1999) or even up to 95% under optimized operation conditions (de Godos et al. 2014). Whatever the CO_2 source, two major boundary conditions must be accomplished: (1) to supply enough carbon to satisfy the demand of the culture, thus avoiding carbon limitation conditions and (2) to optimize the CO_2 supply system to avoid CO_2 losses to the atmosphere that would increase the biomass production cost. In this chapter, these major aspects are taken into account to provide effectively and economically the necessary supply of CO_2 to photobioreactors producing microalgae.

10.2 PHYSICAL CHEMISTRY OF CO₂ ABSORPTION

The amount of CO_2 that can be dissolved from a gas phase into an aqueous phase is a function of the composition of the gas and liquid phases, in addition to operational conditions (pressure and temperature). This behavior is defined by Henry's law (Equation 10.1) that determines the solubility of CO_2 in water, $[CO_2^*]$, which is proportional to the molar fraction of CO_2 in the gas phase in equilibrium with the aqueous phase y_{CO_2}, the total pressure P, with the proportionality constant being Henry's law constant for CO_2, H_{CO_2}:

$$\left[CO_2^*\right] = H_{CO_2} \cdot y_{CO_2} \cdot P \tag{10.1}$$

Henry's law is only valid for ideal gases (pure gases at moderate pressure and temperature), but in the case of CO_2, only small deviations are observed including when CO_2 is in a mixture with other gases. Henry's law constant for CO_2, H_{CO_2}, is usually considered as constant, a value of $3.4 \cdot 10^{-2}$ mol/L·atm being reported for pure water at standard conditions (25°C, or 298 K, and 1 atm: 0.1 MPa). Consequently, under standard conditions, the solubility of CO_2 in water varies from $1.36 \cdot 10^{-5}$ mol/L when in equilibrium with air ($y_{CO_2} = 0.0004$) to $3.40 \cdot 10^{-2}$ mol/L when in equilibrium with pure CO_2 ($y_{CO_2} = 1.00$). However, Henry's law constant varies as a function of temperature and to a lesser extent with the ionic strength of the aqueous phase. Although a rigorous analysis of these variations can be performed from the thermodynamic point of view, several empirical equations have been proposed that have been demonstrated to be valid under moderate temperature and pressure conditions. Thus, a simple solution to determine Henry's law constant for CO_2 at different temperatures other than 298 K ($H_{CO_2,T}$) can be obtained from the van't Hoff equation (Equation 10.2), where T is the temperature in Kelvin. According to this equation, the solubility of CO_2 in equilibrium with air varies from $1.97 \cdot 10^{-5}$ to $1.13 \cdot 10^{-5}$ mol/L for temperatures ranging from 10°C to 30°C that are usually found in microalgal cultures. Alternatively, more precise equations have been reported from experimental data (Equation 10.3), valid for pressures up to 1 MPa and temperature from 273 to 433 K (Carroll et al. 1991). In this equation, the last factor of 5.556 has been included to obtain the value of $H_{CO_2,T}$ as mol/L·atm:

$$H_{CO_2,T} = H_{CO_2,298} \exp\left[2400\left(\frac{1}{T} - \frac{1}{298}\right)\right] \tag{10.2}$$

$$Ln\left(H_{CO_2,T}\right) = \left(-6.8346 + \frac{1.2817 \cdot 10^4}{T} - \frac{3.7668 \cdot 10^6}{T^2} + \frac{2.997 \cdot 10^8}{T^3}\right) \cdot 5.556 \tag{10.3}$$

Regarding the ionic strength I, this is a measurement of the concentration of ions present in a solution and is calculated as the sum of ion concentrations c_i multiplied by the square of the charge of each one z_i, according to Equation 10.4. The relationship between Henry's law constant for water $H_{CO_2,w}$ and a solution of defined ionic strength $H_{CO_2,I}$ was previously defined by Danckwerts to be logarithmic with an empirical parameter b being defined (Equation 10.5). The value of the empirical parameter b varies with temperature but in the range of 0°C–50°C, a mean value of 0.10 can be used. According to this phenomenon, the solubility of CO_2 in seawater containing 35 g/kg of NaCl ($I = 0.72$ mol/kg) is lowered to $1.13 \cdot 10^{-5}$ mol/L, comparing with the value of $1.36 \cdot 10^{-5}$ mol/L for pure water at 25°C:

$$I = \frac{1}{2} \sum_{i=1}^{n} c_i \cdot z_i^2 \tag{10.4}$$

$$\log\left(\frac{H_{CO_2,W}}{H_{CO_2,I}}\right) = b \cdot I \tag{10.5}$$

Once the CO_2 is dissolved into the aqueous phase as carbonic acid H_2CO_3, it can react with other compounds thus modifying the real dissolved CO_2 concentration. In the case of microalgal cultures, the most typical case is the reaction of CO_2 with hydroxyl ions (OH^-) present in the water to produce bicarbonate (HCO_3^-) that can react with more hydroxyl ions to produce carbonate (CO_3^{2-}). These are reversible reactions whose reaction rates r are determined by the concentration of each compound and the reaction rate constants k_{11}, k_{12}, k_{21}, and k_{22}:

$$CO_2 + OH^- \underset{k_{12}}{\overset{k_{11}}{\longleftrightarrow}} HCO_3^-$$

$$HCO_3^- + OH^- \underset{k_{22}}{\overset{k_{21}}{\longleftrightarrow}} CO_3^{2-} + H_2O$$

$$r_{11} = k_{11} \cdot [CO_2] \cdot [OH^-]; \quad r_{12} = k_{12} \cdot [HCO_3^-] \tag{10.6}$$

$$r_{21} = k_{21} \cdot [HCO_3^-] \cdot [OH^-]; \quad r_{22} = k_{22} \cdot [CO_3^{2-}] \tag{10.7}$$

Due to the importance of CO_2 absorption for several industrial and environmental uses, the values of these reaction rate constants have been widely studied for a variety of conditions; these being mainly determined by the composition and temperature of the liquid phase (Cents et al. 2005). The reaction rate is a function of the concentration of each one of the carbonated species and the hydroxyl concentration, so it is a function of pH. With regard to the first reaction, the reaction rate constant for the forward reaction (k_{11}) is high and increases with the value of pH. Meanwhile, the reaction rate constant for the reverse reaction (k_{12}) is low and shows little variation with the value of pH (González-López et al. 2012). Regarding the second reaction, the reaction rate constant for the forward reaction (k_{21}) is an order of magnitude higher than that corresponding to the reverse reaction (k_{22}), and it increases with pH, the value of k_{22} not being influenced by pH variation (González-López et al. 2012).

Under the normal conditions used for the production of microalgae, the reaction rate values are much higher than the CO_2 consumption and transfer rate. Thus, the reactions are in equilibrium (i.e., forward and backward reaction rates are equal), with the concentration of each species being determined by the equilibrium constant K_1 and K_2 (Equations 10.8 and 10.9). The values of these equilibrium constants at standard conditions are widely reported:

$$r_{11} = r_{12} \rightarrow k_{11} \cdot \left[CO_2\right] \cdot \left[OH^-\right] = k_{12} \cdot \left[HCO_3^-\right]$$

$$K_1 = \frac{k_{11}}{k_{12}} = \frac{\left[HCO_3^-\right]}{\left[CO_2\right]\left[OH^-\right]} = 3.2 \cdot 10^7 \text{ L/mol} \tag{10.8}$$

$$r_{21} = r_{22} \rightarrow k_{21} \cdot \left[HCO_3^-\right] \cdot \left[OH^-\right] = k_{22} \cdot \left[CO_3^{2-}\right]$$

$$K_2 = \frac{\left[CO_3^{2-}\right]}{\left[HCO_3^-\right]\left[OH^-\right]} = 3.5 \cdot 10^3 \text{ L/mol} \tag{10.9}$$

The existence of these reactions/equilibriums implies that the real concentration of CO_2 (i.e., the nonionized form) in the liquid is not only defined by Henry's law but by the chemical reactions taking place in the bicarbonate buffer. Thus, the total inorganic carbon (TIC) concentration is the sum of the concentrations of the three carbon species (Equation 10.10), whereas the hydroxyl ion concentration is a function of the proton concentration (pH) and the equilibrium constant for water (Equation 10.11). Accordingly, the real concentration of CO_2 (nonionized form) in the aqueous phase can be calculated as a function of the TIC, and the pH of the aqueous phase (Equation 10.12), and the concentrations of bicarbonate and carbonate can be calculated via Equations 10.13 and 10.14, respectively:

$$\left[TIC\right] = \left[CO_2\right] + \left[HCO_3^-\right] + \left[CO_3^{2-}\right] \tag{10.10}$$

$$K_w = \left[H^+\right] \cdot \left[OH^-\right] = 10^{-14} \tag{10.11}$$

$$\left[CO_2\right] = \frac{\left[TIC\right]}{1 + \dfrac{K_1 \cdot K_w}{\left[H^+\right]} + \dfrac{K_1 \cdot K_2 \cdot K_w \cdot K_w}{\left[H^+\right]^2}} \tag{10.12}$$

$$\left[HCO_3^-\right] = K_1 \cdot \left[CO_2\right] \cdot \frac{K_w}{\left[H^+\right]} \tag{10.13}$$

$$\left[CO_3^{2-}\right] = K_1 \cdot K_2 \cdot \left[CO_2\right] \cdot \left(\frac{K_w}{\left[H^+\right]}\right)^2 \tag{10.14}$$

In conclusion, the pH of the aqueous phase determines the relative concentration of the carbon species, CO_2 being the main carbon compound at acidic pH, whereas at the alkaline pH usually used in microalgae cultures (pH = 7.5–8.5), the main carbon compound is bicarbonate (Figure 10.1). The concentration of each species is determined by the total inorganic carbon concentration (Equations 10.12 through 10.14), which determines if the aqueous phase is CO_2-saturated or not with respect to the gaseous phase. Figure 10.2 shows the variation of CO_2 concentration (expressed as $pCO_2 = -\log([CO_2])$) as a function of pH for water containing 40 mg/L of total inorganic carbon (the typical value for freshwater) and expected equilibrium values with air and pure CO_2. It is observed that when the pH increases, the pCO_2 increases. More importantly, with acid pH, the concentration of CO_2 in equilibrium with air is higher than the concentration of CO_2 in the aqueous phase, $pCO_2 < pCO_2^*$-air, but it becomes the opposite with alkaline pH, the system being in equilibrium at pH = 7.8. Under standard conditions, freshwater containing inorganic carbon is in equilibrium with air at pH 7.8, whereas CO_2 is desorbed to air below this pH and absorbed from air above this pH. Furthermore, if pure CO_2 is used, the concentration of CO_2 in equilibrium with this gas is much higher than the CO_2 concentration in the aqueous phase at any pH, $pCO_2 < pCO_2^*$-pure CO_2, with net absorption of CO_2 taking place at any pH.

The difference between the concentration of CO_2 in the gas phase and the actual concentration of CO_2 in the liquid phase ($[CO_2^*]$-$[CO_2]$) is defined as the driving force for mass transfer. Accordingly, the net amount of CO_2 transferred from a gas

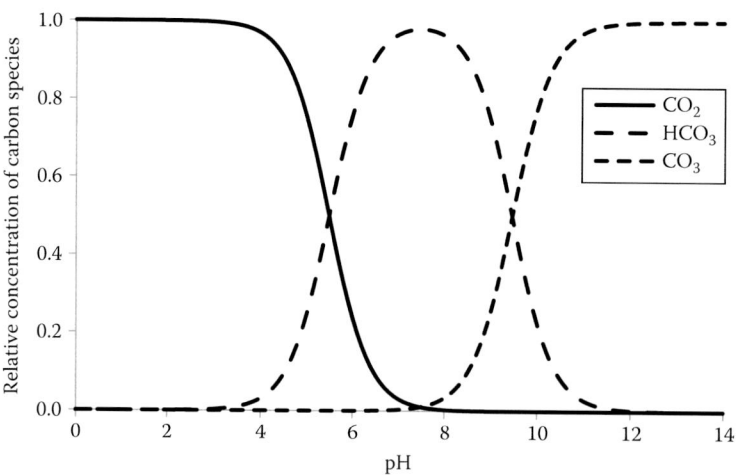

FIGURE 10.1 Influence of pH on the relative concentration of different carbon species for water under standard conditions (25°C, 1 atm).

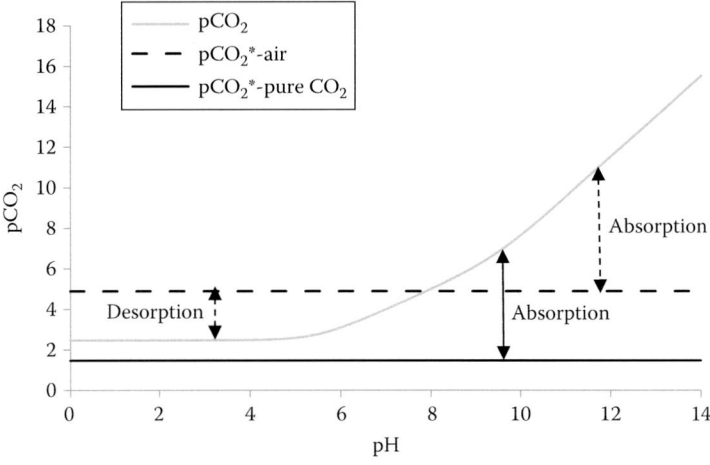

FIGURE 10.2 Variation of CO_2 concentration as a function of pH for water containing 40 mg/L of total inorganic carbon under standard conditions. Comparison with saturation CO_2 concentration in equilibrium with air (y_{CO_2} = 0.0004) (dotted line) and pure CO_2 (y_{CO_2} = 1.00) (continuous line). Values are expressed as pCO_2 = $-\log([CO_2])$.

phase to an aqueous phase N_{CO_2} is defined by the following equation, where K_1a_{1,CO_2} is the volumetric mass transfer coefficient for CO_2 into the system, which takes into account the contact area between the liquid and gas phase:

$$N_{CO_2} = K_1a_{1,CO_2} \cdot \left(\left[CO_2{}^* \right] - \left[CO_2 \right] \right) \qquad (10.15)$$

The value of K_1a_{1,CO_2} is largely dependent on the design and operational conditions of the reactor (P, T, flow rate, etc.). Thus, for channels of raceway reactors, values from 0.2 to 0.9 L/h have been reported (Weissman et al. 1988, 1998; de Godos et al. 2014), whereas for cultures circulating in tubes, values ranging from 1.0 to 3.6 L/h have been measured (Camacho et al. 1999). In thin-layer reactors, volumetric mass transfer coefficients from 5.0 to 7.9 L/h have been reported (Lívanský and Doucha 1996). In aerated systems such as airlift pumps, the value of K_1a_{1,CO_2} ranged from 70 to 110 L/h (Camacho et al. 1999). In bubble columns, the volumetric mass transfer coefficient has been reported to vary from 4 to 27 L/h (Hulatt and Thomas 2011; Fernández et al. 2012), although values up to 350 L/h have been reported (Carvalho et al. 2006). Although the value of K_1a_{1,CO_2} can be estimated from theoretical equations considering the diffusion of compounds through different layers and the variation of interfacial area as a function of the properties of the phases, it is preferable to determine the value of K_1a_{1,CO_2} empirically by various different methods (e.g., dynamic, mass balances) (Chisti and Jauregui-Haza 2002). The potential correlations of K_1a_{1,CO_2} with variables relating to operational conditions, such as the superficial gas velocity U_g or power supply per unit volume P/V, are usually established within their working range using empirical coefficients (a, b, c, d) in order to take into account the variation of K_1a_{1,CO_2} with these variables (Equation 10.16). It is usual to determine the mass transfer coefficient for oxygen and to calculate the corresponding value for CO_2 by multiplying to the square root of the ratio of diffusivities, which is equal to 0.91:

$$K_l a_{l,CO_2} = a \cdot Ug^b = c \cdot \left(\frac{P}{V}\right)^d \tag{10.16}$$

$$K_l a_{l,CO_2} = K_l a_{l,O_2} \cdot \sqrt{\frac{D_{CO_2}}{D_{O_2}}} = 0.91 \cdot K_l a_{l,O_2} \tag{10.17}$$

Finally, it is important to take into account that in microalgal photobioreactors, the injection transfer of CO_2 is not usually performed in the entire volume of the culture, whereas biological demand RCO_2 is a function of total culture volume. Thus, volumes for CO_2 supply $V_{CO_2,supply}$ and for CO_2 demand $V_{CO_2,demand}$ can be different (Equation 10.18), usually the volume used for CO_2 supply being lower than total culture volume. The mass transfer capacity of volume used for CO_2 supply must be enough to supply the amount of CO_2 demanded by the culture in the total culture volume (Equation 10.19 derived from Equations 10.15 and 10.18). Only in bubble columns, flat panels, and similar systems is the entire volume used for both CO_2 supply and demand:

$$N_{CO_2} \cdot V_{CO_2,supply} = RCO_2 \cdot V_{CO_2,demand} \tag{10.18}$$

$$K_l a_{l,CO_2} \cdot \left(\left[CO_2^*\right] - \left[CO_2\right]\right) = RCO_2 \cdot \frac{V_{CO_2,demand}}{V_{CO_2,supply}} \tag{10.19}$$

Thus far, the factors determining the amount of CO_2 that can be transferred from a gas phase to an aqueous phase have been summarized. The key physical parameters determining CO_2 transfer have been identified, and in addition, actual values of these main parameters have been determined under normal microalgal culture conditions. However, to optimize the supply of CO_2, it is necessary to know also the biological demand for CO_2 and the influence of culture parameters on this demand.

10.3 BIOLOGICAL DEMAND FOR CO_2 BY MICROALGAE

CO_2 is consumed by microalgae to perform photosynthesis. Although it has been reported that microalgae are also able to use bicarbonate as a carbon source, the main carbon source for producing microalgae is CO_2, because CO_2 can easily cross the cell membrane, with the driving force being the concentration gradient near the cell surface (Kurano et al. 1998). The uptake of CO_2 by microalgae is proportional to the photosynthetic rate, which for a given microalgae strain is a function of biomass concentration, light availability, and culture conditions (pH, temperature, dissolved oxygen, nutrient availability, etc.). Moreover, photosynthetic rates have characteristic response times of seconds to minutes, and any culture parameter that is variable on these timescales changes the photosynthesis rate and, consequently, the biological demand of CO$_2$ by the culture. In outdoor conditions, the rate of photosynthesis is influenced by diurnal and other weather-related variability in sunlight and temperature, and also nutrient concentration. Although average values can be calculated, the biological demand for CO$_2$ is not constant and changes during the daylight period and this must be taken into account.

Photosynthesis has been extensively studied and rigorous biological and chemical models have been established. However, from a practical perspective, simple models applying enzymatic kinetic concepts can be used (Eilers and Peeters 1988; Camacho et al. 2003). The influence of culture conditions on the photosynthetic rate of diverse microalgal strains has been extensively reported as photosynthesis versus irradiance, that is, P–I curves. According to a recent detailed study (Costache et al. 2013), the rate of photosynthesis is mainly determined by the irradiance to which the cells are exposed in a hyperbolic relationship. This function can be modelled by normalized factors, taking into account the deviation of culture conditions from optimal values (temperature, pH, dissolved oxygen, etc.) (Figure 10.3) (Equation 10.20) (Costache et al. 2013). In this function, there are two terms for temperature and pH in order to take into account both the adverse effects of temperature and pH extremes as well as more optimal values. If the temperature and pH of the culture are controlled in a narrow range, the factors corresponding to these parameters, $RO_{2,T}$ and $RO_{2,pH}$, are constant, whereas if the dissolved oxygen concentration remains below 250% of air saturation, the factor corresponding to dissolved oxygen is equal to one. Therefore, in these conditions, the photosynthesis rate is only determined by the light availability or average irradiance inside the culture (Equation 10.21) (Figure 10.3). The CO$_2$ demand of the culture can be determined from the photosynthesis rate via multiplication by the coefficient yield Y_{CO_2/O_2}, which can be estimated either from the fundamental photosynthesis reaction or based on the biomass composition or determined experimentally. A Y_{CO_2/O_2} value of 1.004 molCO$_2$/molO$_2$ (1.38 mgCO$_2$/mgO$_2$) has been reported (Equation 10.22) (Rebolloso et al. 1999), and it can be used as a generally appropriate value:

$$RO_2 = \left(\frac{RO_{2,max} \cdot I^n}{K_I^n + I^n} \right) \left(A_1 \exp\left(\frac{-Ea_1}{RT} \right) - A_2 \exp\left(\frac{-Ea_2}{RT} \right) \right)$$

$$\left(B_1 \exp\left(\frac{-C_1}{pH} \right) - B_2 \exp\left(\frac{-C_2}{pH} \right) \right) \left(1 - \left(\frac{DO_2}{K_{O2}} \right)^m \right) \tag{10.20}$$

$$RO_2 = \left(\frac{RO_{2,max} \cdot I^n}{K_I^n + I^n} \right) \cdot RO_{2,T} \cdot RO_{2,pH} \tag{10.21}$$

$$RCO_2 = Y_{CO_2/O_2} \cdot RO_2 \tag{10.22}$$

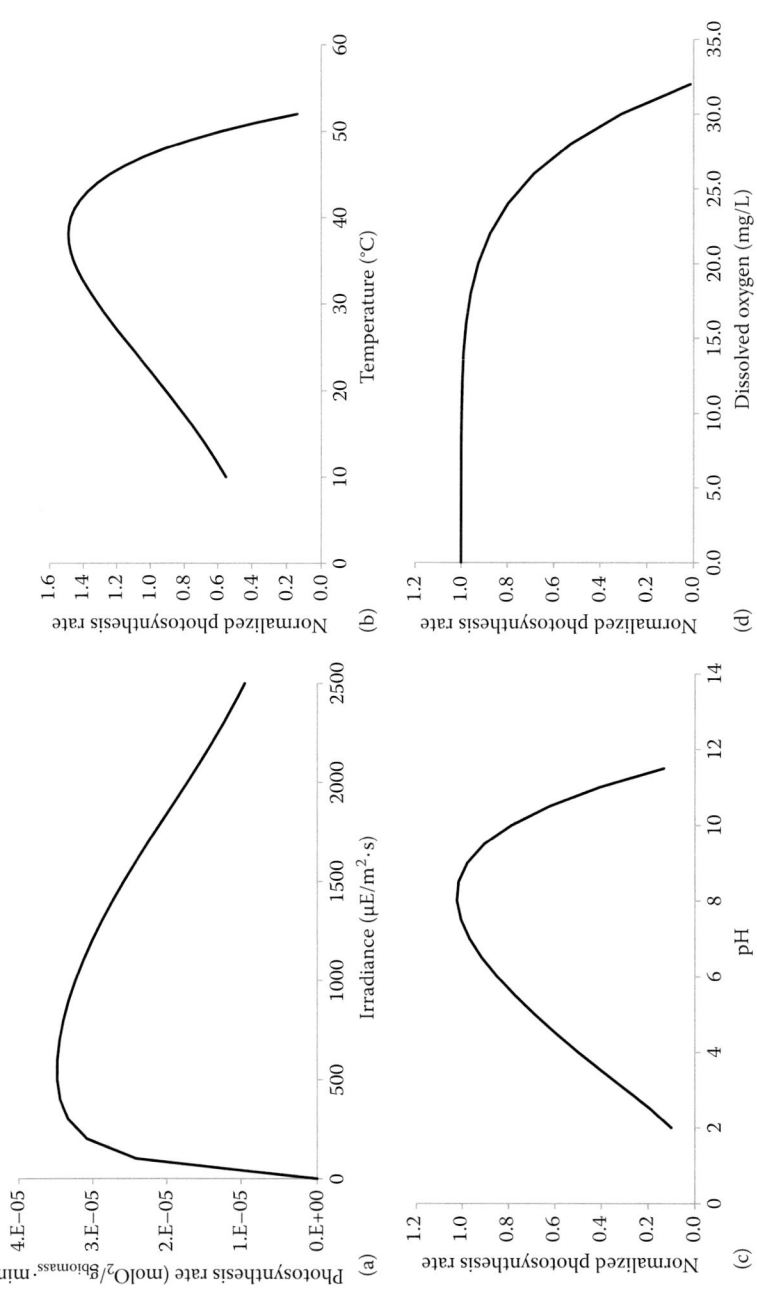

FIGURE 10.3 Variation of photosynthesis rate as a function of major parameters at which the cells are exposed for *Scenedesmus almeriensis* at 20°C and pH 8.0, under nutrient saturation conditions. (a) Irradiance, (b) temperature, (c) pH, and (d) dissolved oxygen. (Values obtained from Costache, T. A. et al., *Appl. Microbiol. Biotechnol.*, 97, 7627, 2013.)

It should be noted that at high cell irradiance, photoinhibition occurs leading to a reduction of photosynthetic rate, for instance, at >500 $\mu E/m^2 \cdot s$, as shown in Figure 10.3a. However, in the model, photoinhibition is not considered, even though under outdoor conditions, the solar irradiance can be higher than 1000 $\mu E/m^2 \cdot s$. This is because the average irradiance at which the cultures are exposed to is much lower, ranging from 20 to 250 $\mu E/m^2 \cdot s$ due to self-shading of the cells and light attenuation (Molina et al. 1996). This is also evident in the data for an outdoor raceway reactor shown in Figure 10.4, where solar irradiance and cell irradiance are compared.

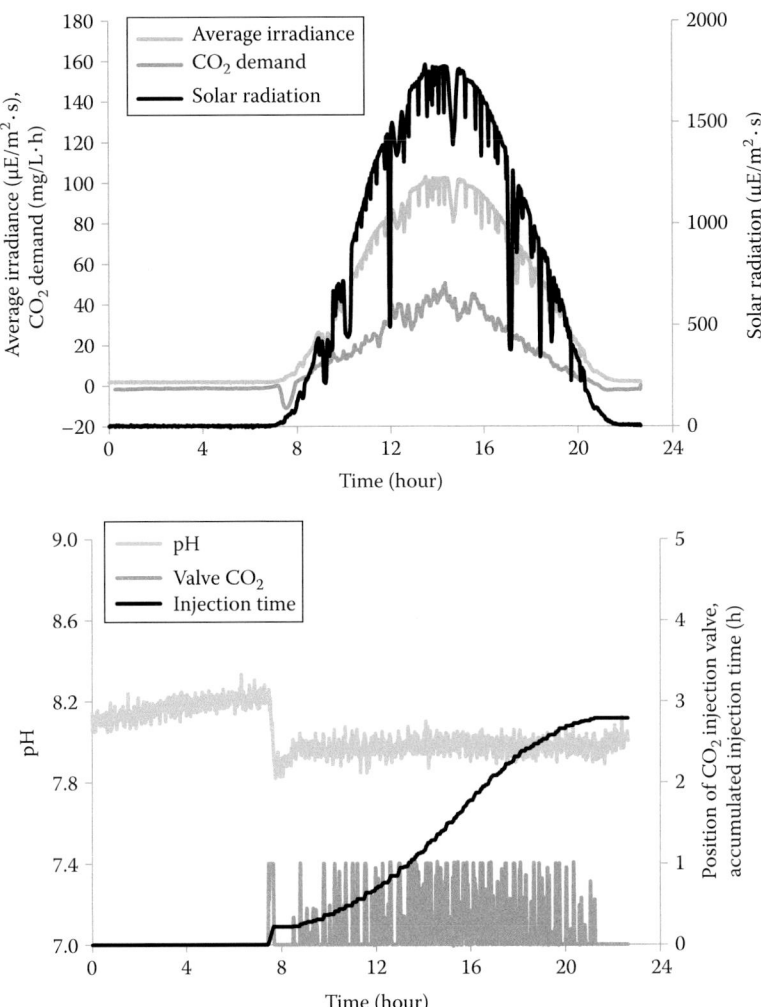

FIGURE 10.4 Daily variation of solar radiation and average irradiance inside the culture, in addition to pH and CO_2 injection, in a raceway reactor of 7.0 m² and 0.20 m depth located in Almería, Spain, with *Scenedesmus* sp. at a biomass concentration of 0.8 g/L. The temperature of the culture was maintained at 24°C ± 3°C and dissolved oxygen was below 250% sat.

The productivity model (RO_2 as a function of irradiance, pH, temperature, and dissolved oxygen) has been demonstrated to fit empirical data obtained from a 0.2 m deep outdoor raceway producing *Scenedesmus* sp. at a dilution rate of 0.2 L/day, using measured parametric values (RO_{2max}, n, K_1, A_1, A_2, Ea_1, Ea_2, B_1, B_2, C_1, C_2, K_{O2}) determined under laboratory conditions (Figure 10.4). This figure shows the daily variation of solar irradiance and average irradiance on the culture; the CO_2 demand and pH of the culture, in addition to the switch on of the CO_2 valve; and accumulated CO_2 injection time. The average irradiance that was experienced by the algal cells changed during daylight period but never reached values above 100 $\mu E/m^2 \cdot s$, even when the solar radiation reached values up to 1850 $\mu E/m^2 \cdot s$. The CO_2 demand rose with the average irradiance from zero during the night period up to 50 mg/L·h at noon, corresponding to a biomass productivity of 5.6 $g/m^2 \cdot h$. As uptake of CO_2 by photosynthesis increased the pH of the culture, CO_2 was injected to reduce the pH to maintain the supply of inorganic carbon to the algal cells. Figure 10.4 shows the pH to have been well controlled, such that it only rose higher than the set point of 8.0 during the night, when the injection of CO_2 was stopped to avoid losses. The CO_2 valve was switched on and off during the daylight period to maintain the pH in the range of 8.0 ± 0.1. The opening frequency for the CO_2 valve during the daylight period was higher at noon, when the CO_2 demand was also highest; hence, the slope of accumulated injection time was maximal at this time. The CO_2 supply was higher than the CO_2 demand, despite the on-demand injection system (Figure 10.5), which supplied pure CO_2 at a fixed flow rate of 3 L/min when the pH was higher than the set point. However, only a fraction of this CO_2 was efficiently transferred to the aqueous phase, a large fraction being lost to the atmosphere. From 1100 g of CO_2 injected, only 414 g of CO_2 was consumed to produce biomass, that is, 62% of the injected CO_2 was lost. The only way to optimize the CO_2 utilization efficiency is (1) by increasing mass transfer capacity into the system to avoid injected CO_2 being transferred out of the culture and (2) by optimizing the control system to better match the injection of CO_2 to the actual demand of the culture. Both tasks are specific to the photobioreactor design and must be performed according to the type of photobioreactor and operational conditions.

10.4   SUPPLY OF CO_2 TO CLOSED PHOTOBIOREACTORS

Closed photobioreactors are mainly used to produce valuable products such as nutraceuticals or other valuable biomass. Consequently, pure CO_2 is mainly used in these reactors as the carbon source although in some cases flue gases can also be used. In these reactors, there is no direct contact between the culture and the atmosphere; thus, the supplied liquid and gases are in contact with each other in the system for long periods. Typical closed photobioreactors are tubular or flat panels, although the utilization of tubular photobioreactors is more extensive, with large variations in their dimensions, designs, and materials. Regardless of type, the capacity for CO_2 supply must satisfy the CO_2 demand by the culture (Equation 10.18).

To optimize the supply of CO_2 in tubular photobioreactors, the first task is to define the volume of the culture to be used for CO_2 transfer, with two major sources of variation being the bubble column and the solar loop (Figure 10.6). The gas

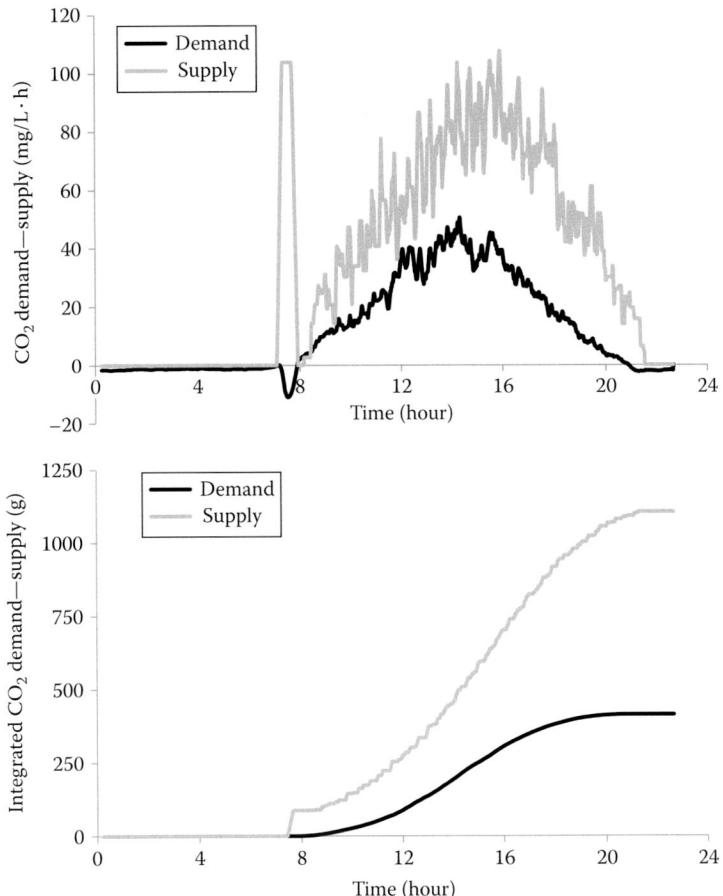

FIGURE 10.5 Daily variation of CO_2 supply and demand, showing in addition the integrated values for a raceway reactor of 7.0 m² and 0.20 m depth located in Almería, Spain. This was producing *Scenedesmus* sp. at a dilution rate of 0.2 L/day and a biomass concentration of 0.8 g/L. The temperature of the culture was maintained at 24°C ± 3°C and dissolved oxygen was below 250% sat.

volume of the bubble column in tubular photobioreactors is only a small fraction of their entire volume (10%–15%); thus, to supply the net amount of CO_2 required by the culture into the entire reactor, the volumetric mass transfer rate N_{CO_2} into the bubble column must be much higher than the volumetric mass transfer rate into the solar loop. Furthermore, because air is sparged into the bubble column to remove oxygen produced by photosynthesis, the injected CO_2 will be diluted with this air, which reduces the concentration of CO_2 and thus the driving force for the absorption of CO_2. Moreover, reduction of the driving force requires a higher value for the volumetric mass transfer coefficient $K_l a_{l,CO_2}$, to satisfy the demand of the culture, which imposes significant design and scale constraints. An alternative is to supply CO_2 in the solar loop (Figure 10.6). The volume of the solar loop is close

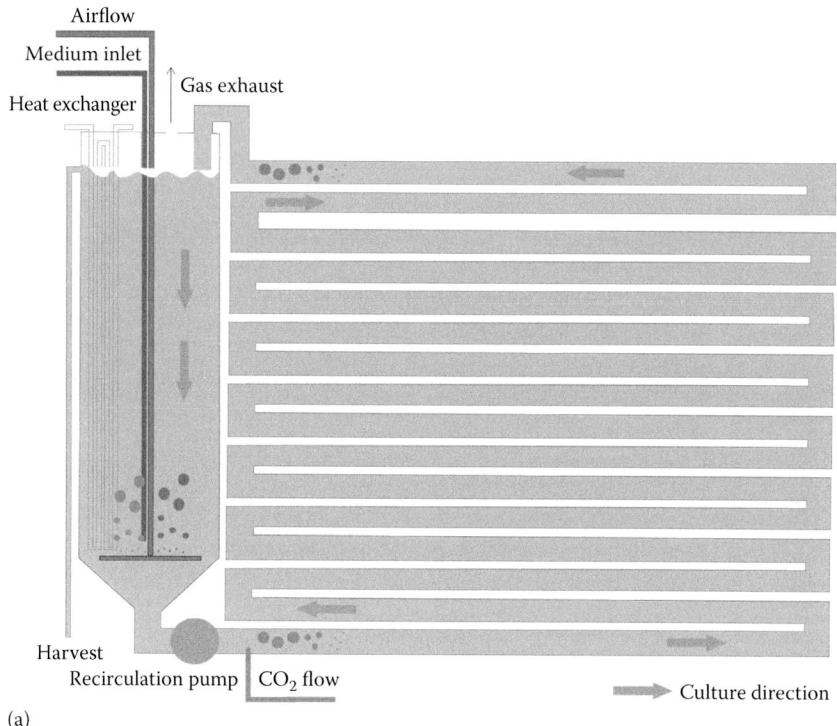

(a)

(b)

FIGURE 10.6 (a) Schematic diagram of a tubular photobioreactor and (b) actual view of tubular photobioreactors at Las Palmerillas Research Center from Fundación Cajamar (Almería, Spain).

to that of the entire volume of the reactor (85%–90%), and in this section, air is not usually supplied. Consequently, the CO$_2$ molar fraction in the injected gas will determine the driving force in the system, with pure CO$_2$ resulting in the largest possible driving force (Figure 10.2). The remaining parameter to consider is the volumetric mass transfer coefficient (K$_l$a$_{l,CO_2}$) itself. To increase this parameter, the contact between the liquid and gas phase must be enhanced as much as possible. CO$_2$ gas injected at the beginning of the solar loop remains as large bubbles when reaching the upper part of the tubes, resulting in low values for K$_l$a$_{l,CO_2}$, of 1.0 L/h (Camacho et al. 1999), thus limiting the mass transfer capacity of the system. To enhance the contact between liquid and gas phase, the CO$_2$ gas can be injected ahead of the centrifugal pump, if used to force the culture through the reactor (Fernández et al. 2012), or membranes or diffusers can be used to decrease the bubble size, thus increasing the contact area between the liquid and gas phases (Cheng et al. 2006). Alternatively, static mixers can be incorporated into the solar loop to enhance the contact between liquid and gas phase, but the pressure drop of these systems is excessive (Zhang et al. 2013; Zirrahi et al. 2013).

Once the injection point is defined, the gas flow to be supplied must be calculated. The gas flow rate must satisfy the demand of the culture at all times, requiring determination of the maximal expected demand of CO$_2$ by the culture. This can be done using models of photosynthetic rate as a function of light availability (Costache et al. 2013). Maximal CO$_2$ demand must be determined at noon under optimal conditions. Figure 10.7 shows the daily variation of CO$_2$ demand as a function of expected biomass productivity. It was observed that the maximal CO$_2$ demand

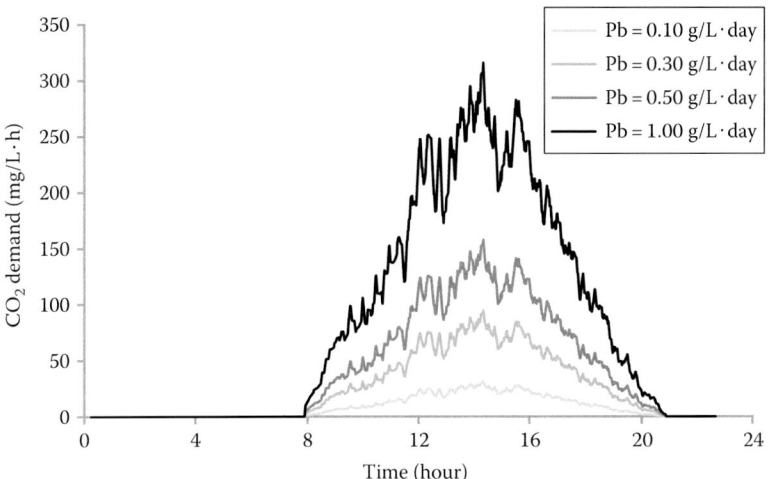

FIGURE 10.7 Estimated daily variation of CO$_2$ demand according to the expected biomass productivity (Pb) of a given system is shown. These data were obtained using the model of Costache et al. (2013) from real experimental data of solar radiation availability in Almería, Spain.

at noon varied from 30 mg/L·h for a biomass productivity of 0.1 $g_{biomass}$/L·day to 300 mg/L·h at a biomass productivity of 1.0 $g_{biomass}$/L·day. For a mean biomass productivity of 0.5 $g_{biomass}$/L·day, the maximal CO_2 demand was 150 mg/L·h; thus, the minimum CO_2 flow rate was $1.1 \cdot 10^{-3}$ v/v/min. This flow rate takes into account the net demand of CO_2, and higher gas flow rates are required if diluted CO_2 or flue gases are used. In any case, choice of the minimum CO_2 flow rate is only valid if no losses of CO_2 take place; otherwise, the actual CO_2 flow rate that is used must be higher. Once the CO_2 gas flow is defined, the efficiency of its use is determined by the contact between the gas and liquid phase. Gas is usually injected at the beginning of the loop, with CO_2 being transferred to the aqueous phase along the tube, with concomitant reduction of pH and removal of evolved O_2 to the gas phase. However, this dilutes the CO_2 in the gas phase and reduces the driving force for CO_2 transfer. Several models of these phenomena based on differential equations have been produced, determining the existence of gradients into the aqueous phase along the tube and quantifying the CO_2 losses as a function of operational conditions (Camacho et al. 1999; Fernández et al. 2012). It was concluded that a compromise between the net transfer of CO_2 and CO_2 loss exists. Figure 10.8 shows a scheme of a tube into which CO_2 is injected and how the composition of gas and liquid phases is modified. Because composition changes along the tube, a logarithmic driving force must be used to determine the net amount of CO_2 transferred to the culture in every section (Equation 10.23, where 'ml' refers to mean logarithmic). The CO_2 lost is a function of the ratio between the CO_2 content of inlet and outlet gas (Equation 10.24). These equations show that to minimize CO_2 losses the $y_{CO_2,out}$ must be as low as possible. However, this implies that the gas phase achieves equilibrium with the liquid phase, the driving force at the end of the tube being minimal: $([CO_2^*]-[CO_2])_{out} \approx 0$. Under these conditions, the denominator of the fraction increases and the logarithmic driving force increases (Equation 10.23). It is not possible to achieve sufficient mass transfer with a finite value for the mass transfer

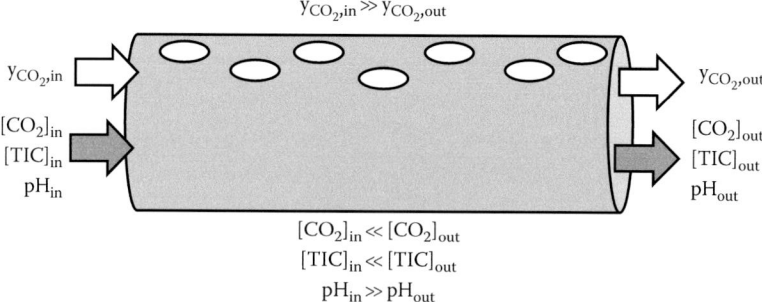

FIGURE 10.8 Schematic diagram of a section of a tubular photobioreactor. The variation in the CO_2 molar fraction of the gas phase, in addition to the CO_2 concentration, total inorganic carbon, and pH in the liquid phase, when CO_2 is injected is shown. Arrows indicate the flow of gas and liquid phases.

coefficient, and thus, an optimal compromise between the mass transfer capacity and carbon losses must be found:

$$N_{CO_2} = K_l a_{l,CO_2} \cdot \left(\left[CO_2^* \right] - \left[CO_2 \right] \right)_{ml}$$

$$= K_l a_{l,CO_2} \cdot \frac{\left(\left[CO_2^* \right] - \left[CO_2^* \right] \right)_{in} - \left(\left[CO_2^* \right] - \left[CO_2^* \right] \right)_{out}}{\ln \dfrac{\left(\left[CO_2^* \right] - \left[CO_2 \right] \right)_{in}}{\left(\left[CO_2^* \right] - \left[CO_2 \right] \right)_{out}}} \qquad (10.23)$$

$$L_{CO_2} = \frac{y_{CO_2,out}}{y_{CO_2,in}} \qquad (10.24)$$

In tubular photobioreactors, the carbon losses have been reported to vary from 10% at noon to 100% at night, with a mean value of 20% (Camacho et al. 1999; García et al. 2003). The utilization of on–off control systems is the most frequent strategy for pH control in microalgae cultures. In these systems, just by switching off the injection of CO_2 during the night reduces the carbon losses by 60% (Camacho et al. 1999). In addition, the reduction of the CO_2 flow in the first hours of the daylight period also reduces carbon losses by an additional 17% (Mazzuca et al. 2000), although this requires special flow meters or valves that allow a variable flow rate to be implemented. However, the dynamics of the system (delay times, mixing times, etc.) mean that the carbon uptake efficiency cannot be increased over this value without modification of the pH control strategy. To ensure sufficient carbon in a long tubular bioreactor while minimizing carbon losses would require multiple gas injection points along the length of the tube, a severe drawback due to the high cost involved. Also, the distance between the injection points would depend on several factors, including the flow velocity of the liquid, rate of photosynthesis, gas–liquid mass transfer coefficient, and rate of CO_2 injection. Alternatively, advanced control strategies capable of adjusting the injection of CO_2 to match the mass transfer capacity and ultimately the demand of the culture can be used (García et al. 2003).

Thus, once the mass transfer capacity and gas flow have been optimized to satisfy the CO_2 demand of the culture, the only option for further improvement is the utilization of advanced control strategies such as model-based predictive control (MPC). This methodology makes use of an algorithm to predict the process output at future time points (horizons), calculates the control actions required to optimize a defined objective function, and implements the strategy so that at each sampling instant the horizon is shifted toward the future, which involves application of the first control signal of the sequence that is calculated at each step (Camacho and Bordons 2007). MPC controllers have several advantages: (1) they can be used to control different processes from physical to biological, (2) they have compensation for dead times, and (3) they introduce feed-forward control to compensate for measurable disturbances. The main task to apply MPC strategies is to develop a model of the system, which can be obtained (1) empirically from pulse-response experiments requiring experimental studies for each reactor design and range of culture conditions (García et al. 2003), or (2) using first-principle concepts to obtain a model that is applicable to different reactors and culture conditions (Fernández et al. 2012). In the first case, utilization of empirical models reduced CO_2 losses in tubular photobioreactors from 19.8% to 15.6% and even down to 5.5% if the feed-forward effect of irradiance is included (García et al. 2003). More importantly, the biomass productivity increased by 16%, as a consequence of better pH control and minimization of pH gradients along the reactor (García et al. 2003). On the other hand, models based on the first principles do not necessarily enhance the quality of control or performance of a system when compared with empirical models, but these models can be applied in a wider range of scenarios. Moreover, models based on the first principles are also useful to study the influence of operational parameters or environmental conditions in the performance of the system, additionally allowing identification of local gradients inside the reactor and determining culture parameters as virtual sensors (Fernández et al. 2012). Figure 10.9 shows the daily variation of major culture parameters in a pilot-scale (3 m^3) tubular photobioreactor and values obtained using a model based on the proposed first principles (Fernández et al. 2012). The model fits experimental data during the day and allows optimization of pH control by using an MPC strategy. The pH is controlled within a narrow set point range (7.8 ± 0.1), but, more importantly, the CO_2 losses are lower than 15% throughout the day, despite using an on–off valve that is only capable of supplying CO_2 at a fixed flow rate.

As an alternative to tubular photobioreactors, flat panels have been proposed as a promising technology for the production of high-quality microalgal biomass and wastewater treatment (Quinn et al. 2012; Ruiz et al. 2013). The major advantage of these reactors is the potential to increase the surface exposed to solar radiation, and their behavior is similar to bubble columns; thus, mixing and mass transfer is high (Sierra et al. 2008). Regarding CO_2 supply in these reactors, the same concepts as previously discussed for tubular photobioreactors can be applied (CO_2 demand, CO_2 mass flow supplied, mass transfer capacity, etc.). In flat panels, the entire volume is used for mass transfer because different sections do not usually exist in these reactors. In these reactors, mixing is performed by aeration at flow rates from 0.1 to 0.3 v/v/min. Thus, if CO_2 is injected into the airflow entering the reactor, it is diluted by the air and the driving force for CO_2 absorption is reduced. To solve this problem, it is recommended to use separate pipes to supply CO_2 and air or, better still, to switch off the air supply when injecting CO_2 using three-way valves. This is feasible if flue gases are used because with pure CO_2 the necessary gas flow is not sufficient to maintain the mixing in the system. It is important to note that whatever the design of the photobioreactor the CO_2 injection flow rate is determined by the CO_2 demand of the culture, which is a function of the culturing conditions in the reactor. As previously noted, for a mean biomass productivity of 0.5 $g_{biomass}$/L · day, the maximal CO_2 demand is 150 mg/L · h; thus, the required minimum CO_2 flow rate is $1.1 \cdot 10^{-3}$ v/v/min, much lower than the 0.1 v/v/min aeration rate used for mixing. If flue gas containing 10% CO_2 is used, the required gas flow rate is 0.01 v/v/min,

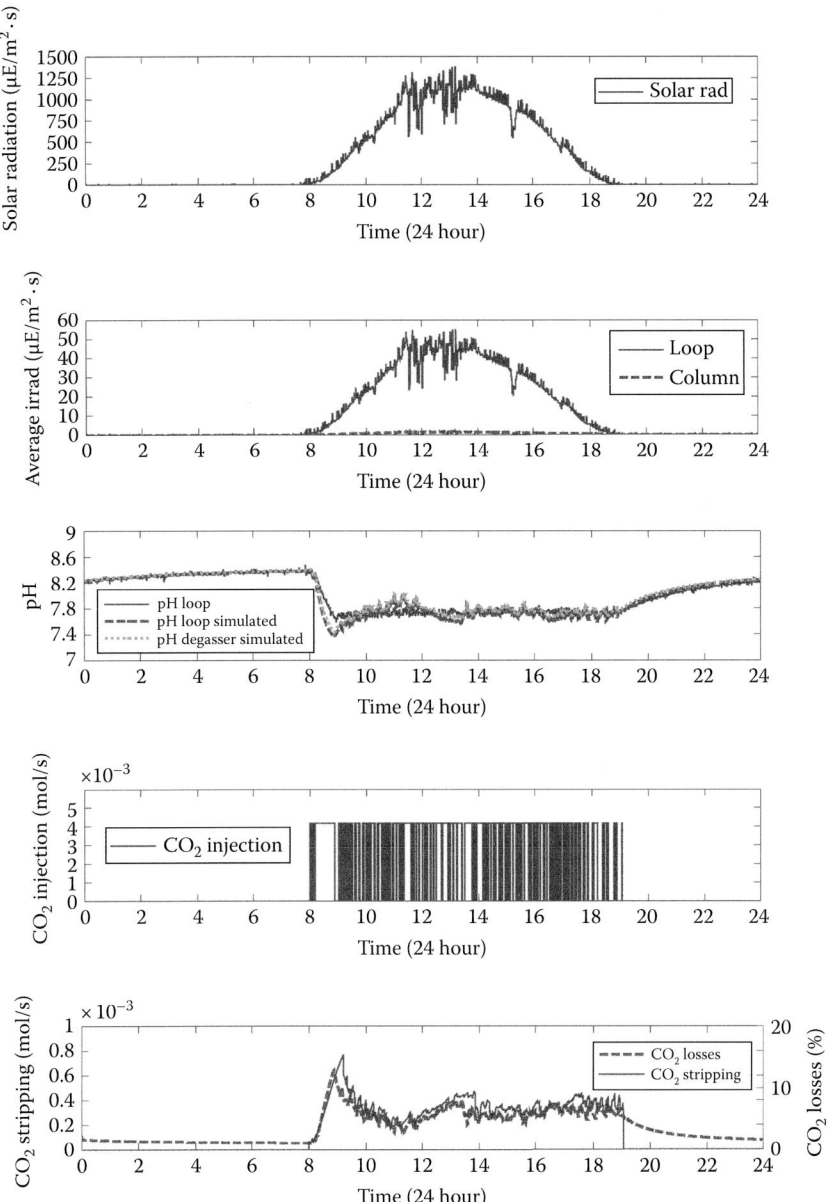

FIGURE 10.9 A comparison of experimental and simulated values of the major culture parameters in pilot scale (3 m³) tubular photobioreactors during the day. Simulated values were obtained using a model based on the first principles proposed by Fernández et al. (2012).

which approaches the aeration rate used for mixing. Regarding mass transfer capacity, in these reactors the volumetric mass transfer coefficient increases with the aeration rate, maximal values of 25 L/h being reported at 0.3 v/v/min, whereas this value reduces to 7.2 L/h at 0.1 v/v/min (Sierra et al. 2008).

10.5   SUPPLY OF CO_2 TO OPEN REACTORS

Open reactors include open unmixed ponds, raceway reactors, and thin-layer reactors among other designs. Although in open reactors the culture is in direct contact with the atmosphere and CO_2 can be absorbed from air, the cultures become carbon limited because only a maximum of 5% of the carbon required by the culture can be transferred from the atmosphere (Stepan et al. 2002). Thus, additional CO_2 must be supplied. Unmixed open ponds have so little control of culture conditions that the supply of CO_2 is never considered, this being only practical when using raceway or thin-layer reactors. In these systems, the same considerations about the supply and demand of CO_2 for microalgal cultures apply as for tubular photobioreactors, although with additional considerations due to the large area of culture that is in contact with the atmosphere and to which the CO_2 can be lost, along with oxygen.

Raceway reactors are the most widely used technology, mainly to produce biomass for nutraceuticals or as food and feed, with pure CO_2 being mainly used as the carbon source, although flue gases can be also utilized. Raceway reactors can be also used for wastewater treatment, where the supply of flue gases enhances the process efficiency by reducing the carbon limitations and by stabilizing the pH (Park and Craggs 2010). Being cheap and with minimal power consumption, raceway reactors have been proposed as the most promising technology to produce biofuels from microalgae, especially in combination with wastewater treatment (Benemann 1997; Jorquera et al. 2010; Benemann 2013; Singh and Ahluwalia 2013). The use of CO_2 from flue gas for growing algae to produce biofuels would generate energy more sustainably than processes that capture and sequester CO_2 from flue gases using chemical methods that consume up to 4 MJ/kg_{CO_2} (González-López et al. 2012). It is also possible to capture CO_2 from flue gases using water enriched by a carbonate–bicarbonate buffer in an optimized contact unit, the resulting liquid phase being depurated in a biological photobioreactor with microalgae and the regenerated phase being used again for recarbonation. However, this technology is only useful for cyanobacteria growing in highly alkaline conditions, such as *Anabaena* or *Spirulina* (González-López et al. 2012).

The simplest method to supply CO_2 is to inject the flue gases into the raceways directly, though the low molar fraction of CO_2 in flue gases (from 4% in natural gas combined cycles to 15% in coal power stations and 30% in cement kilns) means that the driving force for absorption is lower than using pure CO_2. Furthermore, the gas flow will be high and will need to be transferred from the power plant with blowers, requiring power, and the high temperature flue gas will require cooling. Flue gases contain SO_x and NO_x that are also transferred to the aqueous phase, making the culture more acidic than expected due to CO_2 transfer alone. The control systems must take this into account to avoid, for example, carbon limitation due to lack of gas injection in pH-controlled systems. Finally, flue gases can contain dust or heavy metals that can be absorbed by the biomass thus limiting its use for animal or human consumption.

Several studies have investigated the tolerance of microalgae to flue gases. Injection of flue gases containing 15% CO_2 has been reported to reduce the pH of microalgae cultures to 6.2–7.2 (Li et al. 2013). A mutant strain of *Chlorella* was shown to tolerate the continuous bubbling of real flue gases from a coke oven, containing NO_x and SO_x (Chiu et al. 2011). Alternatively, strains tolerant to high temperature and low pH have also been studied to enhance CO_2 removal efficiency from flue gases, including the removal of metals from the gas phase (Hsueh et al. 2007). Strains of the genus *Scenedesmus* have also been widely reported to be suitable for large-scale outdoor production because of their tolerance to wide-ranging culture conditions and contamination. Strains from this genus have been reported to be capable of converting 15%–25% of atmospheric CO_2 into biodiesel for transportation fuel (Mandal and Mallick 2009; Ho et al. 2010).

Whatever the carbon source used, the water depth in raceway ponds is low, from 0.10 to 0.40 m; thus, if gas is bubbled directly from the bottom of the channel, the contact time between gas and liquid phase is limited, the mass transfer capacity is very low, and most of supplied CO_2 is lost to the atmosphere, up to 80%–90% (Weissman et al. 1988; Richmond 2004). This level of CO_2 loss increases the cost of algal production; therefore, the volume of CO_2 leaving the raceway pond needs to be minimized (Acién et al. 2012b). The alternative is to include in the design of the raceway a specifically designed carbonation system. Thus, several designs have been proposed to improve the efficiency of carbon supply into raceway reactors, from external bubble columns to surfaces to maintain the injected gas in contact with the aqueous phase and also inclusion of sumps into the design of the raceway pond that also circulate the liquid along the system (Weissman et al. 1988; Park and Craggs 2010; Putt et al. 2011; Ketheesan and Nirmalakhandan 2012). The use of sumps is the simplest way to enhance carbonation in raceways because they can be easily constructed and do not need an external energy supply. Sumps can also be provided with a baffle, which allows the injection of CO_2 in countercurrent conditions to increase the rate of mass transfer (Weissman et al. 1988). Their design and operation must be capable of supplying the CO_2 demand by the culture (Equation 10.18), but at the same time, its design must minimize their cost and power consumption. A relevant analysis of the influence of sump configuration in a pilot-scale (100 m^2) raceway pond has been published (Mendoza et al. 2013a,b). According to these studies, the use of baffles to force the liquid to move up and down in an orderly way along the sump imposes an excessive pressure drop in the system, which is not justified by the low enhancement of mass transfer capacity obtained. An efficient mass transfer capacity into the sump can be achieved instead by using diffusers. It is recommended to use membranes, resulting in a low-pressure drop but producing small-sized bubbles, which enhances the mass transfer capacity into the sump (Mendoza et al. 2013b).

The utilization of flue gases as a carbon source in raceway ponds has been recently studied in a raceway reactor of 20 m^3 (100 m^2) with a sump of 0.7 m^3, in which flue gas from a diesel boiler containing 10.6% CO_2 is injected (de Godos et al. 2014). The results demonstrated that under alkaline conditions (pH = 8), most of the CO_2 is

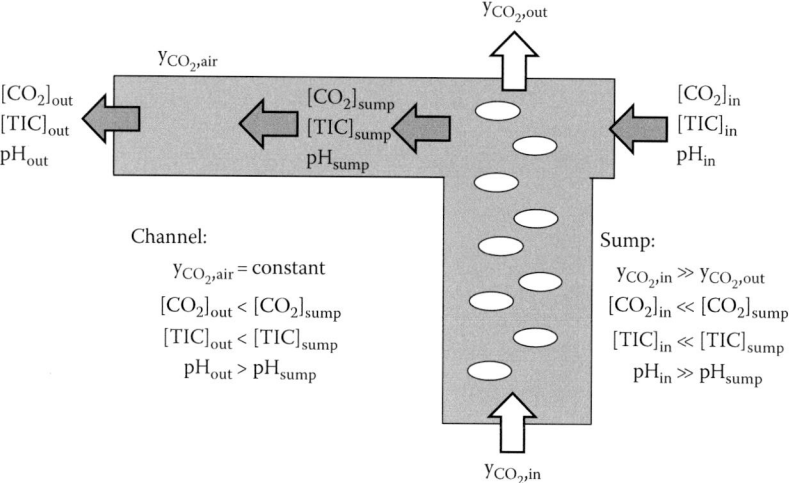

FIGURE 10.10 Schematic diagram of a section of a raceway pond channel containing a sump. The variation of CO_2 molar fraction in the gas phase and CO_2 concentration, total inorganic carbon, and pH in the liquid phase, when CO_2 is injected is indicated. Arrows indicate the flow of gas and liquid phases.

efficiently absorbed, being accumulated in the aqueous phase, whereas if the pH is acidic (pH = 6.5), the CO_2 absorption is reduced. More importantly, at pH 6.5 most of absorbed CO_2 is stripped to atmosphere into the channels. Figure 10.10 shows a scheme of CO_2 transfer processes taking place in the channel of a raceway pond utilizing a sump. Gas containing a large molar fraction of CO_2 is injected into the sump, the CO_2 being transferred to the liquid, increasing the carbon content and decreasing the pH of the liquid phase. Gas leaving the sump contains a lower content of CO_2. Liquid oversaturated with CO_2 in the sump will be stripped of CO_2 when moving into the channel, and this is released to the atmosphere. As in tubular photobioreactors, the CO_2 mass transfer into the sump is a function of driving force (Equation 10.23), whereas the loss of CO_2 is a function of the efficiency of CO_2 absorption into the sump (CO_2 in injected and exhausting gas; Equation 10.24) and the loss of CO_2 in the channels to the atmosphere, calculated as the mass transfer from the liquid to the air (Equation 10.25). To minimize the CO_2 losses, CO_2 oversaturation of the culture in the sump must be minimized, in addition to minimizing the contact between liquid and air in the channels (so the sump is located after the paddle wheels and bends hence avoiding turbulence):

$$N_{CO_2,channel} = K_l a_{l,CO_2,channel} \cdot \left(\left[CO_2^* \right]_{air} - \left[CO_2 \right] \right) \qquad (10.25)$$

Under the normal operating conditions of microalgal cultures (pH = 8), the driving force for CO_2 absorption into the sump when using flue gas containing 10% CO_2 is only 37 mg/L, and the volumetric mass transfer coefficient is 65 L/h, whereas the driving force for CO_2 desorption into the channel is 0.7 mg/L and the volumetric mass transfer coefficient is 0.9 L/h (de Godos et al. 2014). Thus, the net CO_2 absorption rate into the sump is 33.6 g/min, whereas the CO_2 stripping rate in the channel to the air is 0.2 g/min (de Godos et al. 2014). Therefore, the mass transfer capacity in the sump is much higher than in the channel. It has been reported that the use of a sump with a middle baffle with countercurrent injection of gases enhances the mass transfer capacity (Weissman et al. 1998; Craggs et al. 2012). However, the use of the baffle increases the power consumption, but does not sufficiently increase the mass transfer capacity to justify the higher power consumption (Mendoza et al. 2013b). Efficiencies of carbon absorption up to 82% have been reported when using a 3.1 m high bubble column, much higher than the value of 36% when using a sump in the same reactor (Putt et al. 2011). However, with a sump of only 1.0 m depth but optimized in its design and operational use, the carbon absorption efficiency can increase up to 95%, this value being higher than that reported for closed photobioreactors (de Godos et al. 2014).

The major parameter determining the mass transfer capacity into the sump is the volumetric ratio between the liquid and gas phases circulating through the sump: L/G (de Godos et al. 2014). Because the liquid flow rate is determined by the liquid velocity in the channels (0.2–0.3 m/s), the remaining parameter is the gas flow rate. Operating at low gas flow rate maximizes the efficiency of CO_2 absorption, but the net amount of CO_2 transferred to the culture is reduced, whereas operating at high gas flow rates the net amount of CO_2 supplied to the culture increases, but CO_2 is also lost to the atmosphere from the sump. Removal of CO_2 above 95% for a flue gas with 10.6% CO_2 was achieved when operating at an L/G ratio in the sump of above 25. However, under these conditions, the CO_2 transfer capacity was lower than 20 gCO_2/min, limiting biomass productivity to 25 $g_{biomass}$/m²·day (de Godos et al. 2014). In this study, it was demonstrated that CO_2 losses in the sump and channels can be low, only 4% and 6%, respectively, if the system is adequately managed. Here, the major source of CO_2 loss becomes that discharged from the harvested culture medium, representing up to 24% of inlet carbon (Figure 10.11). The final percentage of CO_2 fixed into the biomass can be as high as 66% (de Godos et al. 2014). This value is close to the 64% reported for a fully enclosed raceway system covered by plastic (Li et al. 2013). In laboratory-scale closed raceways, the maximum recovery was 46% when using *Chlorella* and 35% when using *Spirulina* while maintaining 10% of CO_2 in the atmosphere over the culture (Ramanan et al. 2010). In all these studies, no differences in performance of the cultures were observed when using flue gases instead of pure CO_2–air mixtures. Additionally, it is also possible to incorporate advanced control strategies for the operation of raceway reactors, which enhances the quality of control, minimizing not only CO_2 losses but also pH gradients and thus enhancing the biomass productivity of the culture and the overall performance of the reactor (Pawlowski et al. 2014).

Finally, thin-layer reactors have been proposed for the production of microalgae biomass as an improved design versus raceway reactors. In thin-layer reactors, the culture is pumped from a stirred tank to flow continuously on inclined lanes to return to the tank, with the slope of the inclined surfaces determining the depth of the culture, typically from 0.6 to 0.8 mm (Doucha and Lívanský 2009). In relation to the unit of cultivation area, the suspension volume is 50× lower and the density of algae

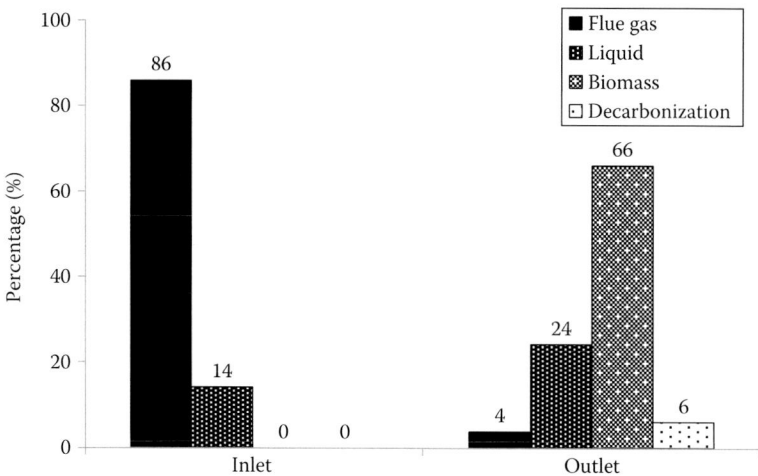

FIGURE 10.11 Carbon mass balance for a culture of *Scenedesmus* sp. operated at a 0.3 L/day dilution rate in a 100 m² raceway pond in which the pH is controlled by on-demand injection of flue gas from a diesel boiler containing 10.6% CO_2. (Data from de Godos, I. et al., *Bioresour. Technol.*, 153, 307, 2014.)

at harvest is 100× higher compared with raceway reactors. Such higher yields and cheaper cultivation/processing technology reduce the costs of biomass production to about 15%–20% of that associated with raceway reactors (Doucha and Lívanský 2009). In these reactors, large amounts of CO_2 must be supplied to satisfy the high demand imposed by the high volumetric biomass productivity. Because of the shallow culture depth, the CO_2 must be injected into the stirred tank, equivalent in function to the sump in raceways, with CO_2 losses occurring in the sump and inclined lines. Because the mass transfer in these reactors has not been optimized, the yield of CO_2 absorption into the stirred tank ranges from 10% to 50% using flue gas containing 6%–8% of CO_2 (Doucha et al. 2005). Under optimal growth conditions, 51% of the supplied CO_2 is lost to the atmosphere, and an additional 10% is lost in the circulation of the culture along the inclined lines, the overall maximum efficiency use of CO_2 being 38% (Doucha et al. 2005).

10.6 CONCLUSIONS

The efficient and cost-effective supply of CO_2 to microalgae cultures is central to the development of microalgal production processes, the relevance of this concern being greater when the scale of the process increases. As a result of considerable research in recent and past years, we have a good understanding of the fundamental aspects of the CO_2 supply requirements, of CO_2 transfer into photobioreactors of different types, and of associated CO_2 losses. Improvements in photobioreactor design and biomass productivity will require more efficient CO_2 supply systems and control strategies. Still, much remains to be studied, learned, and applied as microalgal production processes are scaled up and move from R&D to commercialization.

REFERENCES

Acién, F., C. V. González-López, J. M. Fernández, and E. Molina. 2012a. Conversion of CO_2 into biomass by microalgae: How realistic a contribution may it be to significant CO_2 removal? *Appl. Microbiol. Biotechnol.* 96:577–586.

Acién, F. G., J. M. Fernández, J. J. Magán, and E. Molina. 2012b. Production cost of a real microalgae production plant and strategies to reduce it. *Biotechnol. Adv.* 30:1344–1353.

Benemann, J. 2013. Microalgae for biofuels and animal feeds. *Energies* 6:5869–5886.

Benemann, J. R. 1997. CO_2 mitigation with microalgae systems. *Energy Convers. Manag.* 38:S475–S479.

Borowitzka, M. A. 2013. High-value products from microalgae—Their development and commercialisation. *J. Appl. Phycol.* 25:743–756.

Camacho, E. F. and C. Bordons. 2007. Nonlinear model predictive control: An introductory review. *Lect. Notes Contr. Inf. Sci.* 358:1–16.

Camacho, F., F. G. Acién, J. A. Sánchez, F. García, and E. Molina. 1999. Prediction of dissolved oxygen and carbon dioxide concentration profiles in tubular photobioreactors for microalgal culture. *Biotechnol. Bioeng.* 62:71–86.

Camacho, F., F. García, J. M. Fernández, Y. Chisti, and E. Molina. 2003. A mechanistic model of photosynthesis in microalgae. *Biotechnol. Bioeng.* 81:459–473.

Carroll, J. J., J. D. Slupsky, and A. E. Mather. 1991. The solubility of carbon dioxide in water at low pressure. *J. Phys. Chem. Ref. Data* 20:1201–1209.

Carvalho, A. P., L. A. Meireles, and F. X. Malcata. 2006. Microalgal reactors: A review of enclosed system designs and performances. *Biotechnol. Prog.* 22:1490–1506.

Cents, A. H. G., D. W. F. Brilman, and G. F. Versteeg. 2005. CO_2 absorption in carbonate/bicarbonate solutions: The Danckwerts-criterion revisited. *Chem. Eng. Sci.* 60:5830–5835.

Cheng, L., L. Zhang, H. Chen, and C. Gao. 2006. Carbon dioxide removal from air by microalgae cultured in a membrane-photobioreactor. *Sep. Purif. Technol.* 50:324–329.

Chisti, Y. and U. J. Jauregui-Haza. 2002. Oxygen transfer and mixing in mechanically agitated airlift bioreactors. *Biochem. Eng. J.* 10:143–153.

Chiu, S., C. Kao, T. Huang et al. 2011. Microalgal biomass production and on-site bioremediation of carbon dioxide, nitrogen oxide and sulfur dioxide from flue gas using *Chlorella* sp. cultures. *Bioresour. Technol.* 102:9135–9142.

Costache, T. A., F. Acién, M. M. Morales, J. M. Fernández-Sevilla, I. Stamatin, and E. Molina. 2013. Comprehensive model of microalgae photosynthesis rate as a function of culture conditions in photobioreactors. *Appl. Microbiol. Biotechnol.* 97: 7627–7637.

Craggs, R., D. Sutherland, and H. Campbell. 2012. Hectare-scale demonstration of high rate algal ponds for enhanced wastewater treatment and biofuel production. *J. Appl. Phycol.* 24:329–337.

de Godos, I., J. L. Mendoza, F. G. Acién, E. Molina, C. J. Banks, S. Heaven, and F. Rogalla. 2014. Evaluation of carbon dioxide mass transfer in raceway reactors for microalgae culture using flue gases. *Bioresour. Technol.* 153:307–314.

Doucha, J. and K. Lívanský. 2009. Outdoor open thin-layer microalgal photobioreactor: Potential productivity. *J. Appl. Phycol.* 21:111–117.

Doucha, J., F. Straka, and K. Lívanský. 2005. Utilization of flue gas for cultivation of microalgae (*Chlorella* sp.) in an outdoor open thin-layer photobioreactor. *J. Appl. Phycol.* 17:403–412.

Eilers, P. H. C. and J. C. H. Peeters. 1988. A model for the relationship between light intensity and the rate of photosynthesis in phytoplankton. *Ecol. Modell.* 42:199–215.

Fernández, I., F. G. Acién, J. M. Fernández, J. L. Guzmán, J. J. Magán, and M. Berenguel. 2012. Dynamic model of microalgal production in tubular photobioreactors. *Bioresour. Technol.* 126:172–181.

García, J. L., M. Berenguel, F. Rodríguez, J. M. Fernández, C. Brindley, and F. G. Acién. 2003. Minimization of carbon losses in pilot-scale outdoor photobioreactors by model-based predictive control. *Biotechnol. Bioeng.* 84:533–543.

González-López, C. V., F. G. Acién, J. M. Fernández-Sevilla, J. F. Sánchez, and E. Molina. 2012. Development of a process for efficient use of CO_2 from flue gases in the production of photosynthetic microorganisms. *Biotechnol. Bioeng.* 109:1637–1650.

Ho, S., W. Chen, and J. Chang. 2010. *Scenedesmus obliquus* CNW-N as a potential candidate for CO_2 mitigation and biodiesel production. *Bioresour. Technol.* 101:8725–8730.

Hsueh, H. T., H. Chu, and C. C. Chang. 2007. Identification and characteristics of a cyanobacterium isolated from a hot spring with dissolved inorganic carbon. *Environ. Sci. Technol.* 41:1909–1914.

Hu, Q., N. Kurano, M. Kawachi, I. Iwasaki, and S. Miyachi. 1998. Ultrahigh-cell-density culture of a marine green alga *Chlorococcum littorale* in a flat-plate photobioreactor. *Appl. Microbiol. Biotechnol.* 49:655–662.

Hulatt, C. J. and D. N. Thomas. 2011. Productivity, carbon dioxide uptake and net energy return of microalgal bubble column photobioreactors. *Bioresour. Technol.* 102:5775–5787.

Jorquera, O., A. Kiperstok, E. A. Sales, M. Embiruçu, and M. L. Ghirardi. 2010. Comparative energy life-cycle analyses of microalgal biomass production in open ponds and photobioreactors. *Bioresour. Technol.* 101:1406–1413.

Ketheesan, B. and N. Nirmalakhandan. 2012. Feasibility of microalgal cultivation in a pilot-scale airlift-driven raceway reactor. *Bioresour. Technol.* 108:196–202.

Kurano, N., T. Sasaki, and S. Miyachi. 1998. Carbon dioxide and microalgae. *Stud. Surf. Sci. Catal.* 114:55–63.

Li, S., S. Luo, and R. Guo. 2013. Efficiency of CO_2 fixation by microalgae in a closed raceway pond. *Bioresour. Technol.* 136:267–272.

Lívanský, K. and J. Doucha. 1996. CO_2 and O_2 gas exchange in outdoor thin-layer high density microalgal cultures. *J. Appl. Phycol.* 8:353–358.

Mandal, S. and N. Mallick. 2009. Microalga *Scenedesmus obliquus* as a potential source for biodiesel production. *Appl. Microbiol. Biotechnol.* 84:281–291.

Mazzuca, T., F. Garcia, F. Camacho, F. G. Acién, and E. Molina. 2000. Carbon dioxide uptake efficiency by outdoor microalgal cultures in tubular airlift photobioreactors. *Biotechnol. Bioeng.* 67:465–475.

Mendoza, J. L., M. R. Granados, I. de Godos et al. 2013a. Fluid-dynamic characterization of real-scale raceway reactors for microalgae production. *Biomass Bioenergy* 54:267–275.

Mendoza, J. L., M. R. Granados, I. de Godos et al. 2013b. Oxygen transfer and evolution in microalgal culture in open raceways. *Bioresour. Technol.* 137:188–195.

Molina, E., J. M. Fernández, J. A. Sánchez, and F. García. 1996. A study on simultaneous photolimitation and photoinhibition in dense microalgal cultures taking into account incident and averaged irradiances. *J. Biotechnol.* 45:59–69.

Oswald, W. J. and C. G. Golueke. 1960. Biological transformation of solar energy. *Adv. Appl. Microbiol.* 2:223–262.

Otsuki, T. 2001. A study for the biological CO_2 fixation and utilization system. *Sci. Total Environ.* 277:21–25.

Park, J. B. K. and R. J. Craggs. 2010. Wastewater treatment and algal production in high rate algal ponds with carbon dioxide addition. *Water Sci. Technol.* 61:633–639.

Pawlowski, A., J. L. Mendoza, J. L. Guzmán, M. Berenguel, F. G. Acién, and S. Dormido. 2014. Effective utilization of flue gases in raceway reactor with event-based pH control for microalgae culture. *Bioresour. Technol.* 170:1–9.

Putt, R., M. Singh, S. Chinnasamy, and K. C. Das. 2011. An efficient system for carbonation of high-rate algae pond water to enhance CO_2 mass transfer. *Bioresour. Technol.* 102:3240–3245.

Quinn, J. C., C. W. Turner, and T. H. Bradley. 2012. Scale-Up of flat plate photobioreactors considering diffuse and direct light characteristics. *Biotechnol. Bioeng.* 109:363–370.

Ramanan, R., K. Kannan, A. Deshkar, R. Yadav, and T. Chakrabarti. 2010. Enhanced algal CO_2 sequestration through calcite deposition by *Chlorella* sp. and *Spirulina platensis* in a mini-raceway pond. *Bioresour. Technol.* 101:2616–2622.

Rebolloso, M. M., J. L. García, J. M. Fernández, F. G. Acién, J. A. Sánchez, and E. Molina. 1999. Outdoor continuous culture of *Porphyridium cruentum* in a tubular photobioreactor: Quantitative analysis of the daily cyclic variation of culture parameters. *J. Biotechnol.* 70:271–288.

Richmond, A. 2004. Principles for attaining maximal microalgal productivity in photobioreactors: An overview. *Hydrobiologia* 512:33–37.

Ruiz, J., P. D. Álvarez-Díaz, Z. Arbib, C. Garrido-Pérez, J. Barragán, and J. A. Perales. 2013. Performance of a flat panel reactor in the continuous culture of microalgae in urban wastewater: Prediction from a batch experiment. *Bioresour. Technol.* 127:456–463.

Sierra, E., F. G. Acién, J. M. Fernández, J. L. García, C. González, and E. Molina. 2008. Characterization of a flat plate photobioreactor for the production of microalgae. *Chem. Eng. J.* 138:136–147.

Singh, U. B. and A. S. Ahluwalia. 2013. Microalgae: A promising tool for carbon sequestration. *Mitig. Adapt. Strat. Global Change* 18:73–95.

Stepan, D., R. Shockey, T. Moe, and R. Dorn. 2002. Carbon dioxide sequestering using microalgae systems. U.S. Department of Energy, Pittsburgh, PA.

Weissman, J. C., R. P. Goebel, and J. R. Benemann. 1988. Photobioreactor design: Mixing, carbon utilization, and oxygen accumulation. *Biotechnol. Bioeng.* 31:336–344.

Weissman, J. C., J. C. Radway, E. W. Wilde, and J. R. Benemann. 1998. Growth and production of thermophilic cyanobacteria in a simulated thermal mitigation process. *Bioresour. Technol.* 65:87–95.

Zhang, K., S. Miyachi, and N. Kurano. 2001. Photosynthetic performance of a cyanobacterium in a vertical flat-plate photobioreactor for outdoor microalgal production and fixation of CO_2. *Biotechnol. Lett.* 23:21–26.

Zhang, Q., X. Wu, S. Xue, K. Liang, and W. Cong. 2013. Study of hydrodynamic characteristics in tubular photobioreactors. *Bioprocess Biosyst. Eng.* 36:143–150.

Zirrahi, M., H. Hassanzadeh, and J. Abedi. 2013. Modeling of CO_2 dissolution by static mixers using back flow mixing approach with application to geological storage. *Chem. Eng. Sci.* 104:10–16.

11 Harvesting and Dewatering of High-Productivity Bulk Microalgae Systems

Navid R. Moheimani, Mark P. McHenry,
Koenraad Muylaert, and Patrick V. Brady

CONTENTS

11.1 INTRODUCTION

Energy-efficient and cost-effective microalgae dewatering, nutrient recycling, and wastewater treatment are some of the major challenges facing large-scale industrial-scale microalgae production (Benemann 2013; Borowitzka and Moheimani 2013; Milledge and Heaven 2013). Biomass concentrations in microalgal cultures are generally limited from a few hundred mg dry weight L^{-1} in open raceway ponds to a few g L^{-1} in typical photobioreactors. This low density of biomass of phototrophic algal cultures necessitates that a large volume of water must be processed to harvest the biomass. For example, to harvest 1 ton of dry biomass from an outdoor pond culture with 0.5 g L^{-1} dry matter content, 2000 m^3 of the culture medium needs to be processed. Filtration through fine screens is not practical as they clog rapidly, while the use of wider mesh screens, such as microstrainers, is only feasible with filamentous species such as *Spirulina* (*Arthrospira*). Filtration through fine screens may not be practical in many situations due to fouling. Micro-/ultra-/nanofiltration membrane selection is dependent on the culture medium and level of contamination with various macro- and microsized molecules, elements, and components. However, harvesting microalgal cells from their culture medium by simple gravity sedimentation is generally not feasible due to the small size of the individual cells (typically just under 3 to about 50 μm) and the small difference in density with the medium. Only larger colonies, clusters, or aggregates ("flocs") will readily settle. Although, microalgal cell suspensions are stabilized by the negative surface charge of the cells; thus, they generally do not spontaneously coagulate to form larger aggregates that sediment rapidly enough by gravity alone. One exception is *Haematococcus pluvialis*, commercially produced for its astaxanthin content, which typically produces relatively large and immotile cysts with a high astaxanthin content that settle well (Kobayashi et al. 2001; see also Chapter 12, this volume). Another is *Pediastrum*, a large colonial green alga, which can dominate in high-rate algal wastewater treatment ponds and which has a high sedimentation rate (Park et al. 2011). For commercial *Chlorella* production, centrifugation is used to harvest the biomass, but is too expensive for all but such high-value products (e.g., nutritional supplements). The development and application of microalgae harvesting and dewatering technologies capable of processing large volumes of culture medium at a minimal cost are thus essential for large-scale and low-cost production of lower-cost commodities such as feeds and fuels (Greenwell et al. 2010; Uduman et al. 2010; Christenson and Sims 2011).

The objective of harvesting and dewatering is to raise the concentration of the microalgal biomass by more than two orders of magnitude to reach over 10% solids, sufficiently concentrated for subsequent processing or drying. It is widely believed that this is best achieved using a combination of technologies in a two-stage process (Benemann et al. 1982; Shelef et al. 1984; Vandamme et al. 2013) such as flocculation followed by centrifugation. For example, the first step would be to increase the pond culture biomass concentration from about 0.05% (0.5 g L^{-1}) dry matter content to about 2% dry matter slurry, a 40-fold concentration, with centrifugation in the second step, achieving a paste with a 20% dry matter content (a further 10 × up-concentration).

The harvesting and dewatering process selection often interacts with other process steps in microalgae production such as strain selection, cultivation system design, growth medium composition, biomass fractionation, and water or nutrient recycling (Wijffels et al. 2010; de Boer et al. 2012). Performance of any harvesting process will be influenced by properties of microalgae such as cell size, cell fragility, cell surface charge—properties specific to the species and strains—and also influenced by the culture medium used and culture conditions (Henderson et al. 2008). Microalgae can excrete considerable amounts of organic matter in the culture medium (Hulatt and Thomas 2010), particularly when stressed (Zhang et al. 2012), and this may interfere with harvesting and dewatering and may also result in potential contamination.

One consideration is that the harvesting–dewatering technology should not impair the quality of microalgal biomass such as by contamination with coagulants (i.e., alum, iron, and synthetic organic flocculants). This is particularly important for feed applications (Wijffels et al. 2010). The harvesting method should also allow for recycling of the culture medium in order to reduce the total system water demand (Park et al. 2011).

11.2   HARVESTING FREQUENCY AND PRODUCTIVITY

The main operational modes in microalgal culture are batch, semibatch, semicontinuous, and continuous (Moheimani and Borowitzka 2006). In batch cultures, a single inoculation (typically from 2% to 20% of final biomass concentration) is made into a prepared culture system, followed by several days of algal growth until cells reach the desired cell density, followed by harvest of the entire culture. Batch cultivation is used where there is major danger of contamination from invading algae, grazers, or infections, and where a high cell density is desired if, for instance, centrifugation is used for harvesting. This is the case for commercial *Chlorella* production, where both high cell densities are desired and contamination is a constant problem. In such a batch process, algal cultures are transferred starting from small laboratory cultures to increasingly larger culture volumes until the final production pond or photobioreactor is reached, where the culture is allowed to attain maximum density and then harvested. In semibatch cultivation, harvesting would take place at an earlier point, and a fraction of the culture is retained serving as inoculum for the following batch, this being repeated as often as contamination allows. In semicontinuous dilution, a fraction of the algal pond is harvested daily, typically between 20% and 40% of the culture volume that accordingly results in maximum productivity. Continuous cultures are similar except that the culture is constantly diluted and harvested (at least during the daytime). A variant of continuous culture is the turbidostat, operating at a fixed density rather than a fixed dilution. Clearly, these operational modes blend into each other and are optimized for balancing productivity (higher for semicontinuous or continuous processes) and harvesting cost and/or contamination risk (lower for batch and semibatch cultures).

As an example of these cultivation modes, Moheimani (2013a,b) investigated long-term growth and productivity of *Dunaliella tertiolecta*, *Chlorella* sp., and *Tetraselmis suecica* grown in outdoor bag photobioreactors: with and without the use of external inorganic carbon sources under batch, semibatch (harvested once every 7, 5, and 3 days), semicontinuous (harvested daily), and continuous (turbidostat) cultures (Figure 11.1). Changing the harvest regime from batch mode to any of the semibatch or semicontinuous operations resulted in significant increases in the specific growth rate and biomass productivity for all three species. The *T. suecica* culture achieved the highest specific growth rate and biomass productivity when diluted every day, but lower productivity in a continuous culture. *Chlorella* sp. and *D. tertiolecta* showed no difference when harvested every 3 days or daily. Continuous harvesting

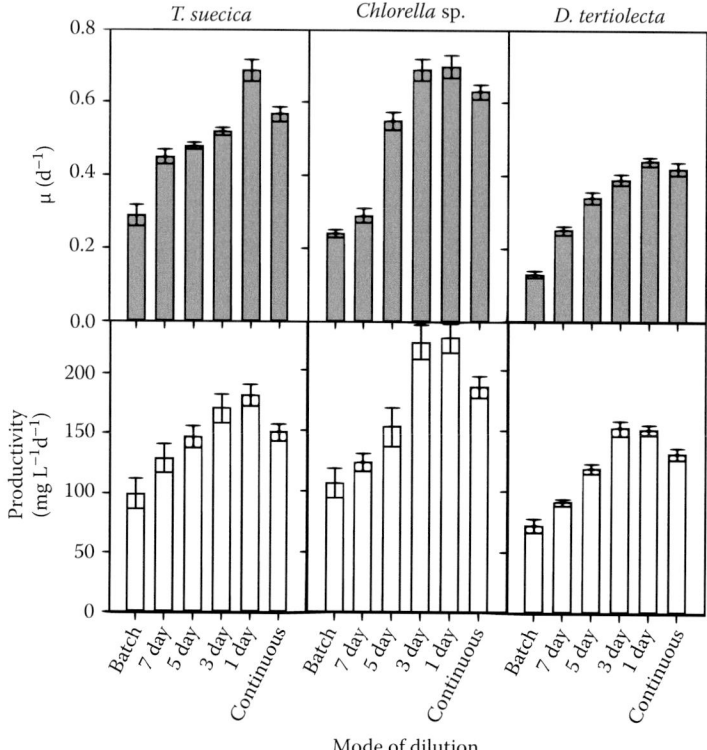

FIGURE 11.1 Effect of different modes of dilution on the specific growth rate and biomass productivity of *Tetraselmis suecica*, *Chlorella* sp., and *Dunaliella tertiolecta* (data are mean ± SE, n = 4, 7, 18, 36, 73, and 73 for cultures grown under batch, diluted every 7 days, 5 days, 3 days, 1 day, or continuously).

resulted in less growth rate and biomass productivity. These results demonstrated the importance of the cultivation mode and its effect on biomass productivity of the particular microalgae species being cultivated.

11.3 HARVESTING OF MICROALGAE

After the growth phase, the next step is to harvest the microalgae, that is, to separate the solids (algal biomass) from the liquid medium (Mohn 1988). The microalgae can be harvested by constraining either the liquid or the biomass, as shown in Figure 11.2. Liquid-constrained (particles are free to move) methods include flotation (i.e., dispersed or dissolved air), sedimentation, and centrifugation. Algae-constrained processes in which the biomass is fixed but the liquid is mobile include filtration and screening. Both types of harvesting methods might be implemented at different stages of the harvesting process (Figure 11.2). Important microalgae properties that can influence their separation are shape (filamentous, rods, sphere, colonial, etc.), size, specific gravity, and surface charge (usually negative).

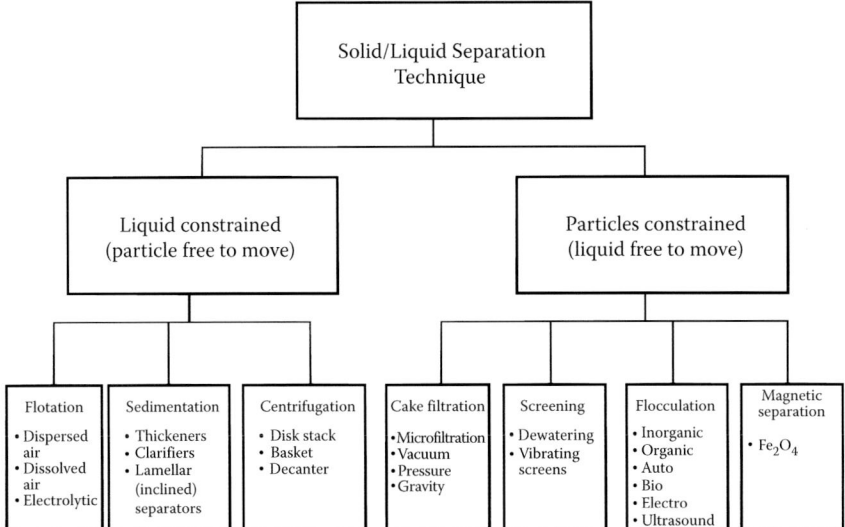

FIGURE 11.2 Classification of common industrial solid–liquid separation techniques. (Modified and redrawn from Shelef, G. et al., Microalgae harvesting and processing: A literature review, SERI/STR-231-2396, Technion Research & Development Foundation Ltd., Haifa, Israel, http://www.nrel.gov/docs/legosti/old/2396.pdf, 1984, last accessed April 1, 2015.)

11.3.1 Flocculation

Due to the small size and low specific gravity of microalgae particles, most of the solid–liquid separation techniques shown in Figure 11.2 are not very effective or economical, or both. Separations based on sedimentation or flotation are generally improved by inducing flocculation, that is, the aggregation of small individual cells into larger aggregates. This can be achieved by addition of chemical flocculants, positively charged metal salts (e.g., alum, ferric salts), or polymers that neutralize the negative surface charge of the algal cells and by doing so initiate flocculation (Lee et al. 2013a). Flocculation followed by sedimentation can also be the result of autoflocculation (Sukenik and Shelef 1984), in which the mineral that precipitates in the culture medium act as flocculants, or bioflocculation (Benemann et al. 1980; Lee et al. 2013a) in which the algae themselves or, more commonly, associated bacteria produce or act as flocculants. However, the most commonly used harvesting technology for microalgae, which is already applied commercially to wastewater treatment ponds, is chemical flocculation (Lee et al. 2013a).

11.3.1.1 Chemical Flocculation

A wide range of inorganic and organic flocculants are used on a large scale in wastewater treatment, drinking water supplies, mining, and other industries. Inorganic flocculants include iron and aluminum salts, which, although effective, are required in relatively high doses, which is both expensive and results in a high content of these flocculants in the biomass as well as the culture medium. Synthetic organic flocculants such as polyacrylamide-based polymers are also used in wastewater treatment including the separation of algae from sewage treatment pond effluents. For animal feed applications, these flocculants would not be acceptable due to potential toxicity. Therefore, flocculants based on natural polymers are preferred with modified natural polymers such as chitosan (deacetylated chitin) or cationically modified starch being currently used. However, as for the inorganic metal salts or polyacrylamide polymers, their cost is too high for use in low-cost production of microalgae for feeds or fuels (Uduman et al. 2010; Vandamme et al. 2013).

11.3.1.2   Physical Flocculation

Standing ultrasound waves have been recently investigated to flocculate microalgae, but this technique is energy-demanding and difficult to upscale (Bosma et al. 2003). Electroflocculation (Danquah et al. 2009), another process for harvesting microalgae, uses electrical energy to dissolve iron or aluminum ions from an electrode (anode), which then act as flocculants. Due to the high conductivity of seawater, power consumption of electroflocculation is substantially lower in seawater than freshwater, making this a potentially attractive method for harvesting marine microalgae (Lee et al. 2013a; Shuman et al. 2014). As for other flocculants, electroflocculation results in contamination of the biomass as well as the culture medium, in this case with iron or aluminum.

11.3.1.3   Bioflocculation

Bioflocculation generally refers to flocculation induced by the algae themselves or by other microorganisms growing with algae or separately, principally bacteria (Benemann et al. 1982). Bacterial flocculation is a well-known phenomenon in wastewater treatment and used, for example, in the activated sludge process to remove the biomass produced by gravity settling (sedimentation). Such bacteria can also induce flocculation of microalgae (Lee et al. 2009), presumably through the production of extracellular polymers that act as flocculants (Lee et al. 2013b). Bioflocculating bacteria (such as the ones produced in the activated sludge processes) can be cultured separately and added to a microalgae culture to induce flocculation. Alternatively, consortia of microalgae and bacteria can be cultivated together as the so-called microalgae bacterial flocs (Van Den Hende et al. 2011). Bacteria-mediated bioflocculation/sedimentation is an algae-harvesting method in wastewater treatment where the organic matter supports bacterial growth. In microalgae-based wastewater treatment systems, bioflocculation indeed often occurs spontaneously (Benemann et al. 1982; Craggs et al. 2012).

Some studies have shown that some fungi may also induce flocculation of microalgal suspensions, plausibly due to positive charges on the fungal hyphae that interact with the negative charges on the microalgal cell walls (Zhang and Hu 2012). Finally, many microalgae will flocculate spontaneously due to production of extracellular substances or the formation of larger cell colonies. In high-rate (paddle wheel mixed raceway) ponds for algal wastewater treatment, sedimentation of the culture is relatively efficient when the microalgal community is dominated by the large species such as *Pediastrum* (Park et al. 2011). Bioflocculating microalgae may also be used to induce flocculation of a nonflocculating species of microalgae (Salim et al. 2011).

11.3.1.4   Autoflocculation

Given that bioflocculation, either with or without bacteria, is typically difficult to control and thus unreliable, and considering that chemical flocculants are expensive and can interfere with further processing or use of the biomass, alternative methods are still needed. One such alternative is autoflocculation—the tendency for algae to flocculate in response to shifts in pH and water hardness. Autoflocculation is due to the coprecipitation of algal cells with Mg^{2+}, Ca^{2+}, carbonates, and phosphates, which precipitate at high pH. Seawater, with 52 mM Mg^{2+}, 10 mM Ca^{2+}, and 2.2 mM carbonate, will precipitate Mg and Ca carbonates when the pH is raised from roughly pH 8 to above pH 10 due to formation of Mg hydroxide and calcite (Ayoub et al. 1986; Vandamme et al. 2012a). Autoflocculation by calcium phosphate requires both Ca and phosphate and a pH above ~8 (Vandamme et al. 2012b; Beuckels et al. 2013). A key goal is the engineering of pH, Mg, Ca, and P levels to achieve maximum autoflocculation with minimal chemical inputs. pH rises can be achieved in algal cultures by cutting off the CO_2 supply, letting algal photosynthesis consume the residual CO_2 and bicarbonate until a high pH is reached, typically in about an hour on a sunny day. The addition of lime can be used to accelerate the process. Better understanding of the mechanistic link between pH, precipitate formation, and charge neutralization on the algal cell surface should allow the process to be optimized.

11.3.1.5   Factors Affecting Microalgae Flocculation

Algae flocculation is affected by extracellular organic material secreted by the algae, which can aid or hinder flocculation (Bernhardt and Clasen 1991; Henderson et al. 2008). Algae can also modify their surface charge and their morphology, potentially increasing flocculant demand and/or slowing flocculation (Pieterse and Cloot 1997). Physical factors play an important role as the rate of alga–alga interactions increase with algae concentration, with shear mixing, and with differential particle settling velocities (Bratby 1980). Flocculation processes are often species, strain, and cultivation condition specific such that a flocculant type and dose that work well with one microalgal species, or even strain and cultivation condition, is unsuitable for another. The degree of flocculation also depends on the chemical composition of the microalgal growth medium such as the ionic environment, temperature, or pH.

11.3.2   Microalgae-Constrained Harvesting Methods

11.3.2.1   Filtration

Filtration requires a pressure differential and a filter media and can be operated in continuous or batch modes. Filtration is widely used in mining and other industries, with the main advantage being that most filtration devices are "off-the-shelf" items. The main filtration methods proposed for microalgae separation include tangential (cross) flow filtration (Danquah et al. 2009), combined gravity-belt thickener and dewatering (Shelef et al. 1984), rotary press (Grima et al. 2004), automated filter press (Grima et al. 2004), vacuum drum filters (Shelef et al. 1984), combined screw thickener and dewatering (Shelef et al. 1984), and vibrating screens (Vonshak et al. 1983). The main concerns with filtering microalgae at large scale are their small particle sizes and the compressibility of the microalgal cells. These properties typically result in filter blockages, fouling, and low biomass recoveries. The most common means to overcome these issues are by using flocculants, filter aids (e.g., diatomite) such as diatomaceous earth, or a combination of these. As with any separation technology, filtration is very species-specific.

Filtration with large mesh (>25 microns) vibrating, inclined, and/or rotating backwashed screens (microstrainers) can be used for large algal colonies or, in particular, filamentous algae such as *Spirulina*, where such devices are used commercially (Vonshak et al. 1983). However, these cannot be used for unicellular or most colonial microalgal species as a much smaller screen size would be required, resulting in low throughput (e.g., liters per m^2 of filter area).

11.3.3   LIQUID CONSTRAINED

11.3.3.1   Centrifugation

Most microalgae can be harvested by centrifugation, which depends on a density difference between the particles and the culture medium. However, due to the low specific gravity of microalgae, even centrifugation can be difficult and is also very expensive. Types of centrifuges available include a perforated basket, sieve scroll, tubular bowl, multichamber solid bowl, scroll discharge, disc bowl, nozzle bowl, and decanters (Shelef et al. 1984). Each design has advantages and disadvantages, and to date, decanters are generally the most effective centrifuge for separating most microalgae. The main disadvantages of centrifuges are the high capital costs, operating costs (power consumption), and, although a relatively minor issue, the effect on microalgae of associated temperature increases (Moheimani et al. 2013). To overcome or at least ameliorate these issues, centrifugation is often used in combination with other separation techniques.

11.3.3.2   Sedimentation

Gravity separation or sedimentation was discussed earlier in the context of chemical, bacteria-induced, or spontaneous algae flocculation methods and is the simplest and lowest cost method of solid–liquid separation (Shelef et al. 1984; Uduman et al. 2010). Sedimentation depends on settling velocity (which can translate into an overflow velocity for the engineering design). Settling velocity is in turn related linearly to the specific gravity difference with the culture medium and as a square function of the particle size. Thus, a 10-fold increase in a particle size (e.g., going from single cells to colonies of a hundred cells) can increase settling rates a 100-fold. This allows the relatively rapid settling of larger colonies or flocs, assuming a modicum of specific gravity differences. Continuous gravity thickeners and lamellar settlers, are commonly used sedimentation technologies. Sedimentation separates the microalgae medium into a clear liquid supernatant that can be decanted from the sedimented microalgal slurry (with typically about 1%–2% solid content, though higher concentration factors are reported) (Dassey and Theegala 2013). The advantages of sedimentation include: lower energy demand, lower capital costs and installation costs, and also lower operating skill requirements. The main disadvantages of sedimentation are the low solid content and long settling times required, leading to large space requirements. In the particular case where algae need to go through a flocculation phase before actually settling, larger batch quiescent sedimentation tanks may be required. Sedimentation characteristics are very species and even strain-specific and depend on many factors including growth conditions but most critically, as noted earlier, colony or floc size. As mentioned earlier, only a few species can be harvested by sedimentation (e.g., *Pediastrum* or immotile cysts of *Haematococcus*). Due to unpredictable and uncertain control over these factors, sedimentation methods will likely require chemical flocculation as a backup to achieve the overflow velocity design specifications at all times. Sedimentation using inclined settlers is an alternative to settling tanks (Smith and Davis 2013).

11.3.3.3 Flotation

Flotation had its beginnings in mineral (ore) and wastewater processing and has a long history in solid–liquid separation applications using stable froths to selectively separate different minerals from each other (Smith and Davis 2013). Where the small size of algae makes their recovery from media very difficult, even by centrifugation, an improvement can often be achieved by flocculation followed by flotation. A few microalgal species have buoyancy (due to regulation of gas vacuoles, in cyanobacteria), but for most species, assisted flotation techniques can be used in which small gas bubbles attach to the flocs; with the separation dependent on the size of the gas bubbles. Large air bubbles can be continuously pumped into a flotation cell for the so-called froth flotation, though froth flotation generally has low recovery efficiency, as large bubbles cause the bulk of the fine particles to follow the streamlines around the bubbles creating low particle inertia (Pahl et al. 2013). Coward et al. (2015) showed that generating airflows with higher levels of air–particle interface can improve overall harvesting cost-benefits using this method. Coward et al. (2014) also found that flotation using surfactants can specifically increase lipid recovery and improve fatty acid methyl ester characteristics of *Chlorella*. Dissolved air flotation (DAF) units were developed that injected air-supersaturated water under pressure into the flotation cells, with the floating algae skimmed from the surface. The higher density and smaller footprint of this system has resulted in full-scale installations at several wastewater treatment ponds. Some studies suggest the possibility of bioflotation, whereby the photosynthetically generated oxygen would be utilized; however, this is yet to be tested at large scale. The interaction between microalgae and air bubbles depends on the hydrophobicity of the cell surface, which can be enhanced by addition of surfactants (Garg et al. 2012). Flotation can be used as an alternative to sedimentation for separating flocculated microalgae from the medium, and as noted previously, often yields a higher concentration factor than sedimentation (Besson and Guiraud 2013).

11.3.3.4 Magnetic Separation

Removal of suspended particles such as microalgae is possible by using magnetic particles such as Fe_3O_4. The coagulated particles can then be passed through a magnetic field, leaving the product water relatively microalgae-free. Bitton et al. (1975)

reported microalgal removal efficiencies between 55% and 94% from five Florida Lakes by use of a commercial magnetic filter. There are no cost estimates available for magnetic separation of microalgae.

11.4   MULTISTAGED SEPARATION

Biomass concentration in microalgal cultures ranges from less than 0.5 g L^{-1} in open raceway pond systems to about 5 g L^{-1} in photobioreactors, compared to >50 g L^{-1} for industrial fermentations (Pahl et al. 2013). Harvesting requires 40–400 up-concentration of the biomass from a dilute culture with 0.05%–0.5% dry matter content to a microalgal paste with a dry matter content of about 20%, typically required for drying or further processing. To reduce the cost and energy inputs, a multistage process is generally specified for harvesting microalgae from dilute raceway pond cultures with the recovery efficiency and concentration factor of each stage being the key parameters (Pahl et al. 2013). The recovery efficiency is the proportion of the biomass in the suspension that can be recovered during the harvesting stage. The concentration factor is the ratio of the biomass concentration in the original suspension and in the final concentrate obtained in a harvesting stage. As a multistage harvesting process will generally require more time than a single-step harvesting, care should be taken that the biomass quality does not deteriorate due to the associated bacteria, which can occur rapidly. Inclined settlers or lamella separators have much shorter settling distances and a larger surface area for collecting settled particles. They can therefore be more efficient than gravity thickeners, but may require chemical flocculation to increase the particle size and enhance the settling rate of the microalgal cells. If the distance between the lamellae is very small (mm scale), inclined settlers can be used to harvest cells without the need of a coagulant (Smith and Davis 2013). However, lamellar settler performances also critically depend on the flow characteristics of the concentrated biomass.

During thickening of the microalgal suspension into dense slurry, the biomass concentration reaches several tens of g L^{-1}, and the viscosity may increase until it displays non-Newtonian behavior such as shear thinning (Wileman et al. 2012). This may also be important when the biomass slurry is transferred to the next harvesting or processing stage by means of pumping. The final dewatering prior to drying, if any, is best achieved using a mechanical method such as centrifugation or belt filtration. While these are energy-intensive technologies, the volumes that need to be processed are generally relatively low at this stage. Due to the high content of intracellular water of the harvested biomass, residual moisture even after centrifugation will be 80%–85%, requiring removal of between about 4 and 6 kg of water to produce 1 kg of algal biomass (with a 10% residual moisture). Just to evaporate water requires 2.5 MJ/kg, thus, it would theoretically require about 10–15 MJ/kg to dry algae. In practice, it would most likely be 30%–50% more than this, depending on efficiency and on any ancillary electricity requirements of the drying technology. Drying could easily consume all the energy that could be recovered as fuel from the algae. Consequently, it was suggested that the drying step should be completely omitted and other wet processing technologies should be used in biofuels applications (i.e., hydrothermal liquefaction, wet extraction) (de Boer et al. 2012).

11.5   CONCLUSIONS

Cost-effective and nondestructive microalgae harvesting and dewatering techniques have been recognized, since the inception of algal mass culture technologies, as a major challenge to the industry in expanding from current high-value to lower-value commodity products (Benemann 2013; Moheimani et al. 2014). Dewatering options will need to be customized for selected species of microalgae to be mass cultured, and for each application (e.g., high-value products, feeds, or biofuels). High productivities are required alongside efficient harvesting and dewatering technologies to reduce total algal product cost (Borowitzka 1992; Chisti 2007; Griffiths and Harrison 2009). The development of a range of commercially viable and sustainable algal production, harvesting, and dewatering processes will be required to supply mass-produced microalgae in a shorter time frame (Borowitzka 1992; Griffiths and Harrison 2009; Moheimani et al. 2013). The final commercial bioproduct production streams and associated costs will be heavily dependent on the species, product, and final use (Borowitzka 1992), with industrial production system integration likely required to enable cost-effective competitive production of bulk microalgal biomass across a range of final markets.

ACKNOWLEDGMENT

Patrick V. Brady thanks Sandia National Laboratories LDRD office for the support.

REFERENCES

Ayoub, G. M., S.-I. Lee, and B. Koopman. 1986. Seawater induced algal flocculation. *Water Res.* 20:1265–1271.

Benemann, J. 2013. Microalgae for biofuels and animal feeds. *Energies* 6:5869–5886.

Benemann, J. R., R. P. Goebel, J. C. Weissman, and D. C. Augenstein. 1982. Microalgae as a source of liquid fuels. Final technical report. In *Proceedings of the June 1982 SERI Biomass Program Principal Investigators' Review Meeting*, Aquatic Species Program Reports, SERI/CP-231-1808. Washington, DC, pp. 1–16. http://www.nrel.gov/docs/legosti/old/1808.pdf (accessed April 1, 2015).

Benemann, J. R., B. L. Koopman, J. C. Weissman, D. M. Eisenberg, and P. Goebel. 1980. Development of microalgae harvesting and high rate pond technologies in California. In *Algae Biomass: Production and Use*, eds. G. Shelef and C. J. Soeder, pp. 457–496. Amsterdam, the Netherlands: Elsevier.

Bernhardt, H. and J. Clasen. 1991. Flocculation of micro-organism. *J. Water Supply Res. Technol. Aqua* 40:76–87.

Besson, A. and P. Guiraud. 2013. High-pH-induced flocculation–flotation of the hypersaline microalga *Dunaliella salina*. *Bioresour. Technol.* 147:464–470.

Beuckels, A., O. Depraetere, D. Vandamme, I. Foubert, E. Smolders, and K. Muylaert. 2013. Influence of organic matter on flocculation of *Chlorella vulgaris* by calcium phosphate precipitation. *Biomass Bioeng.* 54:107–114.

Bitton, G., J. Fox, and H. Strickland. 1975. Removal of algae from Florida lakes by magnetic filtration. *Appl. Microbiol.* 30:905–908.

Borowitzka, M. A. 1992. Algal biotechnology products and processes—Matching science and economics. *J. Appl. Phycol.* 4:267–279.

Borowitzka, M. A. and N. R. Moheimani. 2013. Sustainable biofuels from algae. *Mitig. Adapt. Strat. Global Change* 18:13–25.

Bosma, R., W. A. van Spronsen, J. Tramper, and R. H. Wijffels. 2003. Ultrasound, a new separation technique to harvest microalgae. *J. Appl. Phycol.* 15:143–152.

Bratby, J. R. 1980. *Coagulation and Flocculation: With an Emphasis on Water and Wastewater Treatment*. Croydon, UK: Uplands Press.

Chisti, Y. 2007. Biodiesel from microalgae. *Biotechnol. Adv.* 25:294–306.

Christenson, L. and R. Sims. 2011. Production and harvesting of microalgae for wastewater treatment, biofuels, and bioproducts. *Biotechnol. Adv.* 29:686–702.

Craggs, R., D. Sutherland, and H. Campbell. 2012. Hectare-scale demonstration of high rate algal ponds for enhanced wastewater treatment and biofuel production. *J. Appl. Phycol.* 24:329–337.

Coward, T., G. J. Lee, and G. S. Caldwell. 2014. Harvesting microalgae by CTAB-aided foam flotation increases lipid recovery and improves fatty acid methyl ester characteristics. *Biomass Bioenerg.* 67:354–362.

Coward, T., J. G. Lee, and G. S. Caldwell. 2015. The effect of bubble size on the efficiency and economics of harvesting microalgae by foam flotation. *J. Appl. Phycol.* 27:733–742.

Danquah, M. K., L. Ang, N. Uduman, N. Moheimani, and G. M. Forde. 2009. Dewatering of microalgal culture for biodiesel production: Exploring polymer flocculation and tangential flow filtration. *J. Chem. Technol. Biotechnol.* 84:1078–1083.

Dassey, A. J. and C. S. Theegala. 2013. Harvesting economics and strategies using centrifugation for cost effective separation of microalgae cells for biodiesel applications. *Bioresour. Technol.* 128:241–245.

de Boer, K., N. Moheimani, M. Borowitzka, and P. Bahri. 2012. Extraction and conversion pathways for microalgae to biodiesel: A review focused on energy consumption. *J. Appl. Phycol.* 24:1681–1698.

Garg, S., Y. Li, L. Wang, and P. M. Schenk. 2012. Flotation of marine microalgae: Effect of algal hydrophobicity. *Bioresour. Technol.* 121:471–474.

Greenwell, H., L. Laurens, R. Shields, R. Lovitt, and K. Flynn, K. 2010. Placing microalgae on the biofuels priority list: A review of the technological challenges. *J. R. Soc. Interface* 7:703–726.

Griffiths, M. J. and S. T. Harrison. 2009. Lipid productivity as a key characteristic for choosing algal species for biodiesel production. *J. Appl. Phycol.* 21:493–507.

Grima, E. M., F. G. A. Fernández, and A. R. Medina. 2004. Downstream processing of cell-mass and products. In *Handbook of Microalgal Culture: Biotechnology and Applied Phycology*, p. 215–251. ed. A. Richmond. Oxford, UK: Blackwell Publishing.

Henderson, R., S. A. Parsons, and B. Jefferson. 2008. The impact of algal properties and pre-oxidation on solid–liquid separation of algae. *Water Res.* 42:1827–1845.

Hulatt, C. J. and D. N. Thomas. 2010. Dissolved organic matter (DOM) in microalgal photobioreactors: A potential loss in solar energy conversion? *Bioresour. Technol.* 101:8690–8697.

Kobayashi, M., T. Katsuragi, and Y. Tani. 2001. Enlarged and astaxanthin-accumulating cyst cells of the green alga *Haematococcus pluvialis*. *J. Biosci. Bioeng.* 92:565–568.

Lee, A. K., D. M. Lewis, and P. J. Ashman. 2009. Microbial flocculation, a potentially low-cost harvesting technique for marine microalgae for the production of biodiesel. *J. Appl. Phycol.* 21:559–567.

Lee, A. K., D. M. Lewis, and P. J. Ashman. 2013a. Harvesting of marine microalgae by electroflocculation: The energetics, plant design, and economics. *Appl. Energy* 108:45–53.

Lee, J., D.-H. Cho, R. Ramanan, B.-H. Kim, H.-M. Oh, and H.-S. Kim. 2013b. Microalgae-associated bacteria play a key role in the flocculation of *Chlorella vulgaris*. *Bioresour. Technol.* 131:195–201.

Milledge, J. J. and S. Heaven. 2013. A review of the harvesting of micro-algae for biofuel production. *Rev. Environ. Sci. Biotechnol.* 12:165–178.

Moheimani, N. and M. Borowitzka. 2006. The long-term culture of the coccolithophore *Pleurochrysis carterae* (Haptophyta) in outdoor raceway ponds. *J. Appl. Phycol.* 18:703–712.

Moheimani, N. R. 2013a. Long-term outdoor growth and lipid productivity of *Tetraselmis suecica*, *Dunaliella tertiolecta*, and *Chlorella* sp. (Chlorophyta) in bag photobioreactors. *J. Appl. Phycol.* 25:166–177.

Moheimani, N. R. 2013b. Inorganic carbon and pH effect on growth and lipid productivity of *Tetraselmis suecica* and *Chlorella* sp. (Chlorophyta) grown outdoors in bag photobioreactors. *J. Appl. Phycol.* 25:167–176.

Moheimani, N. R., R. Cord-Ruwisch, E. Raes, and M. A. Borowitzka. 2013. Non-destructive oil extraction from *Botryococcus braunii* (Chlorophyta). *J. Appl. Phycol.* 25:1653–1661.

Moheimani, N. R., H. Matsuura, N. M. Watanabe, and M. A. Borowitzka. 2014. Non-destructive hydrocarbon extraction from *Botryococcus braunii* BOT-22 (race B). *J. Appl. Phycol.* 26:1453–1463.

Mohn, F. H. 1988. Harvesting of micro-algal biomass. In *Micro-Algal Biotechnology*, eds. M. A. Borowitzka and L. J. Borowitzka, pp. 357–394. Cambridge, UK: Cambridge University Press.

Pahl, S. L., A. K. Lee, T. Kalaitzidis, P. J. Ashman, S. Sathe, and D. M. Lewis. 2013. Harvesting, thickening and dewatering microalgae biomass. In *Algae for Biofuels and Energy*, eds. M. A. Borowitzka and N. R. Moheimani, pp. 165–185. Dordrecht, the Netherlands: Springer.

Park, J., R. Craggs, and A. Shilton. 2011. Recycling algae to improve species control and harvest efficiency from a high rate algal pond. *Water Res.* 45:6637–6649.

Pieterse, A. and A. Cloot. 1997. Algal cells and coagulation, flocculation and sedimentation processes. *Water Sci. Technol.* 36:111–118.

Salim, S., R. Bosma, M. H. Vermuë, and R. H. Wijffels. 2011. Harvesting of microalgae by bio-flocculation. *J. Appl. Phycol.* 23:849–855.

Shelef, G., A. Sukenik, and M. Green. 1984. Microalgae harvesting and processing: A literature review. SERI/STR-231-2396. Technion Research and Development Foundation Ltd., Haifa, Israel. http://www.nrel.gov/docs/legosti/old/2396.pdf (accessed April 1, 2015).

Shuman, T. R., G. Mason, M. D. Marsolek, Y. Lin, D. Reeve, and A. Schacht. 2014. An ultra-low energy method for rapidly pre-concentrating microalgae. *Bioresour. Technol.* 158:217–224.

Smith, B. T. and R. H. Davis. 2013. Particle concentration using inclined sedimentation via sludge accumulation and removal for algae harvesting. *Chem. Eng. Sci.* 91:79–85.

Sukenik, A. and G. Shelef. 1984. Algal autoflocculation—Verification and proposed mechanism. *Biotechnol. Bioeng.* 26:142–147.

Uduman, N., Y. Qi, M. K. Danquah, G. M. Forde, and A. Hoadley. 2010. Dewatering of microalgal cultures: A major bottleneck to algae-based fuels. *J. Renew. Sustain. Energ.* 2:012701.

Van Den Hende, S., H. Vervaeren, H. Saveyn, G. Maes, and N. Boon. 2011. Microalgal bacterial floc properties are improved by a balanced inorganic/organic carbon ratio. *Biotechnol. Bioeng.* 108:549–558.

Vandamme, D., I. Foubert, I. Fraeye, B. Meesschaert, and K. Muylaert. 2012a. Flocculation of *Chlorella vulgaris* induced by high pH: Role of magnesium and calcium and practical implications. *Bioresour. Technol.* 105:114–119.

Vandamme, D., I. Foubert, I. Fraeye, and K. Muylaert. 2012b. Influence of organic matter generated by *Chlorella vulgaris* on five different modes of flocculation. *Bioresour. Technol.* 124:508–511.

Vandamme, D., I. Foubert, and K. Muylaert. 2013. Flocculation as a low-cost method for harvesting microalgae for bulk biomass production. *Trends Biotechnol.* 31:233–239.

Vonshak, A., S. Boussiba, A. Abeliovich, and A. Richmond. 1983. Production of *Spirulina* biomass: Maintenance of monoalgal culture outdoors. *Biotechnol. Bioeng.* 25:341–349.

Wijffels, R. H., M. J. Barbosa, and M. H. Eppink. 2010. Microalgae for the production of bulk chemicals and biofuels. *Biofuels Bioprod. Biorefin.* 4:287–295.

Wileman, A., A. Ozkan, and H. Berberoglu. 2012. Rheological properties of algae slurries for minimizing harvesting energy requirements in biofuel production. *Bioresour. Technol.* 104:432–439.

Zhang, J. and B. Hu. 2012. A novel method to harvest microalgae via co-culture of filamentous fungi to form cell pellets. *Bioresour. Technol.* 114:529–535.

Zhang, X., P. Amendola, J. C. Hewson, M. Sommerfeld, and Q. Hu. 2012. Influence of growth phase on harvesting of *Chlorella zofingiensis* by dissolved air flotation. *Bioresour. Technol.* 116:477–484.

12 Cultivation of *Haematococcus pluvialis* for Astaxanthin Production

Jianguo Liu, John P. van der Meer,
Litao Zhang, and Yong Zhang

CONTENTS

12.1 INTRODUCTION

Haematococcus pluvialis Flotow (hereafter referred to as "*Haematococcus*") is a biflagellate, unicellular, green alga with spherical, ellipsoidal, or pear-shaped cells. The cup-shaped chloroplasts bear multiple pyrenoids. The protoplast is connected to the outer cell wall by numerous strands of cytoplasm that extend through the gelatinous surrounding wall material (Elliot 1934; Santo and Mesquita 1984). Studies of *Haematococcus* go back over a century's history with Flotow's early observations on this alga in 1844. The first extensive description of the life history of *Haematococcus* in English was by Hazen (Hazen 1899). The ability of this alga to accumulate high levels of astaxanthin has long been recognized

267

(Goodwin and Jamikorn 1954; Boussiba 2000). At the time Hazen (1899) described the life history of *Haematococcus*, the chemical nature of this red coloring matter within the alga was unknown but was given the name "haematochrom." Astaxanthin is present in lipid globules outside the chloroplast of *Haematococcus*. Its functions in the cell are complex, at least including protection from photodamage by reducing the amount of light impacting the light-harvesting pigment–protein complex (Boussiba and Vonshak 1991; Yong and Lee 1991). Astaxanthin has broad applications in the aquaculture and poultry industries as a feed additive to improve the coloration of cultured fish and crustaceans, and of egg yolks, and is also a potent bioactive antioxidant with applications in the cosmetic, nutraceutical, and pharmaceutical industries (Benemann 1992; Hussein et al. 2006). The price for 5% natural *Haematococcus* astaxanthin in oil currently sells in bulk for about USD 800 per kg, while synthetic astaxanthin (mainly used in aquaculture) sells for almost 10-fold less, based on market sources (J. Benemann, 2014, pers. comm.).

Laboratory culture of *H. pluvialis*, like that of many other microalgae, is not particularly difficult, requiring only relatively simple techniques. However, the scale-up to a commercial production level is very challenging because cell growth and astaxanthin accumulation are affected by many interactive environmental factors whose efficient control during the culturing process, especially for large-scale, mass cultivation of *H. pluvialis* to obtain optimal astaxanthin production, is a difficult task. Many common and higher-level techniques must be mastered, requiring a deep understanding of the underlying scientific biological principles, along with knowledge of engineering, process control, personnel management, and so on, before a successful mass cultivation of *Haematococcus* for astaxanthin production can be realized.

12.2   CELL CYCLE REGULATION AND TWO-STAGE CULTURING STRATEGY

Haematococcus is widely distributed in nature and can be readily found in many natural settings, from fresh to slightly brackish pools of water in rock depressions or man-made environments, in particular, birdbaths. It also can be found in similar pools inland (Hazen 1899; Droop 1954, 1961). More than 50 strains of *Haematococcus* have been isolated from such natural environments and are available in microalgal culture collections such as the Biological Resource Center, Culture Collection of Algae and Protozoa, Freshwater Algae Culture Collection at the Institute of Hydrobiology, and Sammlung von Algenkulturen. Thus far, 42 wild strains of *H. pluvialis* and 10 UV- and chemical-induced mutants derived from them (Sun et al. 2008) have been obtained in our laboratory. Molecular analysis of their 800 bp rDNA internal transcribed spacer (ITS) sequences including 18S, ITS1, 5.8S, ITS2, and part of 28S showed the genetic distance between any two different wild-type strains of *H. pluvialis* is quite small (Liu et al. 2014). Therefore, all of these strains of *Haematococcus* can be classified into one highly conservative, pan-global species, no matter where the isolate originated (Liu et al. 2014).

Although *H. pluvialis* strains cannot tolerate high salt concentrations, they are rather resistant to desiccation, nutrient deficiencies, and various other environmental stress conditions. The reproduction of algae belonging to this genus is similar to *Chlamydomonas*, having several cell types and cell cycles (Elliot 1934).

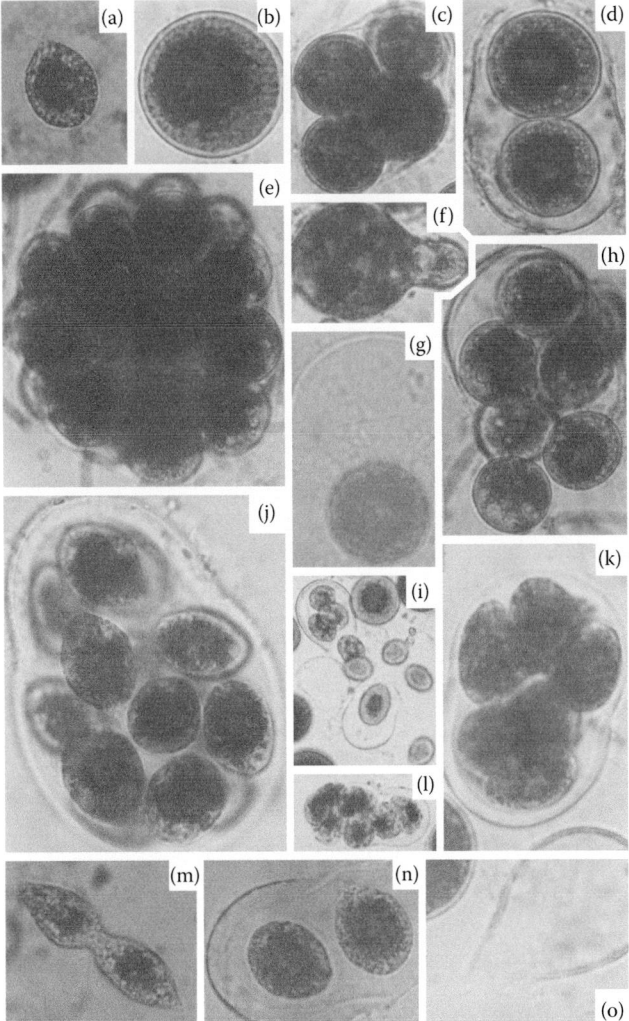

FIGURE 12.1 **(See color insert.)** The cell forms of *Haematococcus pluvialis*: (a) Motile cell. (b) Nonmotile cell. (c, d) Sporangia with 2–4 aplanospores. (e) Sporangia with >20 aplanospores. (f) Vegetative reproduction by cell budding. (g, h) The moment of aplanospore release. (i) The moment of zoospore release. (j–l) The early process of zoospore release. (m) Vegetative reproduction by direct cell division. (n) Sporangia with two zoospores. (o) Theca after spore release.

The complex life history of *Haematococcus* can be divided into two phases: the motile phase and the nonmotile phase (Figure 12.1). The cells can be classified into four forms, namely, motile cells, nonmotile cells, aplanospores, and zoospores (Elliot 1934; Liu et al. 2000). The main cell proliferation of this alga is by asexual reproduction both in the motile phase and the nonmotile phase (Mesquita and Santo 1984). Vegetative reproduction in the motile phase is by direct cell division (Figure 12.1m) and in the nonmotile phase by cell budding (Figure 12.1f) (Liu et al. 2000). Although isogamous sexual reproduction has been described (Elliot 1934; Lee and Ding 1994), the reported evidence remains unconvincing (Liu et al. 2000). Stronger, direct evidence is still needed to make the case.

The motile cells (Figure 12.1a) usually produce two (Figure 12.1n), sometimes four, and exceptionally eight zoospores by asexual reproduction. Under stressed conditions, the fast-growing motile cells gradually lose their flagella and motility and are then transformed into larger, nonmotile cells with thick cell walls (Figure 12.1b). When the nonmotile cells experience unfavorable growth conditions, they are also able to release new cells at a slow rate through the formation of sporangia that release 2, 4, or 8, and occasionally more than 20, aplanospores (Figure 12.1c-e,h) and start to accumulate carotenoids (mostly astaxanthin) in droplets within the cytoplasm. In nature, such cells color rock pools and concrete birdbaths in orange, red, or purple. Once the stresses are relieved, for example, by rainfall, or when cells are transferred into fresh medium in culture, large, nonmotile cells produce 16 zoospores, and the smaller ones produce 8 zoospores (Figure 12.1i-l). Many factors such as nutrient level, illumination level, temperature, pH, and salinity as well as the existence of an autoinhibitor within the cells can affect the process (Boussiba and Vonshak 1991; Kobayashi et al. 1992; Tjahjono et al. 1994; Yin et al. 1998, 2007; Sun et al. 2001; Liu et al. 2002; Li et al. 2006).

The autoinhibitor produced by *H. pluvialis* strongly regulates the cell cycle through a feedback growth-inhibition mechanism (Sun et al. 2001). This autoinhibitor reduces biomass production by inhibiting the growth of the motile cells and inducing them to undergo transformation into nonmotile cells. The inhibition and induction functions of the autoinhibitor are both culture-time and cell-density dependent (Sun et al. 2001). No inhibitory effect is detected in early-stage cultures or in low-density cultures. The longer the algae are cultured, the more autoinhibitor is accumulated and the stronger the cell growth is inhibited. The higher the cell density, the more autoinhibitor is accumulated, which then leads to strong motile cell growth inhibition and rapid cell transformation from motile cells into nonmotile cells (Sun et al. 2001). Interestingly, autoinhibitor is mainly produced by the motile cells of *H. pluvialis*. When motile cells are transformed into nonmotile cells, the concentration of autoinhibitor stops increasing. Cytoplasmic division, rather than DNA replication, is inhibited by the autoinhibitor in the nonmotile cells. Thus, each nonmotile cell may contain several sets of chromosomes (Liu et al. 2004). Once the stressing factor is relieved, the nonmotile cells of *Haematococcus* complete their cytoplasmic divisions and within hours release zoospores. This explains how generally slow-growing *Haematococcus* can successfully dominate the ecosystem of shallow water pools on natural rocks (Liu et al. 2004). The cell growth rate of *Haematococcus* is thus dependent on the cell phase and is influenced by the action of the autoinhibitor and various environmental factors (Boussiba and Vonshak 1991; Kobayashi et al. 1992; Tjahjono et al. 1994; Sun et al. 2001; Liu et al. 2002; Domínguez-Bocanegra et al. 2004; Li et al. 2006; Yin et al. 2007).

The carotenoid content of *H. pluvialis* is also dependent on such factors, including the level of nutrients (N and P concentration in particular) (Sun et al. 2001; Liu et al. 2004), the light density and wavelength, water temperature, pH, and salinity

(Boussiba and Vonshak 1991; Kobayashi et al. 1992; Tjahjono et al. 1994; Yin et al. 1998, 2007; Liu et al. 2002; Fábregas et al. 2003; Li et al. 2006). Parameters favorable for cell growth are generally unfavorable for astaxanthin accumulation (Boussiba and Vonshak 1991; Kobayashi et al. 1992; Tjahjono et al. 1994; Yin et al. 1998, 2007; Liu et al. 2002; Li et al. 2006). A high astaxanthin concentration is only obtained after the fast-growing, motile cells are transformed into slow-growing, nonmotile ones (Yin et al. 1998). Thus, effective control of the culture parameters is needed to manage this alternation of cell growth and astaxanthin accumulation in *H. pluvialis*. Obtaining a good balance between growth and astaxanthin production is crucial for pilot- and industrial-scale mass cultivation for the production of natural astaxanthin. In commercial cultivation of *Haematococcus* for maximal astaxanthin production, a two-stage culture protocol has been developed based on the existing observations and theory (Yin et al. 1998). In the first stage, highly favorable growth conditions are provided to maintain the *Haematococcus* in the fast-growing, motile cell phase to obtain a high biomass, and then, in the second stage of the growth strategy, the motile cells are stressed by N-limitation, P-starvation, and strong illumination to induce them to transform into nonmotile cells, so as to obtain high astaxanthin production (Kobayashi et al. 1992; Sun et al. 2001; Li et al. 2006; Liu et al. 2004).

12.3   CULTURE MODES

The principal aim of *H. pluvialis* mass culture is, of course, to produce a steady supply of high-quality astaxanthin products. Astaxanthin production is significantly dependent on two negatively correlated parameters, the total biomass production and the biomass astaxanthin content (% dry weight). Fast cell growth (increase in biomass) is usually linked to low cellular astaxanthin content. In contrast, high levels of astaxanthin accumulation are generally associated with a large decrease in the cell growth rate. In addition, the production of astaxanthin and its quality are also strongly related to both the strains of *Haematococcus* selected for cultivation (Table 12.1) and the culture mode used during the mass culture. In Table 12.1, eight strains are compared in a large number of batch cultures, for their ability to accumulate astaxanthin. The ability of strains H3, H6, H10, and H11, to accumulate astaxanthin, appears to be significantly higher than strains H0, H2, H5, and H9. Many studies have demonstrated that *Haematococcus* is capable of photoautotrophic, heterotrophic, and mixotrophic growth (Boussiba and Vonshak 1991; Kobayashi et al. 1992; Chen et al. 1997; Boussiba 2000; Orosa et al. 2001; Kang et al. 2005; Li et al. 2006). Each of these growth modes has culturing advantages and disadvantages. The great advantage of using a heterotrophic culture mode is that it results in an accelerated cell proliferation rate for *Haematococcus* and thus boosts biomass productivity (Chen et al. 1997; Kang et al. 2005). It can also be used for indoor laboratory and small-scale cultivation studies of the alga to obtain high cell density or to quickly generate small amounts of biomass. Unfortunately, the advantages of heterotrophic cultivation of *Haematococcus* disappear when cultivation is scaled up to achieve tons of commercial production because astaxanthin synthesis and accumulation are strongly light-induced processes for *Haematococcus* (Steinbrenner and Linden 2000, 2003). With current knowledge and technology (e.g., the available strains), it is essentially

TABLE 12.1

Astaxanthin Content in Various Strains of *Haematococcus pluvialis*, Which Were Batch Cultured on a Large Scale in Open Raceway Ponds at the Yunnan Alphy Biotech Co., Ltd (Chuxiong, Yunnan, China)

Strains	Average Astaxanthin Content ± Standard Error (% of Dry Weight)	Number of Samples	Maximum (%)	Minimum (%)	P	
Total	1.99 ± 0.64	896	4.08	0.3		
H0	1.90 ± 0.63	174	3.59	0.30	0.0436	<0.05
H2	1.90 ± 0.61	309	3.55	0.30	0.014	<0.05
H3	2.24 ± 0.74	102	4.08	0.57	0.0007	<0.01
H5	1.90 ± 0.74	30	3.31	0.57	0.2579	>0.05
H6	2.13 ± 0.48	159	3.50	0.92	0.0008	<0.01
H9	1.56 ± 0.65	23	2.76	0.46	0.0024	<0.01
H10	2.16 ± 0.58	58	3.20	0.54	0.0179	<0.05
H11	2.29 ± 0.67	18	3.99	1.03	0.0385	<0.05

Notes: The samples of *Haematococcus* biomass in this analysis were photoautotrophically produced over a 5-year period (2009–2013). Only one sample was collected from each batch cultivation for analysis. From different large-scale cultivations, 896 samples were collected during that period.

impossible to obtain a high astaxanthin content and high-quality astaxanthin products using heterotrophic metabolism, without subjecting the cells to a follow-up growth period under photosynthetic, photoautotrophic, and/or mixotrophic, cultivation. A further complication is that during such a secondary phototrophic cultivation period for astaxanthin accumulation, especially in large-scale cultivation, it is nearly impossible to maintain an axenic culture of *Haematococcus*. Unwanted microorganisms (including fungi, bacteria, and other species of microalgae) invade the system from various sources and quickly multiply using the residual organic materials remaining after the heterotrophic growth stage, inevitably reducing the quality of the final products. Contamination with algal-grazing zooplankton can even destroy the entire *Haematococcus* culture.

Large-scale mixotrophic culture of *Haematococcus* faces similar shortcomings and problems as those just described for the heterotrophic culture approach. The advantage for mixotrophic culturing, which is to obtain high biomass production and astaxanthin accumulation in small-scale cultures, vanishes on scale-up due to the inevitable culture-contamination issue. For heterotrophic or mixotrophic approaches to be useful for large-scale cultivation of *Haematococcus*, a breakthrough in contamination control will be required. The principal advantage of using the photoautotrophic growth mode for large-scale *Haematococcus* cultivation derives from its steady rate of production and higher astaxanthin content, required for a high-quality astaxanthin product, although specific cell growth rate is slower than that of the other modes. Photoautotrophic cultivation of *Haematococcus* also

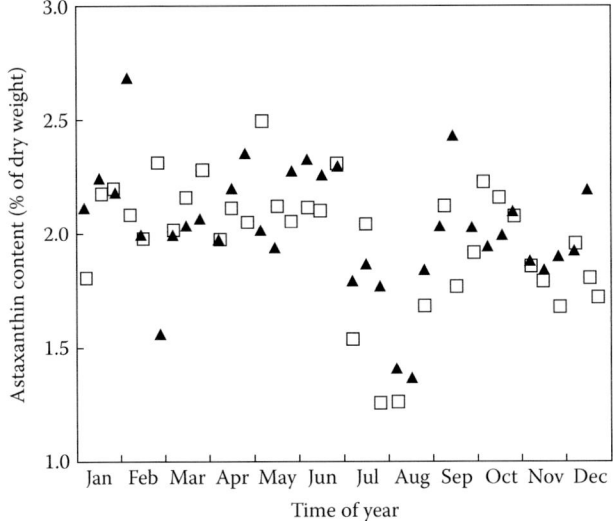

FIGURE 12.2 The annual variation of astaxanthin content in photoautotrophically mass-cultivated *Haematococcus pluvialis*. Each datum represents an average of all samples, no matter which strains were cultivated, taken every 10 days from open raceway ponds both inside greenhouses (squares) and outdoors (triangles).

suffers less from the contamination problems associated with the heterotrophic and mixotrophic cultivation modes. This is because there is little organic matter in the culture throughout the growth cycle, thus allowing extended periods of steady, large-scale industrial production. Currently, essentially all *Haematococcus* cultivation companies worldwide have adapted the photoautotrophic culture mode for commercial-scale production of natural astaxanthin, regardless of the types of cultivation systems, photobioreactors (closed flat, column, or tubular photobioreactors) or open raceway ponds, or strains of *Haematococcus* used.

In the case of our photoautotrophic cultivation of *Haematococcus* in Yunnan, China, the average annual astaxanthin content of the algal powder has fluctuated around 2.0% of dry weight for open raceway ponds (Figure 12.2) and was generally between 3.0% and 4.0%, reaching a high of 4.9% of dry weight for the tubular glass photobioreactors (data not shown). The decrease of astaxanthin content in the algal biomass during the summer rainy season in July–August (Figure 12.2) is mainly correlated with decreased illumination during cloudy weather and heavy precipitation and was also due in part to increased levels of biological contamination. The cultures are easily contaminated both in the open raceway ponds and closed photobioreactors. The variation in astaxanthin content of *Haematococcus* biomass was relatively large, obscuring the effects of strain differences and seasonal weather changes (e.g., light and temperature) (see Table 12.1 and Figure 12.2). The medium composition and its nutrient content mainly affect biomass productivity and the length of the production cycle for astaxanthin. The more nutrients added to the photoautotrophic cultures, the higher the biomass yield, and the longer the astaxanthin

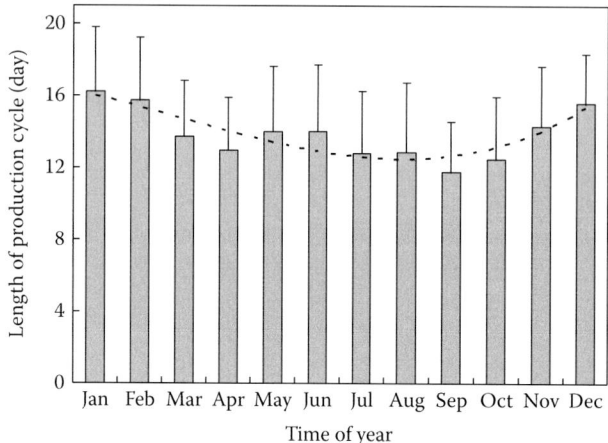

FIGURE 12.3   The annual variation in the length of a production cycle for mass cultivation of *Haematococcus pluvialis* for natural astaxanthin, using open raceway ponds in Yunnan, China. The statistical analysis is based on 5 years of industrial-scale cultivation by the Yunnan Alphy Biotech Co., Ltd.

production cycle. With a fixed amount of nutrients, the length of the production cycle for astaxanthin is relatively constant but somewhat longer in winter than other seasons (Figure 12.3).

12.4   LIEBIG'S LAW OF THE MINIMUM APPLIED TO MICROALGAL CULTURING

Liebig's law of the minimum (originally formulated by Carl Sprengel) states that growth is controlled by the most limiting factor in the system, though it is sometimes difficult to identify this with biological systems. Production of *H. pluvialis* biomass and natural astaxanthin is affected by many physical, chemical, and biological factors, constantly in flux (Boussiba 2000). These factors, including light (intensity, quality, seasonal photoperiod), temperature, pH, CO_2, O_2, and nutrients (nitrogen, phosphate, potassium, calcium, magnesium, iron) and many other microelements and also contaminating microorganisms, are constantly changing over the course of the culture period, especially in outdoor, open pond cultivation systems (Borowitzka et al. 1991; Boussiba and Vonshak 1991; Kobayashi et al. 1992; Tjahjono et al. 1994; Boussiba 2000; Sun et al. 2001; Liu et al. 2002, 2004; Li et al. 2006; Yin et al. 2007). The influences of environmental conditions on cell growth and astaxanthin accumulation are far more complex in commercial-scale farm cultivation under natural conditions than for indoor, laboratory-scale simulations.

Currently, the optimal conditions for cell growth and astaxanthin accumulation are not well established, even though many studies have been conducted, both published and proprietary to commercial companies. Batch cultures are a commonly used method in preliminary studies, but any conclusions from such experiments are only qualitative rather than quantitative. For example, in such experiments, the nutrient level added at the beginning of the experiment is consumed by the growth of the algae, with high astaxanthin accumulation observed usually only at the end of the batch growth experiment, obscuring the real relationship between the actual instantaneous specific growth rate and the astaxanthin accumulation rate (Liu et al. 2002). Accordingly, in commercial cultivation, the growth conditions have to be further optimized. As noted earlier, nitrogen, phosphorus, and other nutrient limitations are favorable for astaxanthin accumulation but not conducive to biomass production. The key level of each nutrient determining the transition from cell growth to astaxanthin accumulation is still quantitatively unclear. Furthermore, in most previous batch culture studies, the effects of a single factor on cell growth or astaxanthin accumulation were examined, missing the potential dynamic influence of multiple factors. Erroneous conclusions from such studies can easily lead to the wrong decisions in mass culture of *Haematococcus*.

A detailed understanding of the complex impacts of environmental factors on cell growth and astaxanthin accumulation in *Haematococcus* is an important prerequisite for culture condition optimization and process control. Maximal cell growth and astaxanthin accumulation are greatly dependent on the limiting factors at any given time, which can change over time, except in cases of gross deficiencies. For commercial cultivation, every effort must be made to identify the factors limiting growth and favoring astaxanthin accumulation, and the complex interactions among these various factors, so that any limitations can be remedied in a timely fashion to sustain a high and stable level of astaxanthin production.

12.5 LIGHT

In photoautotrophic cultures, light plays an important role both in inducing astaxanthin accumulation and in regulating the cell cycles of *Haematococcus*, as well as in supporting biomass accumulation via photosynthetic metabolism. The specific rate of cell growth and astaxanthin accumulation is a function of the photon flux intensity, quality, quantity, and photoperiod to which *Haematococcus* cultures are exposed. In most small-scale laboratory studies, relatively stable illumination of predetermined light intensity, photoperiod, and light wavelength are provided artificially. Natural sunlight, however, always has fluctuations, from annual variations of the irradiation angle and seasonal light/dark period to the daily light cycle from sunrise to sunset, as well as some irregular, unpredictable variations due to cloud conditions. Therefore, both the motile cells and the nonmotile cells in large-scale-cultured *Haematococcus* frequently suffer from both strong photoinhibition because of strong sunlight, and also photolimitation under dim light conditions or darkness at night.

12.5.1 LIGHT INTENSITY

The light reaching the algal cells in an algal culture can be viewed in terms of light intensity averaged over the population density and also in terms of the actual light intensity and its time variations experienced by individual cells. The latter will depend on the culture system (e.g., its geometry), and the cell's movement through the light gradient in the culture, between the more and less illuminated parts of the culture (which depends also on the mixing system). From our studies (Zhang 2004) on the fast-growing motile cells of *Haematococcus*, three completely different relationships exist among individual cells if the culture is well mixed and all cells freely shuttle about in the light gradient and mutually share the benefits of illumination and light shielding. The three relationships are photocompetition, concerted photoadaptation, and photoinhibition (Zhang 2004), which are all mainly determined by the level and amount of light energy received by each individual cell (light per cell).

Photocompetition: Under very low light intensity (<2.6×10^9 photons s^{-1} cell^{-1}), each cell receives insufficient illumination to maintain its basic metabolism, and the *Haematococcus* cells respond by absorbing as much of the available light as possible. The motile cells increase in size and chlorophyll content, while their pigments gradually move to a pericellular distribution, allowing the cells to absorb more light. Culture cell density stops increasing and may decline due to some cells dying under such dim light. Larger cells with a high content of chlorophyll pigments survive through photocompetition because of the larger cross section for absorbing light (Zhang 2004).

12.5.2 PHOTOADAPTATION AND PHOTOINHIBITION

When each motile cell of *Haematococcus* is exposed to illumination ranging between 4.2 and 6.0×10^9 photons s^{-1} cell^{-1}, the cell receives enough light for its basic metabolic demands (Zhang 2004). When the light per cell increases toward the upper end of this range, both the volume and chlorophyll content of the cells increase moderately so as to achieve photoadaptation by mutual cell shading. In contrast, if the light per cell declines toward the lower end of this range (4.2), both the size and chlorophyll content of the cells show a corresponding decrease, allowing more cells to absorb light and survive (Zhang 2004). Within this light range, such concerted changes maximize the light utilization by the cell population.

Strong photoinhibition of the motile cells inevitably occurs when the light per cell is over 6.0×10^9 photons s^{-1} cell^{-1} (Zhang 2004). All cells are exposed to a light-saturated condition, the mutual cell shading loses its effectiveness, and the concerted photoadaptation of cells is disrupted. To reduce light absorption, cells respond by reducing their level of chlorophyll, even bleaching. As a result, the surviving cells encounter even higher light intensity, and photoinhibition becomes even more serious. Ultimately, under these light levels, the whole culture will suffer an inevitable collapse (Zhang 2004).

It needs to be mentioned that the light demands and the ability for photoadaptation not only differ among different strains of *Haematococcus* but also vary significantly among the various types of cells in the life cycle. High-light-tolerant strains of *Haematococcus* can be selected for industrial production of astaxanthin in cultivation regions that have strong illumination, especially during the sunny seasons. In contrast, low-light-preferring strains can be used in regions with more moderate illumination, such as during the rainy seasons.

The light requirement for photosynthesis in motile cells is usually higher than that in the nonmotile cells, but interestingly, their photoadaptive capacity is much weaker than the nonmotile cells (Figure 12.4). When *Haematococcus* is exposed

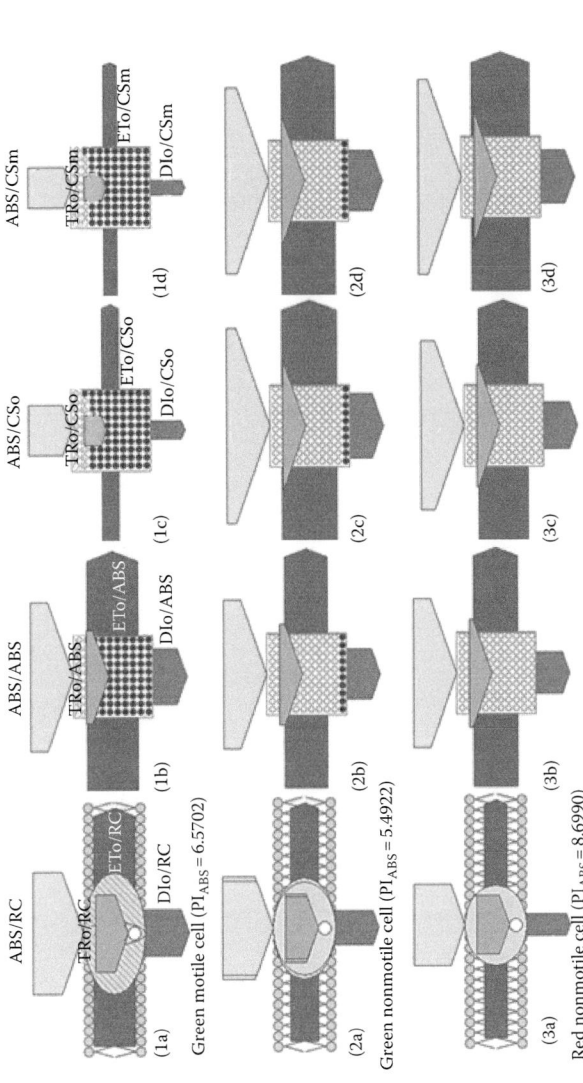

FIGURE 12.4 **(See color insert.)** The photochemical parameters and energy pipeline models of different cell forms of *Haematococcus pluvialis*: (a) specific membrane model (per active reaction center, RCs); (b) phenomenological yield model; (c) phenomenological model (per excited cross section) at time zero; and (d) phenomenological model (per excited cross section, CSm) at the time of reaching the maximal fluorescence. The photochemical parameters were measured with a Handy PEA fluorometer (Hansatech Instruments). All the measurements were done with 10 minute dark-adapted samples at room temperature. The chlorophyll contents in green motile cells, green nonmotile cells, and red nonmotile cells were 0.26, 0.31, and 0.61 μg/10^4 cell, respectively. And the astaxanthin contents in green motile cells, green nonmotile cells, and red nonmotile cells were 0.03, 0.12, and 0.48 μg/10^4 cells, respectively. ABS, TRo, ETo, and DIo represent the energy fluxes for absorption (yellow), trapping (light blue), electron transport (dark blue), and dissipation (red). The relative value of each of the parameters (ABS/CSm, TRo/CSm, ETo/CSm, and DIo/CSm) can be seen from the width of its arrow. Active and inactive RCs are shown as open and solid circles, respectively.

to short periods of strong light, the motile cells adapt by closing most of their PSII (photosystem II) reaction centers (RCs) (Figure 12.4—1b through d) and reducing their activities, rather than reducing light harvested per active reaction center (RC). Only a relatively small amount of the captured energy can be successfully transferred to PSI (Figure 12.4—1c and d). Interestingly, the nonmotile cells show a strong photoadaptation by both reducing the light harvested and light captured per RC (Figure 12.4—2a and 3a) instead of closing their PSII reaction centers (Figure 12.4—2b through d, 3b through d). Major PSII reaction centers of these cells remain activated and can efficiently convert the light energy into chemical energy, which then can be transferred to PSI successfully, thanks to a smooth electron transfer flow (Strasser et al. 2000; Force et al. 2003).

To examine the photoinhibition of PSII in *Haematococcus*, the Fv/Fm and performance index (PI_{ABS}) were measured. Fv/Fm is the maximal quantum yield of primary photochemistry. The performance index (PI_{ABS}) is the most sensitive parameter of the JIP-test because it incorporates several parameters that are evaluated (Strasser et al. 2000). The PI_{ABS} is more sensitive to high light than the Fv/Fm. The Fv/Fm and PI_{ABS} have a significant daily variation in the large-scale outdoor cultures. The variations of Fv/Fm and PI_{ABS} in a tubular photobioreactor are quite similar to those measured in an open raceway pond. However, the rate of decrease of PI_{ABS} in the morning and the rate of increase in the afternoon in open raceway ponds are faster than those measured in the tubular photobioreactor (Figure 12.5). The maximal Fv/Fm and PI_{ABS} usually appear after extended darkness during the night, until dawn. During most of the daylight hours, motile cells of *Haematococcus* experience serious photoinhibition because of the strong sunlight, even in the early morning (Figure 12.5). Comparing with the data before dawn, the Fv/Fm and PI_{ABS} of the motile cells at 8:00 hours are only 78% and 45% of their maxima, respectively. The instantaneous Fv/Fm and PI_{ABS} drop as low as 45% and 3% of their maxima

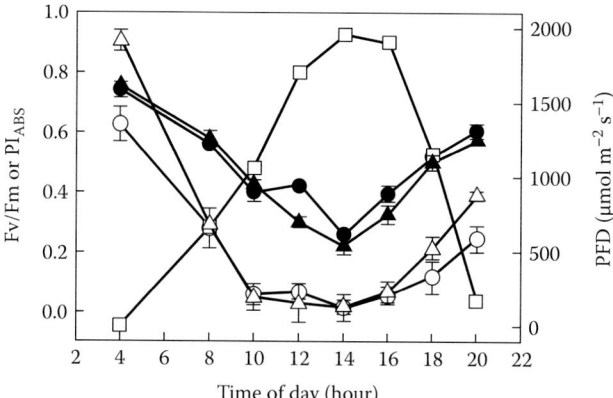

FIGURE 12.5 The diurnal changes in photon flux density (PFD) of sunlight (hollow squares), Fv/Fm (solid circles and triangles), and PI_{ABS} (hollow circles and triangles) of mass-cultivated motile phase cells of *Haematococcus pluvialis* in a tubular photobioreactor (circles) or open raceway pond (triangles).

from 12:00 to 14:00 hours when the most serious photoinhibition occurs in outdoor cultures (Figure 12.5). After that, in the daily cycle, both Fv/Fm and PI_{ABS} start to increase gradually again from 16:00 hours, when the sunshine begins to decrease in intensity, and the Fv/Fm and PI_{ABS} have recovered to 70% and 20%, respectively, by 18:00 hours. The motile cells then fully recover from photoinhibition during the period from dusk to the middle of the night. During the recovery period, the slow recovery of antennae (the efficiency of light absorption) and reaction centers in motile cells are the two processes most limiting photosynthetic recovery.

During a transition period from astaxanthin-free green motile cells to astaxanthin-rich red nonmotile cells (Figure 12.6a), both Fv/Fm and PI_{ABS} change greatly in the tubular photobioreactors (Figure 12.6b). The daily maximal Fv/Fm and PI_{ABS} slightly increase in the first 4 days of cultivation then gradually drop during the following culture period when the motile cells start to transform into nonmotile cells and began to accumulate astaxanthin (Figure 12.6b). The decrease of maximal PI_{ABS} seems much faster than the decrease of maximal Fv/Fm. The daily minimal Fv/Fm

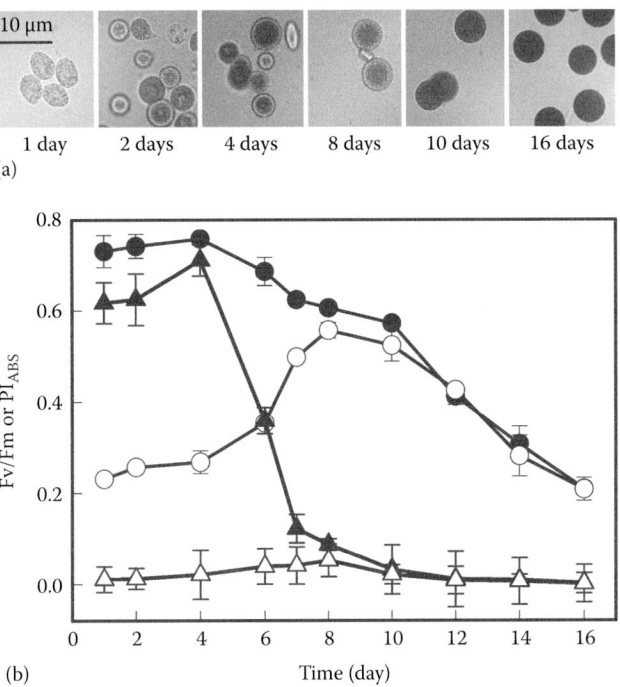

FIGURE 12.6 The variation of cell morphology, the maximal and minimal Fv/Fm and PI_{ABS} in *Haematococcus pluvialis* during a culture period from astaxanthin-free motile cells to astaxanthin-enriched nonmotile cells in a tubular photobioreactor: (a) **(See color insert.)** the cell morphological changes and (b) the maximal and minimal Fv/Fm (circles) and PI_{ABS} (triangles) variation. The samples for maximal Fv/Fm and PI_{ABS} (solid circles and triangles) measurements were collected after extended darkness, during the night, before dawn, and at 4:00 hours, and those for analysis of the minimum (hollow circles and triangles) were obtained at 14:00 hours.

undergoes a significant increase during the first 8 days of cultivation, even reaching a level quite close to its daily maximal Fv/Fm. Then, both of the maximal and minimal Fv/Fm decrease linearly in the following culture period (Figure 12.6b) when it is also apparent that the cell accumulates a large amount of astaxanthin after 8 days of cultivation (Figure 12.6a). With regard to the Fv/Fm and PI_{ABS} of astaxanthin-enriched nonmotile cells (after 8 days of cultivation), no significant daily variations are found either in tubular photobioreactors or in open raceway ponds, which indicates that no photoinhibition occurs in such astaxanthin-containing *Haematococcus* (Figure 12.6). These results also indicate that other stress factors (such as nutrient deprivation), and not just strong light, have inhibitory effects on photosynthesis during the period of astaxanthin accumulation when *Haematococcus* is exposed to strong natural sunlight.

Flashing light has been drawing increasing interest as a potential alternative light source to enhance the efficiency of photosynthesis in indoor cultures and hence increase the system's productivity in generating biomass for algal-derived products (Kim et al. 2006). Flashing low light gave the same final astaxanthin concentration as that obtained under continuous illumination, thus reducing energy consumption per unit of product. In view of these results, flashing light appears to be a promising illumination method for algal cultivation (Kim et al. 2006).

12.5.3 PHOTOPERIOD

From our observations, the number of cell divisions per day appears to be dependent on light quantity (light intensity × net illumination time). When a *Haematococcus* culture is exposed to a short-day light/dark regime, only a fraction of its motile cells produce zoospores, and one mother motile cell divides only once to produce two daughter zoospores within a sporangium (Yin et al. 1998). In contrast, when a *Haematococcus* culture is exposed to a long-day regime, particularly with strong outdoor illumination, most of the motile cells divide the next morning, and some mother cells can produce four, and sometimes even eight, zoospores instead of the usual two. The protoplasm volume of the motile cells before division seems to be directly proportional to the length of the illumination period (Yin et al. 1998). Cytoplasmic division of motile cells usually starts after 7 hours of darkness. Cell division in the natural environment often begins at dawn, and the new daughter cells start to separate in the early morning and finish before noon, usually by 10:00–11:00 hours. Astaxanthin accumulation in *Haematococcus* is also influenced by photoperiod: the longer the day length, the higher the level of astaxanthin accumulation in *Haematococcus*, and *vice versa*. This phenomenon of long-day illumination enhancing astaxanthin accumulation is further evidenced in outdoor mass cultivation. The average astaxanthin concentration in *Haematococcus* in spring (from March to May) is about 7% higher than that in autumn (from September to November; see Figure 12.2). Our early work on static cultures also indicated that prolonging the light period not only accelerates the cell growth rate of motile cells but also induces the cell transformation from motile cells into nonmotile cells and improves the astaxanthin accumulation rate. The results earlier suggest that photoperiod plays an important role in regulating both the cell growth rate and astaxanthin accumulation.

12.5.4 LIGHT WAVELENGTH

The influence of light wavelength on cell growth and astaxanthin accumulation in large-scale outdoor cultivation of *Haematococcus* has long been ignored. Recent indoor experiments have shown that the cell growth rate and cell density of motile cells under red light are much higher than those under other light wavelengths. In addition to naturally diffused sunlight, red LED light is now gradually becoming accepted as accelerating cell growth for extended indoor photoautotrophic cultivation, especially during the night. In contrast, blue light and UV light favor astaxanthin accumulation in the nonmotile cells. Much more work is required to fully understand the effects of light wavelength.

12.6 TEMPERATURE

Over the past decades, much attention has been paid to the effect of temperature on cell growth and astaxanthin accumulation in *Haematococcus* (Lu et al. 1994; Tripathi et al. 2002; Evens et al. 2008; Terence et al. 2008). Overall, there appear to be some differences among strains of *Haematococcus* in terms of the best temperature for cell growth. The most favorable growth temperature for the motile cells varies between 20°C and 25°C. Once the temperature exceeds 28°C, the speed of cell division quickly declines, and the process of cell transformation to nonmotile cells is induced. The nonmotile cells have a relatively high-temperature tolerance compared to the motile cells. The biomass of nonmotile cells continues to increase when the water temperature reaches 33°C or even slightly more. The highest temporary temperature for photosynthesis of nonmotile cells and motile cells is about 37°C and 33°C, respectively. In large-scale outdoor cultivation, algal cultures build up heat from strong sunlight, especially in some tropical, subtropical, and even temperate regions during the hot season. The daily temperature of algal cultures may rise sharply to over 35°C in the morning, then slowly decline in the late afternoon and night. It is usually high water temperature rather than other natural environmental factors that hinders industrial mass cultivation of *Haematococcus* in these regions. Choosing a location near cold, deep seawater or with a mild climate and a narrow annual temperature range for siting a facility for large-scale cultivation of *Haematococcus* is a first step toward success. Cooling systems (e.g., a damp water curtain or sprinklers to spray water intermittently over the external surface of photobioreactors) are essential in some cases for maintaining a suitable temperature for *Haematococcus* cultivation. In these situations, the excessive heat, especially at midday, can be dissipated effectively by the cooling effects of water evaporation and ventilation.

Low temperature also reduces the cell growth rate of both the motile cells and nonmotile cells. However, the motile cells continue to grow when the temperature drops, down to just below 10°C. The lowest temperature for photosynthesis we have measured is about 6.7°C. Astaxanthin formation is significantly affected by culture temperature. When *Haematococcus* is cultivated indoors at 30°C, astaxanthin content per cell is higher than at 20°C. Even at low temperature, *Haematococcus* maintains its high capacity for astaxanthin accumulation. This may be due to the fact that the process of photosynthesis at low temperature is quite slow, even if the illumination is not strong; thus, there is excess light that stimulates the process of astaxanthin accumulation. Moreover, astaxanthin accumulation in the cold winter occurs not only in the nonmotile cells but also in the motile cells.

In nature, temperature variation is usually accompanied by an illumination change. The influence of natural temperature variation on astaxanthin accumulation in large-scale cultivation appears to be linked, at least partly, to illumination; however, this interaction is complex and would likely be nearly impossible to disentangle. The average culture temperature in the greenhouse is constantly higher than that of outdoor mass cultures. At the same time, the daily fluctuation both of culture temperature and the light intensity for outdoor cultivation is greater than that inside the greenhouse. Over the past 5 years of mass cultivation, a modest seasonal fluctuation of astaxanthin content in *H. pluvialis* was noted either in the greenhouse or in the outdoor cultures. However, the average astaxanthin content of greenhouse and outdoor cultures showed no significant difference even with many hundred runs. Under real-life conditions, coexisting factors inevitably lead to considerable variability, which could obscure the effects of the temperature differences. Interestingly, the astaxanthin content (about 2.0%–2.5% of dry biomass) obtained during the cold season (winter and early spring) is usually higher than that in summer and early autumn (about 1.5%–2.0% of dry biomass). More detailed experimental evidence is needed to understand the influences of temperature on astaxanthin accumulation in *H. pluvialis*.

12.7 NUTRIENTS

Nutrients, especially the macro elements, play crucial roles in regulating cell growth and astaxanthin accumulation. Up to now, many culture media (e.g., A9, BG-11, BBM, MCM, PHM-1 and Z-8) (Grung et al. 1992; Tripathi et al. 1999; Liu et al. 2002; Domínguez-Bocanegra et al. 2004), and various modifications of these, have been used for *Haematococcus* cultivation, even though their compositions, including the level of each element, differ significantly. Optimization of culture nutrients for high production of biomass and astaxanthin is dependent not only on the nutrients themselves but also on the strains used and the culture conditions. In general, the more nutrients added, the longer the time required for the algal cells to consume them, the faster the cell growth, and the higher the cell density achieved, concomitantly, however, the lower the astaxanthin content of the biomass. There is a complex, negative, nonlinear correlation between cell growth and astaxanthin accumulation. Therefore, finding a suitable concentration of the major elements supporting both cell growth and astaxanthin accumulation, rather than optimizing the culture medium solely for biomass yield, is crucial in large-scale cultivation of *Haematococcus* for astaxanthin production.

Nitrogen and phosphate, the two major nutrients in microalgal culture, are of course the most critical regulating factors (Borowitzka et al. 1991; Boussiba and Vonshak 1991; Liu et al. 2002; Yin et al. 2007; Göksan et al. 2011). Nitrogen as NO_3^- rather than NH_4^+ or urea has been shown in many studies to be the best nitrogen form for *Haematococcus* cultivation, especially for application in outdoor mass cultivation. Our previous studies have shown that a level of 0.5–10 mM of nitrate is favorable for cell growth in batch cultures (Liu et al. 2002). The dynamic changes of inorganic nitrogen, cell growth, and transformation as well as astaxanthin accumulation in batch cultures of *Haematococcus* were studied by inoculating astaxanthin-enriched, nonmotile cells into gradients of nitrate (Liu et al. 2002). The astaxanthin/chlorophyll (ast/chl) ratio was always over 0.8 in astaxanthin-enriched, brown and red nonmotile cells but was usually less than 0.5 in the green motile cells or yellow-green nonmotile cells. In general, nitrate continuously decreased during the course of culturing. Concomitantly, measurable amounts of NO_2^- and NH_4^+ were also observed, although there was no addition of external nitrite and ammonia. A nonlinear correlation between the ast/chl ratio (or color) changes, and the levels of NO_3^-, NO_2^-, and NH_4^+ are found in *H. pluvialis* cultures. The change in the ast/chl ratio always co-occurs with a perceptible color change from yellow to brown (or red), along with a change of NO_3^-, NO_2^-, and NH_4^+ levels to around 30, 5, and 5 µM, respectively. About 50 µM of total inorganic N seems to be the crucial concentration for triggering the color changes, cell transformation, and astaxanthin accumulation (Liu et al. 2002).

As for the influence of phosphate on cell growth and astaxanthin accumulation in *Haematococcus*, there are two totally different conclusions. Some consider that the cell growth of *Haematococcus* is not significantly influenced by the level of phosphate and that high phosphate is conducive to astaxanthin accumulation (Borowitzka et al. 1991). However, other studies show that phosphate limitation accelerates astaxanthin accumulation in *Haematococcus* (Boussiba and Vonshak 1991; Yin et al. 2007). Our studies, using cultures depleted for phosphate, indicate that phosphate is also a key nutrient that directly regulates cell growth and the astaxanthin/chlorophyll ratio in *H. pluvialis*. A low phosphate concentration limits cell growth, induces cell transformation from motile cells to nonmotile cells, and accelerates the process of astaxanthin accumulation, which can all be reversed by phosphate addition (Yin et al. 2007). The conclusion that the phosphate concentration had no effect on the cell growth rate and that high phosphate concentration was good for astaxanthin accumulation may be due to an experimental artifact. This artifact is caused by a rapid inorganic nitrogen utilization when the culture is exposed to high phosphate, which then leads to a secondary physiological nitrogen deficiency (lower than 50 µM). It is the secondary nitrogen limitation, rather than the high phosphate level, that decreases cell growth and induces cell transformation and astaxanthin accumulation (Yin et al. 2007).

Based on the complex, even contradictory, influence of nutrients on cell growth and astaxanthin accumulation, the two-stage culture strategy for the mass cultivation of *Haematococcus* is a good choice for stable, high astaxanthin production (Yin et al. 1998). In the first stage, highly favorable growth conditions, including sufficient levels of the nutrients required for cell growth, are provided so as to maintain the *Haematococcus* cells in the fast-growing, motile cell phase to obtain a large amount of biomass. Then, in the second culture stage, the motile cells are stressed by nitrate and/or phosphate limitation, to induce the process of motile cell transformation into nonmotile cells for rapid astaxanthin accumulation (Yin et al. 1998).

12.8   CONTROL OF BIOLOGICAL CONTAMINATION

Sustained production of natural astaxanthin is dependent not only on the technologies promoting cell growth and accelerating astaxanthin accumulation but also on the effectiveness of protecting the culture from biological contamination. Usually, biological contamination is not a serious problem in small-scale laboratory studies. However, during the scale-up of cultures, especially for outdoor mass cultivation of *Haematococcus*, biological contaminants inevitably infect the cultures through various routes, whether in open raceway ponds or in closed photobioreactors. Most of the contaminating organisms generally become much more active and multiply rapidly at a moderately high temperature; therefore, biological contamination is a particularly serious problem during the hot and rainy seasons. Biological contamination reduces the cell growth rate and decreases the astaxanthin content in the biomass (Figure 12.2). The biological contaminants found in mass cultures include many species of microorganism from bacteria and fungi to species of fast-growing microalgae, as well as grazers (rotifers, amoebae, protozoa, etc.). These contaminants can enter and infect cultures through nutrient and water addition, gas (air and CO_2) bubbling, incompletely sterilized culture devices, insufficient care during culture manipulation, and even a contaminated inoculum. At present, methods of physical prevention (thermal sterilization, micropore filtration, UV radiation, etc.) are the best options. Chemical control agents (some reagents and synthetic chemical pesticides) are only feasible in small-scale cultures of *Haematococcus*; none of these is sufficiently safe, selective, and effective to deal with biological contamination in large-scale cultures. Once the culture is contaminated by unwanted microorganisms, it is nearly impossible to exterminate them. The contaminated culture may soon collapse due to either the aggressive feeding of grazers or the overgrowth of the culture by fast-growing, contaminating microalgae. Even when the culture manages to survive the contamination, it may still fail economically because of a drop in product quality. Accordingly, biological contamination remains a constant problem that has to be managed effectively for successful, industrial-scale cultivation.

To date, comprehensive methods to avoid biological contamination of cultures during the process of biomass expansion remain the most crucial technology (perhaps exceeding the importance of all other culture aspects) in commercial production of *Haematococcus* for astaxanthin. These integrated technical methods include thorough sterilization or disinfection of culture devices; strict aseptic technique; well-trained process control and management personnel; minimization of contamination opportunities, particularly during the early culture stage; effective amplification of the axenic monoculture volume by using closed photobioreactors; and skilled control of parameters that enhance cell density and keep the culture in a fast, logarithmic, cell growth phase and provide high-quality algal cells for the inoculum. In addition, timely detection and efficient elimination of any biological contaminants that do get into the cultures are primary tasks for daily management and process control. Early detection and treatment reduce the probability of a rapid and extensive proliferation of the biological contaminants, which otherwise might lead to more widespread contamination of the surrounding environment and initiate a vicious cycle of contaminations. Drugs that inhibit biological contaminants without damaging the nontarget microalgae are the preferred treatment choices when physical control measures fail. Biopesticides are a good alternative, owing to their effectiveness and relatively safe properties (Huang et al. 2014a,b). Finding and developing some efficient and highly selective natural pesticides that can take the place of broad-spectrum pesticides that also have low toxicity to *Haematococcus* cells, and degrade rapidly without affecting the quality of products, is another important subject for ongoing studies. Successful selection of such substances is a major challenge for phycologists but has a great potential for industrial application. Then, the development of various, effective, closed photobioreactors, and using them in conjunction with appropriate natural pesticides, should be an effective approach for controlling contaminants in mass cultivation of *Haematococcus*.

12.9 PHOTOBIOREACTORS

Just as the fermenter is central to industrial microbiology, the photobioreactor plays an important role in the exploitation of microalgae. Many types of photobioreactors (Figure 12.7) have been developed during the past decades. The most widely used closed photobioreactors include flat, column, and tubular designs. Those used

(a)   (b)

(c)   (d)

FIGURE 12.7   (See color insert.) Photobioreactors used for industrial-scale production of *Haematococcus pluvialis*: (a) Open raceway pond inside a greenhouse; (b) open raceway pond outdoors; (c) closed column photobioreactors inside a greenhouse; and (d) closed tubular photobioreactors outdoors.

for outdoor mass cultivation, rather than indoor basic scientific studies, have the advantage of using free solar energy to drive the growth of photoautotrophic cultures, which saves energy input and lowers production costs. Photobioreactors used for industrial-scale *Haematococcus* cultivation are quite different from fermenters, being usually constructed using transparent glass and various plastics (e.g., plexiglass). These materials have two inherent shortcomings: the material can be so sensitive to heat such that thermal sterilization is impossible and that they may not tolerate high pressure. Therefore, the volume of a photobioreactor cannot easily be scaled up by increasing its height, which increases water pressure. In addition, how to maintain the ratio of surface area/volume as photobioreactors increase in size, so as to let the external light effectively penetrate into the deepest parts of the culture, has been a driving question. The ratio of surface area/volume, an important parameter for evaluating the efficiency of a photobioreactor, is mainly determined by its shape and thickness. The surface area/volume ratio for flat panel and spherical photobioreactors can be calculated by the equations $2(L + W)/L \cdot W$ and $3/R$, respectively, where L and W are the length and width of a flat panel photobioreactor, and R is the radius of a spherical photobioreactor. The surface area/volume ratio either for a vertical column photobioreactor or a horizontal tubular photobioreactor has a negative correlation with its radius and can be calculated as $2/R$. Any increase in the thickness or diameter of the photobioreactor will obviously expand the culture volume, but inevitably decreases the surface area/volume ratio, which in turn reduces the efficiency of light penetration into the photobioreactor. Another choice for scale-up of a culture without decreasing the photobioreactor's surface/volume ratio is to increase the length of the culture tubes, as has been done for cultivation of *Haematococcus* (Pulz 2001; Boussiba and Aflalo 2005). However, this approach, especially when taken for tube lengths over 50 m, is accompanied by other problems. In long tubular bioreactors the exchange of gases has emerged as an increasingly serious problem (Mirón et al. 1999; Molina et al. 2001). First, supplying CO_2, to compensate for its consumption by the algae, becomes increasingly difficult with length and can limit photosynthesis in regions of long tubes distant from the CO_2 injection point, and that inevitably decreases the cell growth rate and astaxanthin accumulation level. Second, excess oxygen released by photosynthesis can build up quickly in the photobioreactor. At times, O_2 saturation can reach as high as 300% and more (Mirón et al. 1999). The excess O_2 saturation further decreases cell growth and astaxanthin accumulation (Posten 2009) in *Haematococcus*. Solving these problems in photobioreactors of increasing size requires finding an overall good design balance for height, length, and radius of the culture vessel, along with effective technological solutions for enhancing gas distribution and exchange (Posten 2009).

Another common phenomenon encountered in mass culture is that both motile cells and the nonmotile cells of *Haematococcus* can adhere to a greater or lesser extent to the inner wall of the photobioreactor, or settle in "dead areas" (i.e., areas with poor circulation) of the connected tube systems. These adherent and settled algal cells reduce light penetration, which also decreases the cell growth rate and astaxanthin accumulation. In addition, these adherent and settled cells are easily bleached by strong sunlight, and then these dead cells can cause biofouling. It is necessary to remove adherent cells from the walls of the photobioreactor and resuspend them into the culture medium in a timely and efficient manner in industrial-scale cultivation. Not only is the traditional cleaning method of taking apart the photobioreactor for cleaning after harvest and then reassembling time- and labor-consuming, but such downtime also reduces the annual yield from the photobioreactor. An existing patented technology (Liu et al. 2010) that allows continuous, automatic cleaning is in great demand for photobioreactor-based, large-scale cultivation. It would be desirable to have more such technologies available.

Up to now, no single photobioreactor, no matter its type, possesses all the characteristics desired for industrial cultivation of *Haematococcus*. Each type of photobioreactor has its own advantages and disadvantages. For example, precise-control photobioreactors, in which most of the culture parameters can be monitored and controlled, can maintain a fast cell growth and high cell density. However, their shortcomings are that they are expensive and usually of small size. Thus, such a precise-control photobioreactor could only be used for laboratory studies or for indoor cultivation to obtain a pure, dense inoculum for initiating cultivation in larger vessels. Selecting an appropriate mix of photobioreactors that have various strengths and can compensate for each other's weaknesses is an important issue in large-scale cultivation, as this approach can result in a reliable and smooth scale-up of the starter culture to industrial scale. Comparative studies based on actual situations should be performed to get information that would allow a balanced decision about such infrastructure. Closed column photobioreactors and tubular photobioreactors, as well as open raceway ponds, have been selected for use in our two-stage cultivation of *Haematococcus* for astaxanthin production in Yunnan Province of the P.R. China (Figure 12.7).

12.10 HARVESTING AND DOWNSTREAM PROCESSING

Haematococcus cells can be harvested by many methods, including gravity sedimentation, flocculation, membrane filtration, and centrifugation. Because of the relatively small cell size of *Haematococcus* (only 10–60 μm in diameter) and low culture biomass density (only about 0.5% dry weight), fast, efficient, and economical separation of the biomass from the liquid culture medium is not an easy task. Progressive concentration by using a combination of separation techniques appears to be the most feasible for large-scale harvesting of *Haematococcus*. Flocculation, a common method used in industrial separations (Rashid et al. 2013; Dong et al. 2014; Letelier-Gordo et al. 2014), is not the first choice for large-scale harvesting of *Haematococcus*, as the addition of any flocculant will potentially cause product contamination, particularly as it is nearly impossible to remove the flocculant in the subsequent processing steps. Physical separation technologies especially flotation, gravity sedimentation, and filtration are gentle, easy to scale up, and noncontaminating (Chen et al. 2011), thereby ensuring that the natural properties of *Haematococcus* astaxanthin are retained in the products. In general, physical methods are always superior to chemical and biological techniques not only for biomass harvesting but also for cell wall cracking and subsequent astaxanthin extraction.

In large-scale cultivation of *Haematococcus*, most of the astaxanthin is accumulated within the nonmotile cells rather than in the motile ones. The astaxanthin-enriched nonmotile cells are much larger in size and heavier than the motile cells and readily sink to the bottom of the raceway ponds or the lower part of the photobioreactor if they are undisturbed by external agitation. Carefully draining off the upper part of the culture supernatant removes most of the original liquid (80%–90%), thereby giving a 5×–10× biomass concentration. Centrifugation shares the advantage of being chemical-reagent-free, but has high capital (equipment) and operational (electricity) costs. With an eye to the economic feasibility of large-scale *Haematococcus* separation, natural sedimentation rather than centrifugation is commonly used for the primary concentration steps. However, after repeated sedimentations, the biomass concentration reaches a point where no further progress can be made by this method. At this point, filtration and centrifugation are required for any further biomass concentration as well as for the subsequent washing steps for cleaning the harvested biomass.

Haematococcus regulates the size, thickness, and structure of its cell wall under various environmental conditions. The nonmotile cells of *Haematococcus* usually possess a thick, strong, secondary cell wall, whereas its motile cells have fragile, thin cell walls. As a cell transforms from a motile into a nonmotile cell, the cell wall undergoes continuous, dynamic modifications from a gelatinous extracellular matrix to a primary cell wall and eventually to a thick, impermeable secondary cell wall (Hagen et al. 2002). Both the thickness and hardness of the cell wall continue to increase during the entire period of astaxanthin accumulation. The final solid cell wall is composed of an outer primary wall, a trilaminar sheath, a secondary wall, and a tertiary wall. Chemical characterization of the external trilaminar sheath of the wall shows that it is made of algaenan. Mannose and cellulose are present in the secondary and tertiary walls (Damiani et al. 2006). The wall is highly resistant to chemical, enzymatic, and mechanical disruption. Even so, this solid barrier must be successfully ruptured before the raw biomass is ready for commercial utilization. Otherwise, the bioavailability of the cell's astaxanthin would likely be too low, because the thick, impermeable cell walls are difficult to digest.

Traditional physical methods such as ultrasound, grinding, and freeze–thaw techniques are unable to split the solid cell wall of *Haematococcus* effectively. In addition, considerable heat is generated in the former two treatments, which results in substantial losses of astaxanthin. Thus, these traditional technologies are not appropriate for cracking the cell walls, especially in industrial-scale operations. Ultralow temperature thermal shock and fast physical grinding methods are commonly used to break open these thick cell walls. Of the physical methods, ultrasonic flow grinding is used for splitting cells in the dried algal powder and high-pressure homogenization for cracking cells of fresh algal "mud" (biomass paste). In the ultrasonic flow grinder, the dried algal materials are carried into a cylindrical chamber and accelerated continuously by strong jetted gases until they reach ultrasonic speeds (about 3× the velocity of sound). At such ultrasonic speeds, violent collisions between cells are inevitable and strong enough to crack open the cells. In high-pressure homogenization, high pressure (1000–1200 bars), generated by strong pumps, is applied to the wet *Haematococcus* mud followed by an almost instantaneous pressure drop to atmospheric pressure. During this depressurizing process, the *Haematococcus* cells are ejected through a tiny aperture at a very high speed and collide with solid sharp surfaces, resulting in the cell walls of

Haematococcus being quickly broken into micron, submicron, and even nanosized fragments by the joint effects of the shearing, cavitation, and crushing forces.

After *Haematococcus* cell cracking, a further astaxanthin extraction from the algal biomass is often necessary to obtain the desired products. Many traditional organic solvents can be used to extract astaxanthin from *Haematococcus*. However, some of the organic solvents are potentially toxic and unacceptable for the food and pharmaceutical industries. Supercritical fluid extraction using CO_2 (Krichnavaruk et al. 2008; Reyes et al. 2014) is an efficient alternative to organic solvents for the extraction of natural astaxanthin due to its ideal extractive properties, such as high compressibility, liquid-like density, low viscosity, and high diffusivity, and is now used worldwide for extraction of natural astaxanthin. Supercritical CO_2 has a greater ability to diffuse through the ultrafine, complex matrix than conventional organic solvents and can be easily separated from the products in the subsequent depressurizing process. Furthermore, the low temperature of supercritical CO_2 also means that the extraction process can be operated at a lower temperature, so as to avoid heat-induced degradation of astaxanthin. As a result, the astaxanthin product obtained is pure and of high quality, and thus is safe for use as a nutritional additive and for pharmaceutical applications.

12.11   FARM SITING

Locating the algae farm on nonarable land is another strategy being applied in the commercial cultivation of *H. pluvialis* (Figure 12.8). This not only saves arable lands for agriculture but also reduces the risk of biological contamination, improves production efficiency, and ultimately reduces production cost, thus increasing the market competitiveness of products. This barren-hills cultivation model includes building water storage tanks on the top of the hill, digging deep wells at the bottom of the hills, and building multilevel terraces on which closed column and tubular photobioreactors and open raceway ponds can be constructed. The clean water from the deep wells is first pumped to the water tanks in which fresh culture medium

FIGURE 12.8   A model of a large-scale, microalgal cultivation farm constructed on hilly, nonarable land.

is prepared and then filtered, before inoculation and subsequent algal cultivation. During the scaled-up cultivation stage, the culture and fresh medium can automatically flow to the column photobioreactors, tubular photobioreactors, and/or open raceway ponds by gravity. Through such use of the terrain and gravity, the costs for water pumps and electricity and for culture transportation are reduced. Meanwhile, culture transportation between tanks and devices can easily be done in closed pipes so that opportunities for biological contamination caused by necessary manipulations can at least be considerably reduced. Moreover, should biological contamination occur, the downstream contaminated culture can be quickly drained by opening a valve. The contaminated culture then automatically flows to the water dams at the foot of the hills, away from the clean algal cultures upstream. Such a system enhances the feasibility of mass cultivation as the problem of biological contamination can be efficiently minimized. In addition, the culture effluents can be collected behind the water dam where the water slowly sinks through the deep soil and returns to the groundwater reservoir. During the farming process, wastewater can be purified and then recycled by filtration through layers of soil. Because of these advantages, the nonarable hill culture model is strongly recommended for *Haematococcus* cultivation as well as other industrial-scale microalgal cultivation for biomass, bioactive substances, and biofuels.

In conclusion, commercial farming of the green alga, *H. pluvialis*, an excellent source of natural astaxanthin, shows considerable promise with commercial-scale production already ongoing in several countries, including China, Chile, Israel, and the United States. Although there are remaining challenges of increasing outputs and reducing costs, these can be managed by paying conscientious attention to the algae and the cultivation systems. Optimal culture strategies and conditions, nutrition, and contamination control as well as harvesting, processing, and astaxanthin extraction are actively being improved through production experience and ongoing research. Mass cultivation of *Haematococcus* is already physically and economically feasible and profitable, and this industry is bound to expand in the near future.

REFERENCES

Benemann, J. R. 1992. Microalgae aquaculture feeds. *J. Appl. Phycol.* 4:233–245.

Borowitzka, M. A., J. M. Huisman, and A. Osborn. 1991. Culture of the astaxanthin producing green alga *Haematococcus pluvialis*. 1. Effects of nutrients on growth and cell type. *J. Appl. Phycol.* 3:295–304.

Boussiba, S. 2000. Carotenogenesis in the green alga *Haematococcus pluvialis*: Cellular physiology and stress response. *Physiol. Plantarum* 108:111–117.

Boussiba, S. and C. Aflalo. 2005. An insight into the future of microalgae biotechnology. *Innov. Food Technol.* 28:37–39.

Boussiba, S. and A. Vonshak. 1991. Astaxanthin accumulation in the green alga *Haematococcus pluvialis*. *Plant Cell Physiol.* 7:1077–1082.

Chen, C. Y., K. L. Yeh, R. Aisyah, D. J. Lee, and J. S. Chang. 2011. Cultivation, photobioreactor design and harvesting of microalgae for biodiesel production: A critical review. *Bioresour. Technol.* 102:71–81.

Chen, H., F. Chen, and X. D. Dong. 1997. Mixotrophic and heterotrophic growth of *Haematococcus lacustris* and rheological behaviour of the cell suspensions. *Bioresour. Technol.* 1:19–24.

Damiani, M. C., P. I. Leonardi, O. I. Pieroni, and E. J. Caceres. 2006. Ultrastructure of the cyst wall of *Haematococcus pluvialis* (Chlorophyceae): Wall development and behaviour during cyst germination. *Phycologia* 45:616–623.

Domínguez-Bocanegra, A. R., I. G. Legarreta, F. M. Jeronimo, and A. T. Campocosio. 2004. Influence of environmental and nutritional factors in the production of astaxanthin from *Haematococcus pluvialis*. *Bioresour. Technol.* 92:209–214.

Dong, C. L., W. Chen, and C. Liu. 2014. Flocculation of algal cells by amphoteric chitosan-based flocculant. *Bioresour. Technol.* 170:239–247.

Droop, M. R. 1954. Conditions governing haematochrome formation and loss in the alga *Haematococcus pluvialis* Flotow. *Arch. Mikrobiol.* 20:391–397.

Droop, M. R. 1961. *Haematococcus pluvialis* and its allies. III: Organic nutrition. *Rev. Algol. N.S.* 5:247–259.

Elliot, A. M. 1934. Morphology and life history of *Haematococcus pluvialis*. *Arch. Protistenk.* 82:250–272.

Evens, T. J., R. P. Niedz, and G. J. Kirkpatrick. 2008. Temperature and irradiance impacts on the growth, pigmentation and photosystem II quantum yields of *Haematococcus pluvialis* (Chlorophyceae). *J. Appl. Phycol.* 4:411–422.

Fábregas, J., A. Domínguez, A. Maseda, and A. Otero. 2003. Interactions between irradiance and nutrient availability during astaxanthin accumulation and degradation in *Haematococcus pluvialis*. *Appl. Microbiol. Biotechnol.* 61:545–551.

Force, L., C. Critchley, and J. J. S. van Rensen. 2003. New fluorescence parameters for monitoring photosynthesis in plants. *Photosynth. Res.* 78:17–33.

Goodwin, T. W. and M. Jamikorn. 1954. Studies in carotenogenesis. II. Carotenoid synthesis in the alga *Haematococcus pluvialis*. *Biochem. J.* 57:376–381.

Göksan, T., A. K. İlknur, and K. Cenker. 2011. Growth characteristics of the alga *Haematococcus pluvialis* Flotow as affected by nitrogen source, vitamin, light and aeration. *Turk. J. Fish. Aquat. Sci.* 11:377–383.

Grung, M., F. M. L. D'Souza, M. Borowitzka, and S. Liaaen-Jensen. 1992. Algal carotenoids 51. secondary carotenoids 2. *Haematococcus pluvialis* aplanospores as a source of (3S, 3′S)-astaxanthin esters. *J. Appl. Phycol.* 4:165–171.

Hagen, C., S. Siegmund, and W. Braune. 2002. Ultrastructural and chemical changes in the cell wall of *Haematococcus pluvialis* (Volvocales, Chlorophyta) during aplanospore formation. *Eur. J. Phycol.* 37:217–226.

Hazen, T. E. 1899. The life history of *Sphaerella lacustris*. *Mem. Torrey Bot. Club* 3:211–247.

Huang, Y., L. Li, J. G. Liu, and W. Lin. 2014a. Botanical pesticides as potential rotifer control agents in microalgal mass culture. *Algal Res.* 4:62–69.

Huang, Y., J. G. Liu, L. Li, P. Tong, and L. T. Zhang. 2014b. Efficacy of binary combinations of botanical pesticides for rotifer elimination in microalgal cultivation. *Bioresour. Technol.* 154:67–73.

Hussein, G., U. Sankawa, H. Goto, K. Matsumoto, and H. Watanabe. 2006. Astaxanthin, a carotenoid with potential in human health and nutrition. *J. Nat. Prod.* 69:443–449.

Kang, C. D., J. S. Lee, T. H. Park, and S. J. Sim. 2005. Comparison of heterotrophic and photoautotrophic induction on astaxanthin production by *Haematococcus pluvialis*. *Appl. Microbiol. Biotechnol.* 68:237–241.

Kim, Z. H., S. H. Kim, S. H. Lee, and C. G. Lee. 2006. Enhanced production of astaxanthin by flashing light using *Haematococcus pluvialis*. *Enzyme Microb. Technol.* 39:414–419.

Kobayashi, M., T. Kakizono, S. Nishio, and S. Nagai. 1992. Effects of light intensity, light quality and illumination cycle on astaxanthin formation in the green alga *Haematococcus pluvialis*. *Appl. Environ. Microbiol.* 74:61–63.

Krichnavaruk, S., A. Shotipruk, M. Goto, and P. Pavasant. 2008. Supercritical carbon dioxide extraction of astaxanthin from *Haematococcus pluvialis* with vegetable oils as co-solvent. *Bioresour. Technol.* 99:5556–5560.

Lee, Y. K. and S. Y. Ding. 1994. Cell cycle and accumulation of astaxanthin in *H. lacustris* (Chlorophyta). *J. Phycol.* 30:445–449.

Letelier-Gordo, C. O., S. L. Holdt, D. D. Francisci, D. B. Karakashev, and I. Angelidaki. 2014. Effective harvesting of the microalgae *Chlorella protothecoides* via bioflocculation with cationic starch. *Bioresour. Technol.* 167:214–218.

Li, Y. Y., J. G. Liu, W. Lin, X. J. Cui, and Y. B. Xue. 2006. Effects of light intensity on cell transformation, astaxanthin accumulation in three strains of *Haematococcus pluvialis* and their difference. *Mar. Sci.* 30:36–41.

Liu, J. G., Q. Q. Li, Q. Liu et al. 2014. Screening of unicellular microalgae for biofuels and bioactive products and development of a pilot platform. *Algol. Stud.* 145:99–117.

Liu, J. G., Y. N. Sun, M. Y. Yin, W. Liu, and Z. Zhang. 2004. Inorganic carbon and the cell growth regulator in *Haematococcus pluvialis*. *Oceanol. Limnol. Sin.* 35:87–94 (in Chinese with English abstract, English version in *Proc. China Assoc. Sci. Technol.* 2:454–461).

Liu, J. G., M. Y. Yin, J. P. Zhang, W. Liu, and Z. C. Meng. 2002. Dynamic changes of inorganic nitrogen and astaxanthin accumulation in *Haematococcus pluvialis*. *Chinese J. Oceanol. Limnol.* 20:358–364 (in Chinese with English abstract).

Liu, J. G., M. Y. Yin, J. P. Zhang, Z. C. Meng, and W. F. Bourne. 2000. Cell cycle of *Haematococcus pluvialis*. *Oceanol. Limnol. Sin.* 31:145–150 (in Chinese with English abstract).

Liu, J. G., Y. Yuan, L. Li, Y. Huang, and W. Lin. 2010. A method to clear inner wall of tubular photobioreactor. Patent of State Intellectual Property Office of the P.R. China ZL201010500926.9.

Lu, F., A. Vonshak, and S. Boussiba. 1994. Effect of temperature and irradiance on growth of *Haematococcus pluvialis* (Chlorophyceae). *J. Phycol.* 30:829–833.

Mesquita, J. F. and M. F. Santo. 1984. Ultrastructural study of *Haematococcus lacustris* (Girod) Rostafinski (Volvocales) II. Mitosis and cytokinesis. *Cytologia* 49:229–241.

Mirón, S. A., A. C. Gómez, F. G. Camacho, E. M. Grima, and Y. Chisti. 1999. Comparative evaluation of compact photobioreactors for large-scale monoculture of microalgae. *J. Biotechnol.* 70:249–270.

Molina, E., J. Fernández, F. G. Acién, and Y. Chisti. 2001. Tubular photobioreactor design for algal cultures. *J. Biotechnol.* 92:113–131.

Orosa, M., D. Franqueira, A. Cid, and J. Abalde. 2001. Carotenoid accumulation in *Haematococcus pluvialis* in mixotrophic growth. *Biotechnol. Lett.* 23:373–378.

Posten, C. 2009. Design principles of photobioreactors for cultivation of microalgae. *Eng. Life Sci.* 9:165–177.

Pulz, O. 2001. Photobioreactors: Production systems for phototrophic microorganisms. *Appl. Microbiol. Biotechnol.* 57:287–293.

Rashid, N., S. U. Rehman, and J. I. Han. 2013. Rapid harvesting of freshwater microalgae using chitosan. *Process Biochem.* 48:1107–1110.

Reyes, F. A., J. A. Mendiola, E. Ibanez, and J. M. Valle. 2014. Astaxanthin extraction from *Haematococcus pluvialis* using CO_2 expanded ethanol. *J. Supercrit. Fluids* 92:75–83.

Santo, M. F. and J. F. Mesquita. 1984. Ultrastructural study of *Haematococcus lacustris* (Girod) Rostafinski (Volvocales) I. Some aspects of carotenogenesis. *Cytologia* 49:215–228.

Steinbrenner, J. and H. Linden. 2000. Regulation of two carotenoid biosynthesis genes coding for phytoene synthase and carotenoid hydroxylase during stress-induced astaxanthin biosynthesis in the green alga *Haematococcus pluvialis*. *Plant Physiol.* 125:810–817.

Steinbrenner, J. and H. Linden. 2003. Light induction of carotenoid biosynthesis genes in the green alga *Haematococcus pluvialis*: Regulation by photosynthetic redox control. *Plant Mol. Biol.* 52:343–356.

Strasser, R. J., A. Srivastava, and M. Tsimilli-Michael. 2000. The fluorescence transient as a tool to characterize and screen photosynthetic samples. In *Probing Photosynthesis: Mechanism, Regulation and Adaptation*, eds. M. Yunus, U. Pathre, and P. Mohanty, pp. 445–483. London, UK: Taylor & Francis.

Sun, Y. H., J. G. Liu, X. L. Zhang, and W. Lin. 2008. Strain H2-419-4 of *Haematococcus pluvialis* induced by ethyl methanesulphonate and ultraviolet radiation. *Chinese J. Oceanol. Limnol.* 26:152–156.

Sun, Y. N., M. Y. Yin, and J. G. Liu. 2001. Auto-signals in *Haematococcus pluvialis*. *Trans. Oceanol. Limnol.* 89:22–28 (in Chinese with English abstract).

Terence, J. E., R. P. Niedz, and G. J. Kirkpatrick. 2008. Temperature and irradiance impacts on the growth, pigmentation and photosystem II quantum yields of *Haematococcus pluvialis* (Chlorophyceae). *J. Appl. Phycol.* 20:411–422.

Tjahjono, A. E., Y. Hayama, T. Kakizono, Y. Terada, N. Nishio, and S. Nagai. 1994. Hyper accumulation of astaxanthin in a green alga *Haematococcus pluvialis* at elevated temperatures. *Biotechnol. Lett.* 16:133–138.

Tripathi, U., R. Sarada, R. Ramachandra, and G. A. Ravishankar. 1999. Production of astaxanthin in *Haematococcus pluvialis* cultured in various media. *Bioresour. Technol.* 68:197–199.

Tripathi, U., R. Sarada, and G. A. Ravishankar. 2002. Effect of culture conditions on growth of green alga *Haematococcus pluvialis* and astaxanthin production. *Acta Physiol. Plant.* 24:323–329.

Yin, M. Y., J. G. Liu, J. P. Zhang, and Z. C. Meng. 1998. Review study of culture *Haematococcus* and its astaxanthin. *Trans. Ocean. Limnol.* 30:53–62 (in Chinese with English abstract).

Yin, M. Y., J. G. Liu, J. P. Zhang, and Z. C. Meng. 2007. The effect of phosphate on cell growth and astaxanthin accumulation in *Haematococcus pluvialis*. *Oceanol. Limnol. Sin.* 37:249–254 (in Chinese with English abstract).

Yong, Y. Y. R. and Y. K. Lee. 1991. Do carotenoids play a photoprotective role in the cytoplasm of *Haematococcus lacustris* (Chlorophyta)? *Phycologia* 30:257–261.

Zhang, Z. 2004. Biological characteristics of *Haematococcus pluvialis* in a high cell density system. MS dissertation, Institute of Oceanology, University of Chinese Academy of Sciences, Beijing, China.

13 ALGAFARM
A Case Study of Industrial Chlorella *Production*

Diana B. da Fonseca, Luís T. Guerra,
Edgar T. Santos, Sofia H. Mendonça,
Joana G. Silva, Luís A. Costa, and João C. Navalho

CONTENTS

13.1 ALGAFARM: A DESCRIPTION

ALGAFARM is an industrial-scale microalgae production facility for the production of *Chlorella* biomass for food and feed applications. ALGAFARM integrates the entire production process, from small-volume production at a lab scale to final product packaging. It uses a combination of closed photobioreactor (PBR) technologies comprising mainly of large modular arrays of tubular PBRs (Figure 13.1) to cultivate the microalgae. The closed modular PBRs are based on transparent plastic

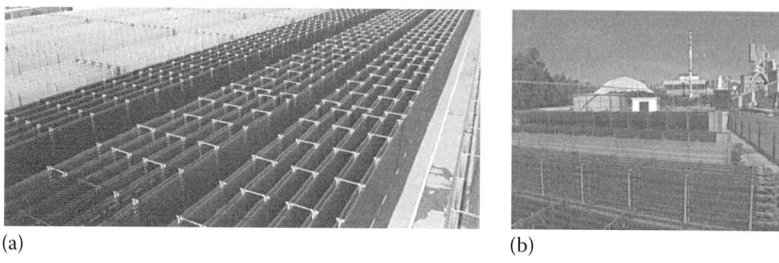

(a)   (b)

FIGURE 13.1 Industrial microalgal production units in a closed system at ALGAFARM located in Pataias, Portugal. (a) View of the outdoor photobioreactors from above and (b) view of the site showing the cement facility in the background.

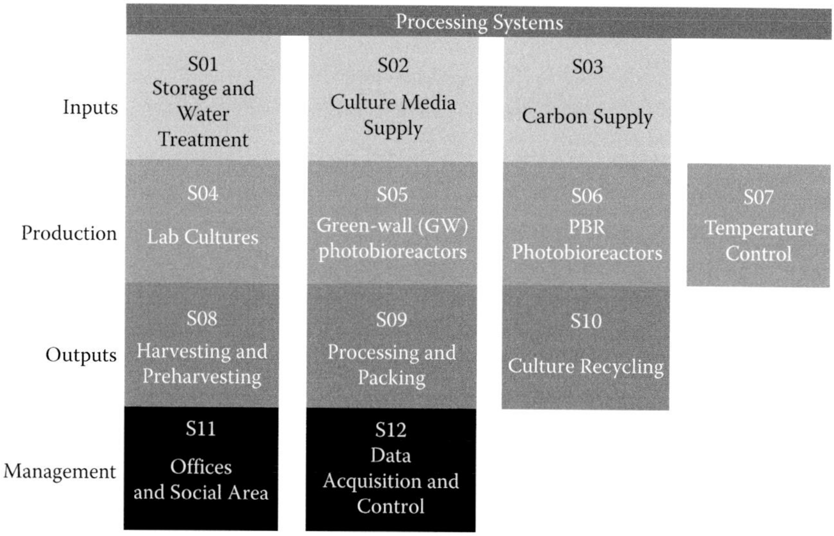

Processing Systems			
Inputs	**S01** Storage and Water Treatment	**S02** Culture Media Supply	**S03** Carbon Supply	
Production	**S04** Lab Cultures	**S05** Green-wall (GW) photobioreactors	**S06** PBR Photobioreactors	**S07** Temperature Control
Outputs	**S08** Harvesting and Preharvesting	**S09** Processing and Packing	**S10** Culture Recycling	
Management	**S11** Offices and Social Area	**S12** Data Acquisition and Control		

FIGURE 13.2 ALGAFARM supply chain systems.

tubes with a total length of around 300 km, arranged in parallel over an area of more than 1 ha and with a volume of more than 1300 m³.

The plant contains 19 tubular PBR modules, the core system, and 12 supply and support chain systems, which are responsible for the specific processes that culminate in the final product, which is dried *Chlorella* (Figure 13.2). These supply systems include water and nutrient supplies, CO_2 delivery and management, several inoculum production stages ranging from the laboratory to the largest PBR, and dewatering, drying, and packaging systems. In addition, there are peripheral systems important for stability and sustainability: water treatment and recirculation, cleaning and disinfection, effluent management, thermoregulation, control and supervision, and laboratory and office support. The entire process and supply chain is certified under several ISO and Occupational Health and Safety Standards (OSHAS). The closed PBR reactor hydraulics have been optimized using CFD modeling, and the complete network of support systems has been optimized regarding production sustainability, safety, quality, energetic, and economic efficiencies, as well as the local environmental and social impacts.

13.2 CULTIVATION OF *CHLORELLA*

Large-scale industrial cultivation of *Chlorella* must provide water, light, fertilizer nutrients, CO_2, and a suitable environment (e.g., temperature, pH, and protection from pests) for algal growth and production, as described later.

13.2.1 SYSTEM S01: STORAGE AND WATER TREATMENT

Large volumes of water are needed for algae biomass production. For example, assuming a biomass concentration of 3 g/L of *Chlorella* (on the high side for autotrophic growth), this requires 333 m^3 of water/ton of biomass produced, assuming that no water is recycled.

This does not include water needed to clean the reactors or for thermal regulation (spray cooling). Thus, water availability plays a large role in determining the location of such a plant, and conservation strategies such as recycling must be implemented to minimize consumption. Industrial *Chlorella* cultivation utilizes freshwater, that is, from surface sources, wells, or municipal water supplies. Except for the latter, a water treatment process is generally required to ensure water quality standards meet those required for algae cultivation. Although *Chlorella* can be cultured on wastewater, this would not be generally suitable for production of higher value human food or animal feed (Ali et al. 2012).

13.2.2 SYSTEM S02: CULTURE MEDIA SUPPLY

Besides water, microalgae have to be supplied with all the essential nutrients required for growth. These can be divided into macro- or micronutrients according to the amounts provided to the cultures. Macronutrients include nitrogen (either nitrate or ammonium or urea), phosphate, potassium, calcium, sulfur, and magnesium. The amount usually provided is based on the content of each nutrient in the algal biomass, ranging from 0.1% (Mg) up to 10% (for N) of the ash-free dry weight. Micronutrients such as iron, copper, cobalt, manganese, boron, molybdenum, and zinc are required in much smaller quantities. Some of these micronutrients may be present in sufficient quantities in the water supply.

In formulating a nutrient mixture for *Chlorella* growth, all constituents should be in a proportion that matches their presence in the biomass, taking into account the excess nutrients that *Chlorella* cells can uptake and store in vacuoles for later usage—so-called luxury consumption. The nutrient mix can be prepared in the solid state or as a concentrated liquid that is then supplied at the desired concentrations and rates to ensure that no inorganic nutrient becomes a limiting factor. To prevent precipitation of nutrients during media preparation, these stocks can be made up separately and/or at reduced pH. Custom-made nutrient media suitable for microalgal growth can be sourced from suppliers, but must be carefully checked for suitability.

pH control is achieved to an extent by the buffering capacity of the medium, with alkali (sodium carbonate) added as needed and by regulating the supply of carbon dioxide, as discussed next.

13.2.3   System S03: Carbon Supply

13.2.3.1   Inorganic Carbon

Supplementation of the culture medium with CO_2 allows for higher growth rates and productivities when compared with cultures that depend solely on air for their CO_2 supply. The CO_2 supplied to a plant such as ALGAFARM must be suitable for algae production of a food product, for example, "food grade." If obtained from an industrial source, such as a smokestack (flue gas) located nearby, it must be free of any deleterious impurities (although NOx and SOx are not a concern, particulates and VOCs would be). Utilization of CO_2 from industrial sources is a useful synergy because it can reduce the costs of CO_2 supply (often a major one) and also benefits the supplying company in providing the use for this otherwise wasted resource. The cost of purchasing pure (food grade) CO_2 must be compared to the investment and operational costs of connecting to a flue gas source and reducing any contaminants of concern. For applications where biomass quality is not as critical (e.g., chemicals or oils), untreated flue gas can be an option, depending on the specific issues of contamination.

13.2.3.2   Organic Carbon

Some strains of *Chlorella* are capable of growing on reduced organic carbon sources such as acetate, glycerol, sugars like glucose, and protein hydrolysates, both in the dark (heterotrophic growth, not discussed further herein) and light (mixotrophic growth) (Heredia-Arroyo et al. 2011). The supplementation with organic substrates in the light allows for a considerable boost in growth rates and productivities since another source of energy is available to the cells (Lin 2005; Kong and Vigil 2013). However, the introduction of a reduced carbon source has some disadvantages: (1) it can lead to proliferation of bacterial and fungal contaminants; (2) the cost of the reduced carbon sources is generally higher than that of CO_2, depending on the source; and (3) some *Chlorella* strains are not capable of utilizing a reduced carbon source.

The major issue for mixotrophic growth regimes is microbial contamination with heterotrophs. Acetic acid is generally used in mixotrophic *Chlorella* growth, as this substrate is not as conducive to contamination as long as supply rates and conditions are carefully managed. Crude glycerol is another potential *low-energy* carbon source for mixotrophic *Chlorella* production, being a by-product of the biodiesel industry and now available at relatively low cost (Thompson and He 2006; Johnson and Taconi 2007).

There are often substantial discrepancies in the results obtained between laboratory experiments and the industrial setting, and this must be considered in any such strategy. In mixotrophic cultivation, fed-batch operations are typically used, in which the organic carbon is supplied incrementally at the same rate as the algae consume it, to keep its concentration low in the medium. The fine-tuning of delivery requires the ability to actually measure carbon concentration in the medium, which is provided independently of the other (inorganic) nutrients. To summarize, in order to implement a mixotrophic regime at an industrial scale, the potential gains in productivity must be carefully compared to the potential disadvantages. In ALGAFARM, a mixotrophic strategy was adopted.

13.2.4 Systems S04 and S05: Laboratory and Green-Wall Cultures

At ALGAFARM, the first step of the scale-up is carried out in the laboratory with strict quality control protocols in place, including water treatment and filtration, culture media, and growth vessel sterilization (autoclaving). The laboratory and inoculation suites are thoroughly cleaned and disinfected weekly. Scale-up starts with seed cultures that are brought in a stepwise fashion to 250 mL, 1 L, and then to 5 L flask cultures. The latter are used to inoculate the 1000 L green-wall (GW) panels (Tredici 2004). In ALGAFARM, it is possible to cultivate 12 GW panels of 1000 L simultaneously. The GW panels are located inside a greenhouse with automatic temperature and sunlight control. Relatively, dense inocula are recommended to enhance the productivity and robustness of the scale-up process.

13.2.5 System S06: Photobioreactors

The most commonly used systems for industrial microalgal cultivation and their advantages and disadvantages are listed in Table 13.1. Choosing the appropriate production technology for a specific product and location is critical to the success of the project, with the main aspects to be taken into consideration being

1. Energy sources, that is, solar and organic
2. CO_2 source (without a free source, poor carbonation efficiencies might incur high CO_2 costs)
3. Strain sensitivity to contamination, which can result in culture crashes (losses)
4. Strain sensitivity to physical stress (e.g., flagellated algae and pumping or mixing)
5. Culture medium (seawater, freshwater, brackish water, etc.)
6. Plant location (climatic conditions, such as cloudy weather, will affect productivity)

The ALGAFARM production unit design was optimized to maximize productivity per unit area and is composed of 19 horizontal tubular PBRs. Designing such a large PBR presents many challenges, requiring expertise and synergy in all fields of science and engineering to achieve the lowest possible capital and operational costs. An optimized PBR design is influenced by number of parameters:

Light exposure: External configuration must maximize solar light utilization.
Mixing: Necessary to maximize light exposure, uniformly distribute nutrients and CO_2, and for thermoregulation.

TABLE 13.1

Cultivation Systems Commonly Used in the Large-Scale Production of Microalgae

Cultivation System		Advantages	Disadvantages	Examples	References
Open systems[a]	Open, circular, raceway, cascade ponds, and open tanks	Cheaper to build and operate Production of low-value products Argued by some as the only economically viable system for large-scale production of algal biomass Currently used to produce over 99% of all algal biomass	Susceptible to contaminants (selective medium and strain dependent) Variation in weather can impact control of nutrients, light intensity, and CO_2 supply Evaporation an issue where freshwater is scarce Requires favorable climate Low productivity rates	*Dunaliella salina* (high salinity) *Arthrospira (Spirulina platensis)* (high alkalinity) *Chlorella vulgaris* *Haematococcus pluvialis*	Hennenberg et al. (2009) Tredici (2010) Hosseini and Shariati (2009)
Closed systems[b]	Large bags, heterotrophic fermenters, horizontal tubular PBRs, vertical tubular PBRs, artificially lit PBRs, flat panel, column, airlift, and green-wall panels	Higher productivities Lower harvesting costs (due to higher concentrations of biomass resulting from high surface-to-volume ratios) Effective culture control Can produce many more algal species than open ponds Contamination much reduced Easier to prevent evaporation	Complex scale-up and operation More expensive (but could design lower-cost systems) Requires further R&D (i.e., low-cost materials, reduced energy requirements)	*Chlorella vulgaris* *Haematococcus pluvialis*	Chaumont (1993) Carvalho et al. (2006) Tredici (2004) Tredici (2010)

[a] Examples of microalgal open cultivation systems are as follows: (1) Circular ponds: Yaeyama, Japan; (2) pilot raceway pond: Necton, Portugal; (3) raceway ponds: Cyanotech (Hawaii).

[b] Examples of microalgal closed cultivation systems are as follows: (1) Green-wall panels and tubular PBRs from the University of Florence; (2) flat panel PBR: Necton, Sociedade Anónima; (3) tubular PBR from ALGAFARM.

CO_2: Maximize utilization, limit losses (critical with purchased CO_2, not with flue gas).

Energy: Essential to minimize power consumption at all phases of the process.

Oxygen: O_2 accumulation requires an efficient degasification process.

Materials: Resistant to chemicals for disinfection/cleaning, UV stable, and inhibits biofilms.

Adopting these criteria should maximize production of the final layout, lead to a low incidence of contamination, and lower capital and operating costs. In addition, the sustainability of the system has to be evaluated. The operational costs are influenced by most of these criteria, considering that high input-use efficiency will reduce the costs of raw materials (nutrients, CO_2, heat/cooling), and the proper choice of material and equipment will reduce maintenance costs.

The best approach when scaling up a PBR system is to do so gradually in small increments. The biggest challenge faced is in designing PBRs that use standardized parts, materials, and equipments. This promotes financial viability in avoiding the added costs of developing, designing, and producing tailor-made parts and fittings. Moreover, the manufacture of larger-scale designs must be outsourced and standard part orders usually have faster delivery times, and this is crucial for efficient operation. The same rationale applies for the consumables required for operation (i.e., fertilizer, CO_2, cleaning chemicals).

13.2.6 SYSTEM S07: TEMPERATURE CONTROL

Temperature is perhaps the most critical of the culture parameters due to its potential for destroying the *Chlorella* culture. For example, when exposed to full sunlight, the culture temperature can reach over 40°C in just 1 hour: sufficient to kill the entire culture. Thus, temperature must be carefully monitored and controlled, particularly in closed PBRs (in ponds evaporation reduces drastic temperature increases).

Two strategies are used for temperature control in closed PBRs: (1) spraying water onto the cultivation system to cool it by evaporation and (2) using heat exchangers to transfer heat into a cooling reservoir (e.g., water, brine). The heat exchanger can be open (using the water from the cooling reservoir) or closed loop (recirculating the same water through the PBR and reservoir). The heat exchanger can be located inside the PBR or the culture can be bypassed through an external heat exchanger. Energy will be required to pump the thermal fluid through the heat exchanger.

In using a heat exchanger network in a closed circuit system, water loss by evaporation is prevented (a major disadvantage of water spraying), but the heat removed from the PBRs will increase the temperature of the water reservoir, which would also increase evaporation—something to also consider. Nevertheless, a heat exchanger may be used both for cooling in summer and for heating in the winter, resulting in a boost in annual productivity in temperate climate regions. In ALGAFARM, internal heat exchangers are used, complemented by a water-spraying backup system. A significant advantage is drawn from its colocation with a cement factory that produces waste heat—useful for heating cultures in the winter.

13.2.7   System S08: Harvesting

Achieving cost-effective harvesting of microalgae is still a challenge. The choice of harvesting strategy and type of downstream processing employed must take into account the equipment cost (CAPEX-capital expenditure) and energy consumption (OPEX-operating expenses), which relates closely to the culture concentration at harvest. The other key factor defining strategy is the mode of cultivation: autotrophic or mixotrophic. In commercial microalgal farming, harvesting should in fact be subordinated and adapted to the cultivation strategy since the latter dictates the yield of a production plant.

In dealing with one of the oldest and most well-known microalgal commercial products, the process of *Chlorella* harvesting has been extensively studied and developed. There are many reports in the literature on algae harvesting by sedimentation, with or without chemical flocculation (Lin 2005; Salim et al. 2011; Vandamme et al. 2012), and mostly for bench or pilot-scale or nonfood applications (e.g., wastewater treatment). These typically achieved low concentration factors of about 10 and required additional concentration steps before drying. Sedimentation tanks have proven scalability and low investment and operational costs (low-energy consumption). However, *Chlorella* requires a chemical flocculant, which would affect the quality of the product in food and feed applications. Thus, flocculants such as Fe^{3+}, Al^{3+}, or polycationic polymers would not be feasible, and even food-grade flocculants, such as chitosan, may be problematic as it changes the composition of the final product, prevents the reutilization of the culture water, and increases production costs. Also, with culture medium recycling, the risk of introducing residual flocculating agent into the cultivation systems requires careful dosing and/or subsequent removal steps.

A strategy employed by most of the *Chlorella* production plants in Asia (for autotrophic or mixotrophic growth) entails batch cultivation for over a week, followed by harvesting of the entire culture and starting again with a fresh inoculum. However, this does not make the best use of *Chlorella*'s biological potential for high growth rates and productivity. It is more advantageous to harvest a fraction of the culture on a daily basis, or every few days, for maximum productivity. The key is maintaining a high-quality culture in continuous production for several weeks, even months, avoiding contamination and then downtime in cleaning, sterilization, and reinoculation. This is a particular challenge in mixotrophic conditions, compared with autotrophic growth, due to the buildup of bacterial contamination from the added carbon source.

If harvesting is more frequent than once per day, the system is essentially in continuous production mode, which is achievable if (1) the dilution or harvesting rate of the system matches the growth rate of the culture, so as to maintain a constant microalgal concentration in the system (turbidostat mode), and (2) the concentration of nutrients (in particular, organic carbon) is maintained as low as possible due to continuous loss via harvesting. This requires continuous monitoring of cell concentration, limiting nutrients, and solar radiation (for both auto- and mixotrophic systems). Automation of harvesting and nutrient supply is extremely useful, sometimes indispensable. Continuous harvesting may be performed exclusively during the photoperiod to avoid excessive dilution of cultures during the night and can be designed to keep the microalgae growing at optimal levels.

Storage of culture harvested from the cultivation system should be carefully considered since microalgae will usually be outside their preferred environment in a dark tank, for instance. This could lead to alterations in cell biochemical composition, for example, pigment content. The time of the day selected for harvesting in a batch or semicontinuous mode should account for differences in biochemical composition associated with the circadian cycles, as noted in *Chlorella* harvested at the end of the day or at the end of a night (Tamiya 1963). It is recommended to minimize the storage period to avoid proliferation of heterotrophic contaminants, such as bacteria and fungi, and also the loss of microalgal biomass via respiration and/or fermentation, which can also lead to the formation of undesirable odor compounds. These phenomena cause reduction in the product's quality; consequently, cooling and illumination of the harvested culture is carried out in some industrial facilities. The general strategy, particularly in mixotrophic production, is to minimize nutrient concentration at the time of harvest in order to prevent microbial contamination, nutrient wastage, and also to meet the effluent discharge regulations without costly additional treatment.

Keeping the stored culture homogenous by aeration, pumped recirculation, or static mixing avoids cell settling, thus ensuring steady operation of the harvesting system. The sizing and selection of harvesting equipment should aim to establish a continuous process, which will reduce the scale required and, thus, the investment and operational costs. The approach that must be taken with continuous harvesting of cultivation systems is self-evident, whereas in the case of batch harvesting strategies, scheduling of production cycles in the various cultivation systems should be organized so as to keep the harvesting step working continuously. It is, however, easier to minimize the storage time of each individual cell outside the cultivation system before harvesting when a continuous or semicontinuous harvesting strategy is implemented, in effect, a just-in-time approach.

Most of the major players in the *Chlorella* market, such as Sun *Chlorella* (Japan), Far East Microalgae Ind. Co. Ltd. (Taiwan), Yaeyama Shokusan Co. Ltd. (Japan), and Roquette Klötze Gmbh & Co. (Germany), among others, use centrifugation to ensure quick cell harvesting with notable efficiency (>95% microalgal cells retained). Despite the high investment costs and energy consumption required (around 1 kWh/m^3 of processed culture), centrifugation in disk-stack bowl centrifuges has become the industry's standard over the past decades due to its reliability and ease of operation, even without highly skilled technicians. Centrifugation is, without a doubt, a mature technology.

Chlorella produced by ALGAFARM, and being commercialized under the brand Allma by AllMicroalgae (Portugal) (see later), is harvested using cross-flow filtration with membranes, a younger and very promising technology—already well established in the water treatment business—which has now been applied successfully to microalgal production at an industrial scale. In addition, it is associated with a high investment cost, this technology is easily scalable due to its modular character. Although requiring more highly skilled operators, membrane filtration has many advantages over centrifugation: (1) using the correct membrane pore size, it yields 100% efficiency in cell harvesting (*Chlorella* and culture contaminants), allowing the culture medium to be reused without further treatment or at least to reduce the requirements of effluent treatment, and (2) the energy consumption is considerably lower (<0.5 kWh/m^3 of processed culture), reducing production costs.

Chlorella is processed into a powder, usually by spray-drying, before being eventually formulated into tablets or capsules. Therefore, for energy-saving purposes, the harvesting process needs only to result in a "pumpable" concentrated *Chlorella* suspension and not necessarily a solid paste.

13.2.8   System S09: Processing and Packing

After harvesting, processing must be carried out in the most energy-efficient and economical way possible, meeting product quality requirements and assurance. Most common biomass slurry (5%–25% solids) processing involves removing the remaining water from the solution or paste, which usually represents up to 75% of the processing costs and 20%–30% of the biomass production cost (Mohn 1978; Shelef et al. 1984; Gudin and Therpenier 1986; Grima et al. 2003). Thus, the choice of drying method depends on the scale of operation and also the intended use of the dried product. Different techniques have been used for microalgal biomass drying, such as spray-drying, drum-drying, sun-drying, and freeze-drying (Table 13.2). The first two use process heat, while the last uses low pressure and low temperature to dry the biomass. Freeze-drying is probably the best system to deliver high-quality product, given that heat is not used, thus preventing biomass degradation. Nevertheless, freeze-drying is very expensive in terms of capital and operational costs and is normally only used for very small operations. Of the drying systems, sun-drying is the oldest and least expensive; however, it is not an effective method for producing biomass for human consumption, and also, it is highly dependent on weather conditions (Shelef et al. 1984). Spray-drying is by far the most commonly used process to dry microalgae and where hot air is freely available (e.g., in the vicinity of a heat-intensive industry), the best choice. Also, spray-drying under continuous operation results in a low residence time, while the low effective drying temperature must be carefully optimized to prevent degradation of compounds and to maintain product stability. Spray-drying involves liquid atomization, gas/droplet mixing, and drying of liquid droplets. The atomized droplets are usually sprayed in a downward movement, concurrent with the gas phase, into a vertical chamber. There, drying is achieved within seconds. Finally, a cyclone separates the gas phase from the solid particles where the dried biomass is collected to be packed in a modified or controlled atmosphere and then stored in the warehouse away from direct light, moisture, and heat. Standard safety measures must be in place in the packaging area that should have restricted access both to ensure the quality of the final product and also to keep ignition sources away from the fine powder produced, which is a fire hazard.

Other processing methods, such as technologies for cell rupture, are briefly described in Table 13.3. These methods are used as a means of increasing digestibility of the biomass and, as such, achieve product differentiation, but are not widespread and still require a consensus among the different players in the field of microalgae.

TABLE 13.2

Processing Technologies Commonly Used in Large-Scale Microalgae Production

Process		Advantages	Disadvantages
Pasteurization		Product contamination control	Requires skilled labor
		Suitable for human consumption	
		Low CAPEX	
Cell wall rupture	Pulsed electric fields	Gentle processing	Recent technology
		Preserves product quality	Industrial-scale proof of concept required
		Low OPEX	Need for skilled labor
			High CAPEX
	Milling	Proven technology	Cell disintegration facilitates contaminant proliferation
		Intracellular contents made completely available	High CAPEX and OPEX
Drying	Spray-drying	Low residence time	Requires close supervision of operating conditions
		Low effective drying temperature	Need for skilled labor
		Continuous operation	High CAPEX and OPEX
		Mature technology	
	Sun-drying	Product considered for feed	Product not suitable for human consumption
		Low CAPEX	Highly dependent on weather conditions
			Low-efficiency process
			Requires area availability
	Drum-drying	Low CAPEX	Low efficiency—heterogeneous temperature
			High effective drying temperature—product degradation
	Freeze-drying	Product bioactivity preservation	Not suitable for large-scale application
		Suitable for high-value/ low-volume products (enzymes)	Batch processing
			Very high CAPEX and OPEX

Additional downstream processing is required when producing biofuels (i.e., lipid extraction or hydrothermal liquefaction) or high-value chemicals (i.e., carotenoid extraction). However, in the case of ALGAFARM, high-quality algae are produced and processed for food and feed applications, taking advantage of the nutritional properties of the entire biomass. Thus, the technology should also provide a product with (1) low microbiological contamination; (2) extended shelf life; and (3) market differentiation, for example, nutrient stability. In order to achieve these goals, the concentrated microalgal solution (5% solids) is pasteurized before being subjected to the spray-drying process. Pasteurization is the biological control point in the system, designed for the destruction of pathogenic bacteria and a reduction in total viable microbial counts per gram of dried cells.

TABLE 13.3

Relative Importance of the Various Characteristics of *Chlorella* sp. Powder for Its Distinct Applications

Area	Characteristic	Feed	Food Ingredient	Dietary Supplements	Cosmetics
Production capacity		3	3	2	1
Nutritional profile	High protein	3	3	3	1
	EPA/DHA[a]	3	2	2	2
	High carotenoids	3	2	2	3
Biomass price		2	2	2	1
Biomass characteristics	Broken cell wall	2	2	2	3
	Enriched with cofactors (e.g., selenium) or DHA	3	1	2	2
	Water solubility	1	3	2	2
	Stability at high temperatures	1	2	1	1
	Stability at extreme pH	1	2	1	1
	Palatability	2	3	2	1
Means of production/ certification	Organic	1	1	2	1
	HACCP[b]	3	3	3	3
	Kosher	1	2	2	1
	Halal	1	2	2	1
	Heavy metals (CA Prop 65)	1	1	2	1
Processing	Aqueous/oil extract	1	2	2	3
Bioactive functionality	Antioxidant	2	2	2	3
	Immunostimulant	3	2	3	1
	Antimicrobial	2	2	1	1
	Improved digestion	3	2	2	1
Biomass origin		2	3	2	1

1, not essential; 2, desirable; and 3, essential.

[a] EPA, eicosapentaenoic acid; DHA, docosahexaenoic acid.

[b] HACCP, hazard analysis and critical control points.

13.2.9 SYSTEM S10: CULTURE MEDIUM RECYCLING, CLEANING, AND EFFLUENT TREATMENT

This system involves recirculation of the culture medium, cleaning of the production unit, and effluent treatment as required by good manufacturing practice (GMP) rules and regulations. Culture medium recirculation is a crucial step in the industrial cultivation of microalgae. It allows a significant reduction in wastage of water and has the advantage of reinstating the nutrients that were not consumed by the algae in the PBRs, thus reducing production costs.

The effluent treatment includes waste resulting from the cleaning of PBR, supporting tubing circuitry and equipment, from tasks performed in the laboratory and automatic cleaning equipment, and so on. Effluents are treated before being discharged into the municipal wastewater treatment system, and special attention is given to physical, chemical, and biological parameters in relation to the imposed limits. The ALGAFARM plant is cleaned periodically and whenever required. Standard procedures are available and each cleaning event is recorded, complying with the implemented quality standards.

13.2.10 SYSTEMS S11 AND S12: OFFICES AND SOCIAL AREA/DATA ACQUISITION AND CONTROL

ALGAFARM works out of several offices, including a social area, and laboratories, at the plant. Careful monitoring of culture parameters is essential for maximizing productivity of *Chlorella* biomass at an industrial scale. The accurate evaluation of the physiological state of the cells, the culture cell density, pH, nutrient availability, temperature, contaminant load, and incident light intensity, among others inform the decision-making process in order to maximize growth. Continuous monitoring can be achieved with probes submerged into the well-mixed areas of the culture medium and the information integrated through software. This is advantageous as it is labor saving, provides the operator with the means to take action, and supplies immediate feedback. Furthermore, inline monitoring allows an automated response to parameter changes. For instance, a routine can be programmed to dispense nutrients when these are nearly depleted or even harvest the culture according to cell concentration.

Although useful, inline monitoring requires a significant initial investment. In addition, it is limited to the probes that are available in the market and to their specificity for the desired parameter or susceptibility to interference from other components of the medium. Otherwise, discrete manual sampling and measurements must be made by the operator. Therefore, inline monitoring systems do not dispense with regular manual sampling and control, and equipment can also fail. Although supervision is still required, such systems greatly improve production by maintaining stable culturing conditions.

13.3 QUALITY CONTROL

Since 2012, ALGAFARM has been certified by the International Standard Guidelines ISO 14000 and OSHAS 18001; those systems were easily implemented since this industrial unit was considered an extension of the integrated management system, which was already implemented in place. In the near future, it is expected that ALGAFARM will add further certification: ISO 9001 and ISO 22000:2005, which describe the quality and the food safety management systems (standard requirements that map out what an organization needs to do to demonstrate its ability to control food safety hazards), respectively. Through ISO 22000:2005, the hazard analysis and critical control points (HACCP) system was created in order to identify specific hazards and actions, to control them, and to ensure food quality and safety. HACCP has become a useful and recognized tool and is now mandatory for the food industry in the EU.

ALGAFARM produces spray-dried *Chlorella vulgaris*. This alga has already been consumed for many years by humans and is recognized and accepted as a food ingredient by the European Commission. Food safety issues are still a challenge for the industry worldwide and, in ALGAFARM's case, were a priority since the beginning of the project. The behavior of European consumers has been gradually changing in relation to nutritional quality, hygiene, health standards, and product/process certification, factors that are taken into account during the purchasing, marketing, and consumption process.

13.4 APPLICATIONS FOR *CHLORELLA*

ALGAFARM aims to produce the premium *C. vulgaris* defined by an exceptional standard of nutritional composition, microbiological quality, and purity levels. This product, being marketed for ALGAFARM by AllMicroalgae, is intended to reach not only the nutraceutical and food supplement markets but also the food ingredients and pet food industries. These last fill two quite distinct market niches, both with demanding requirements concerning biochemical profile and properties (Table 13.3). The competitive advantage in the marketplace depends on having auditable high-quality analyses and process certification. The nutritional value is the prime consideration; some standardization of specific micronutrients is also required. However, according to our experience, this type of product information is difficult to find among the European and American *Chlorella* products available in the market; not only because in most cases the product is imported by the same distributors from different producers in Asia but also due to sparse biochemical analyses.

In terms of applications for aquaculture, dried ingredients are becoming preferable to the traditional microalgal cultures managed in hatcheries (which incur risks, reduced standardization, low production predictability, and high operational costs). The nutritional value of *C. vulgaris* is limited within the aquafeed applications (low fatty acid content, specifically PUFAs). However, it may be considered as an excellent vegetable protein source, contributing also specific minerals and phytochemicals, that is, carotenoids. In some cases, it is used for the "green water technique" in hatcheries (Dâmaso-Rodrigues et al. 2010).

C. vulgaris has been reported as a nutritional supplement for specific animal health conditions and to be capable of stimulating immunity and energy levels (Matsuura et al 1991; Morris et al. 2007; Kwak et al. 2012). Also, its capacity to improve the performance of racehorses and to accelerate their recovery after physical effort (when included as supplement within the feed formulations) is generally advertised (Oilgae Report 2013).

The dietary supplements market for *Chlorella* meal (dried, broken cells) appeals to the consumer through attributes such as (1) increased bioavailability that can be achieved with broken cell biomass (Kay et al. 1991); (2) the means of production (organic/nonorganic); (3) the benefits of *Chlorella* extracts sometimes referenced to as "*Chlorella* growth factor" (Georgiou 2005 and references therein); (4) high protein content; and (5) specific mineral and vitamin contents. Daily suggested intake is defined by the brand manufacturer and most advise between 3 and 6 g/day. Allma is recommending a daily intake of 3 g.

Depending on the food product formulation and on the targeted benefits, *C. vulgaris* may be used as a food ingredient in a dry powder format without having a substantial effect on food texture, appearance, or taste, boosting its nutritional content. The color may be a problem in some cases, but, on the other hand, considering the current food trends (for healthy, natural, and "green" food), this characteristic could represent an innovative opportunity for the food industry leaders and marketers. In fact, attempts have been made to increase *Chlorella* chlorophyll content and coloration for health foods (Nakanishi and Deuchi 2014). It is clear that color and health will become increasingly important, and there is literally no greener food than *Chlorella*.

C. vulgaris' functional characteristics are still being fully assessed, not only from the technical, food bioengineering level but also from the nutritional and bioactive perspectives. Great advances in the industrialization of heterotrophic *Chlorella* production have been made recently, in Europe by Roquette and in the United States by Solazyme, for outstanding examples, leading to new *Chlorella* products on the market, demonstrating the market potential. ALGAFARM's mixotrophic production provides both the advantages of growing on sunlight and in using organic substrates for *Chlorella* production.

When *Chlorella* is produced with the highest quality in a controlled, standardized manner, based on a "natural" process, it can be regarded as a "superfood." It has the potential to revolutionize the healthy, green foods market in the near future. Moreover, it represents a natural, novel, sustainable ingredient for the food industry. Allma's *Chlorella* is produced taking into account the industry's highest standards.

REFERENCES

Ali, S. M., H. S. Nasr, and W. T. Abbas. 2012. Enhancement of *Chlorella* vulgaris growth and bioremediation ability of aquarium wastewater using diazotrophs. *Pak. J. Biol. Sci.* 15:775–820.

Carvalho, A. P., L. A. Meireles, and F. X. Malcata. 2006. Microalgal reactors: A review of enclosed system designs and performances. *Biotechnol. Prog.* 22:1490–1506.

Chaumont, D. 1993. Biotechnology of algal biomass production: A review of systems for outdoor mass culture. *J. Appl. Phycol.* 5:593–604.

Dâmaso-Rodrigues, M. L., P. Pousão-Ferreira, L. Ribeiro et al. 2010. Lack of essential fatty acids in live feed during larval and post-larval rearing: Effect on the performance of juvenile *Solea senegalensis*. *Aquacult. Int.* 18:741–757. dx.doi.org/10.1007/s10499-0099296-9.

Georgiou, G. J. 2005. The discovery of a unique natural heavy metal chelator. *Explore!* 14:1–8. http://www.thebeewellcompany.com/NewFiles/Explore_Article_on_HMD.pdf.

Grima, M. E., E. H. Belarbi, F. G. Acién Fernández et al. 2003. Recovery of microalgal biomass and metabolites: Process options and economics. *Biotechnol. Adv.* 20:491–515.

Gudin, C. and C. Therpenier. 1986. Bioconversion of solar energy into organic chemicals by microalgae. *Adv. Biotechnol. Process.* 6:73–110.

Hennenberg, K. J., U. Fritsche, and R. Herrera. 2009. *Aquatic Biomass: Sustainable Bio-energy from Algae?* Darmstadt, Germany: Öko-Institut.

Heredia-Arroyo, T., W. Wei, R. Ruan, and B. Hu. 2011. Mixotrophic cultivation of *Chlorella vulgaris* and its potential application for the oil accumulation from non-sugar materials. *Biomass Bioenergy* 35:2245–2253.

Hosseini, T. A. and M. Shariati. 2009. *Dunaliella* biotechnology: Methods and applications. *J. Appl. Microbiol.* 107:14–35.

Johnson, D. T. and K. A. Taconi. 2007. The glycerin glut: Options for the value-added conversion of crude glycerol resulting from biodiesel production. *Environ. Prog.* 26:338–348.

Kay, R. A. and L. L. Barton. 1991. Microalgae as food and supplement. *Crit. Rev. Food Sci. Nutr.* 30:555–573.

Kong, B. and R. D. Vigil. 2013. Light-limited continuous culture of *Chlorella vulgaris* in a Taylor vortex reactor. *Environ. Prog. Sustain. Energy* 32:884–890.

Kwak, J. H. et al. 2012. Beneficial immunostimulatory effect of short-term *Chlorella* supplementation: Enhancement of Natural Killer cell activity and early inflammatory response (Randomized, double-blinded, placebo-controlled trial). *Nutr. J.* 11:1–53.

Lin, L.-P. 2005. *Chlorella: Its Ecology, Structure, Cultivation, Bioprocess and Application.* Taipei, Taiwan: Yi Hsien Publishing Company Ltd.

Matsuura, E., T. Nemoto, H., Hozumi et al. 1991. Effect of *Chlorella* on rats with iron deficient anemia. *The Kitasato Archiv. Exp. Med.* 64:193–204.

Mohn, H. F. 1978. Improved technologies for harvesting and processing of microalgae and their impact on production costs. *Arch. Hydrobiol. Bech. Ergebn. Lemnol.* 11:228.

Morris, H. J., O. Carrillo, A. Almarales et al. 2007. Immunostimulant activity of an enzymatic protein hydrolysate from green microalga *Chlorella vulgaris* on undernourished mice. *Enzyme Microb. Tech.* 40:456–460.

Nakanishi, K. and K. Deuchi. 2014. Culture of a high-chlorophyll-producing and halotolerant *Chlorella vulgaris*. *J. Biosci. Bioeng.* 117:617–619.

Oilgae Report. 2013. Energy from algae: Products, market, processes and strategies. http://www.oilgae.com/ref/report/Report_Sample.pdf.

Salim, S., R. Bosma, M. H. Vermuë et al. 2011. Harvesting of microalgae by bio-flocculation. *J. Appl. Phycol.* 23:849–855.

Shelef, G., A. Sukenik and M. Green. 1984. Microalgae harvesting and processing: A literature review. In *A Subcontract Report*. SERI/STR-231-2396. Golden, CO: Solar Energy Research Institute. http://www.osti.gov/scitech/biblio/6204677.

Tamiya, H. 1963. Cell differentiation in *Chlorella*. *Symp. Exp. Biol. Ser.* 17:188–214.

Thompson, J. C. and B. B. He. 2006. Characterization of crude glycerol from biodiesel production from multiple feedstocks. *Appl. Eng. Agric.* 22:261–265.

Tredici, M. R. 2004. Mass production of microalgae: Photobioreactors. In *Handbook of Microalgal Culture: Biotechnology and Applied Phycology*, ed. A Richmond, pp. 178–214. Oxford, UK: Blackwell.

Tredici, M. R. 2010. Photobiology of microalgae mass cultures: Understanding the tools for the next green revolution. *Biofuels* 1:143–162.

Vandamme, D., I. Foubert, I. Fraeye et al. 2012. Flocculation of *Chlorella vulgaris* induced by high pH: Role of magnesium and calcium and practical implications. *Bioresour. Technol.* 105:114–119.

14 Heterotrophic Culturing of Microalgae

Roberto E. Armenta and Zhiyong Sun

CONTENTS

14.1 INTRODUCTION

Fermentation processes to grow heterotrophic microalgae are highly productive in terms of cell biomass, oil, and nonfat products. Although industrial-scale fermentation of microalgae requires a significant capital investment, the economics are greatly improved when considering the high amount of product on a per liter basis (high titer), the short production cycles, and the high productivities not attainable with autotrophic algae.

Currently, only a few heterotrophic microalgal strains are being exploited commercially, and a few more recently discovered ones are being investigated at the R&D and pilot plant levels, aiming to improve current commercial processes or targeting new microalgal products. This chapter focuses on the upstream fermentative culturing of microalgae for the purpose of producing cell biomass, oils, and other products. Figure 14.1 depicts the whole bioprocess for heterotrophic microalgae-derived products. This includes the addition of front-end processing to use low-cost fermentable sugars, fermentation (the "upstream" process) and downstream processes for extracting oils, and, finally, the defatted biomass. In addition, this chapter focuses on the role of culturing or fermentation, discussing research opportunities and the

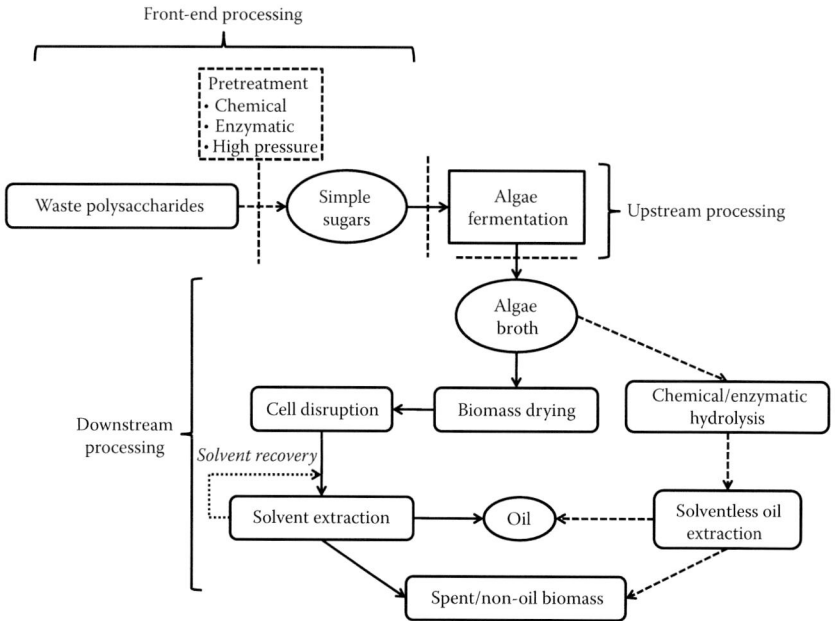

FIGURE 14.1   Processing of heterotrophic algae.

challenges faced when optimizing and scaling up this type of bioprocess. It also aims to provide information about basic algal fermentation metrics that would assist readers in assessing the commercial potential of heterotrophic microalgal technologies.

14.2   PRODUCTS FROM HETEROTROPHIC MICROALGAE

Algal fermentations are currently used worldwide to produce a variety of different products at large scale, including lipids for infant formula, dietary supplements, health food ingredients, cosmetics, and animal feeds. Also, there have been significant R&D and pilot-level efforts globally to achieve industrial-scale heterotrophic production of biofuels and commodity vegetable oils, to optimize current algal fermentation processes and to develop novel products. Table 14.1 shows the type of products and ongoing activities in heterotrophic culturing of microalgae.

The lipid (docosahexaenoic acid [DHA]) is widely used in infant formulas and largely produced commercially in North America using the dinoflagellate microalga *Crypthecodinium cohnii*. Other microalgae, such as some species of thraustochytrids, are most proficient at producing the same lipid (Armenta and Valentine, 2013), and in 2015 a *Schizochytrium* sp. (within the group of thraustochytrids) was authorized for use in infant formula in Europe (Commission Implementing Decision [2015] 2082/F1). Also, the same type of microalgae oil can also be used as an ingredient for infant formula in China. To date, thraustochytrid-derived high-DHA oil is mostly used as a dietary supplement and health food ingredient and also used to fortify premium foods sold in supermarkets, such as juices, milk, and yogurt. Thraustochytrids are related to both fungi and

TABLE 14.1

Heterotrophic Culturing of Microalgae at the Global Stage

Product	North America	South America	Europe	Asia	Oceania	Africa
Lipid for infant formula	☑			☑		
Dietary supplements	☑		☑	☑	✓	
Health food ingredients	☑	✓	☑	☑		✓
Cosmetics	✓		✓	✓		
Animal feed	☑		✓	☑	✓	
Biofuel oil	✓	✓	✓	✓	✓	✓

☑—Industrial fermentation scale.
✓—R&D and/or pilot plant demonstration fermentation.

algae; however, to date, these microorganisms are taxonomically classified as neither fungi nor algae (Yokoyama and Honda 2007; Armenta and Valentine 2013). However, scientific literature on the biotechnological use of thraustochytrids often refers to thraustochytrids as microalgae, and they are included in the literature of algal fermentations. Thus, thraustochytrids are included in this chapter with the microalgae.

In addition to nutritional oils from heterotrophic microalgae, recent new applications in the area of health food ingredients have been proposed, and some of them have recently become available in the market, including algal oils and flour from heterotrophically grown *Chlorella*. This flour is intended to be used in confectionary applications to replace traditional grain-based flour, with applications from bread making to an ingredient in ice cream and smoothies. Also, oil from *Chlorella* has been targeted as a replacement source for palmitic acid (C16:0) (e.g., from palm oil), and production of alternative oils from genetically modified organism (GMO) *Chlorella* strains, is being scaled up. DHA-rich oil from thraustochytrids (Figure 14.2a) has been used for a number of years to fortify a variety of foods that can be found in supermarkets, including juices, yogurt, milk, and sports bars.

For animal feed, cell biomass from both thraustochytrids and *Chlorella* are produced through fermentation mostly in the United States and Japan, respectively. In this case, algal biomass is dried after the fermentation and is sold commercially as whole cell biomass as an ingredient in animal feeds, including fish (i.e., salmon), shrimp, chicken, and pigs. Due to the size and value of this specialty animal market, this is one of the areas with significant growth potential for heterotrophic microalgae.

There has been some effort as well for using both algal oils and nonfat biomass in cosmetic products. A few of these products, mostly from *Chlorella*, have been commercialized, which due to their high value require relatively small pilot plant scale production to make a business case.

Although there have been a few announcements of the imminent industrial large-scale production of biofuel oil with heterotrophic microalgae, these efforts have been in the area of proving the large-scale technical feasibility for these types of fermentation,

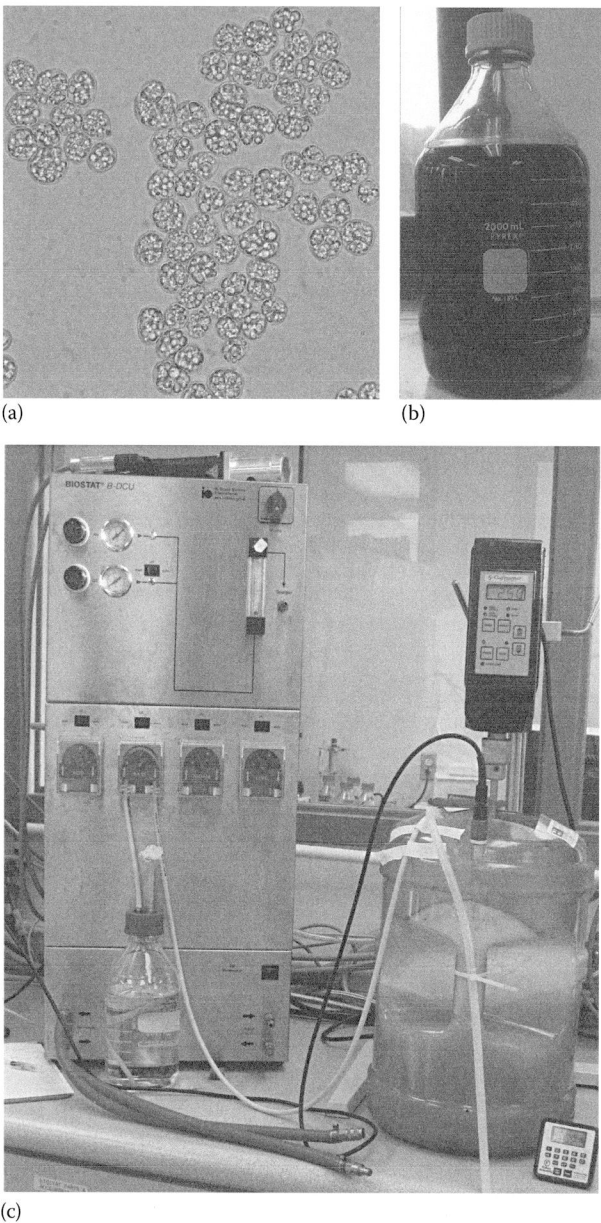

(a)

(b)

(c)

FIGURE 14.2 An oil-producing thraustochytrid (a); unrefined glycerol by-product from the biofuel industry (b); and nonsterile fermentation of heterotrophic algae with minimum operating control (c).

a fundamental step in any process development. Most efforts for producing biofuel oil through algal fermentation remain at the R&D, pilot, or demonstration levels, and economic feasibility can be achievable in the medium to long term. Meanwhile, industrial, large-scale production of heterotrophic microalgal oil for higher-value applications in feeds and specialty products seems to be nearing its realization in large-scale fermentation facilities located now mostly in Brazil, as are many other fermentation projects using yeast and bacteria, due to the abundance of low-cost sugar feedstocks.

Other potential products from microalgal fermentations may be commercialized in the medium to long term, including carotenoids (Sun et al. 2008), squalene (Nakazawa et al. 2013), and vitamin E (Bumbak et al. 2011), though such efforts have been ongoing for some time, for example, for vitamin C (Doncheck et al. 1996; Running et al. 2002). For these, species of *Chlorella*, thraustochytrids, *Euglena*, and others are being investigated, respectively.

14.3   STRAIN DISCOVERY AND DEVELOPMENT

Microalgae can be found in a variety of water environments. *Chlorella* in freshwater and salt water (Shah et al. 2003), the dinoflagellate heterotrophic microalga *C. cohnii* in marine waters (Prabowo et al. 2013), and thraustochytrids have been found around the world in many marine environments such as estuaries, mangroves, bays (Raghukumar 2008), and even Antarctica (Riemann and Schaumann 1993). Overall, heterotrophic microalgae originally surveyed from cold marine waters have a tendency to have a higher proportion of unsaturated fatty acids within the total oil they produce. Meanwhile, those found in tropical marine environments, such as estuaries and mangroves, have a higher proportion of saturated fatty acids within the total oil (Huang et al. 2003; Burja et al. 2006). In general, locations where there is a high probability of finding and collecting heterotrophic algae are those where there are significant inputs of organic matter, as some of these microbes use this material as their main source of nutrients. There is a significant opportunity to discover novel heterotrophic microalgae that could be highly prolific at producing different type of products, including oils and antioxidants.

Promising newly identified heterotrophic strains can be subjected to a targeted laboratory fermentation optimization program to increase biomass and product (i.e., oil) productivities, as discussed further in the following section on fermentation. When the fermentation optimization reaches diminishing returns, additional improvements may be accomplished with genetic approaches. Figure 14.3 depicts strategies for genetic improvement of algal strains. These include classic genetics, metabolic engineering (gene overexpression and knockdown/knockout), and insertion of foreign genes.

Among the classic genetic approaches is exposure to mutagenic agents such as ultraviolet radiation and chemicals. Metabolic engineering is a promising approach for modifying microorganisms in specific genes regulating the biosynthesis of enzymes pathways that perform key metabolic steps. The insertion of a gene, or a set of genes, is another approach to strain improvement (Mercer and Armenta 2011; Armenta and Valentine 2013). For heterotrophic algae, most of these techniques currently focus on achieving strain improvements in carbon source utilization and in achieving specific fatty acid compositions.

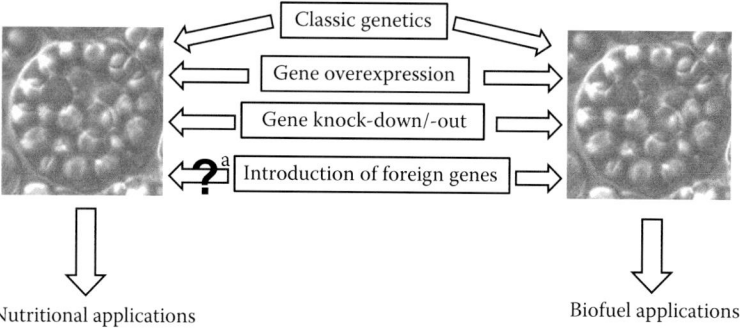

Nutritional applications   Biofuel applications

FIGURE 14.3 Genetic modification strategies for heterotrophic algae. ([a]Consumers' perception and regulatory approval for genetically modified nonfat microalgae by-products must be considered.)

Although genetic modifications can accomplish significant improvements in productivity, which can enhance the economics of microalgal fermentative processes, their usage must be weighed against consumer' perceptions of genetically modified organisms, especially those modified with foreign genes. This may be less of a concern for algal biofuel applications; however, it is an evident concern for products or coproducts intended for human nutritional or even animal feed applications, such as defattened biomass remaining after microalgal oil recovery (Figure 14.3). Therefore, significant effort will be needed to meet regulatory requirements and overcoming market barriers to commercialize such products from genetically modified heterotrophic microalgal strains.

A technical challenge is that when subjected to genetic modifications, the transformed algae may lose robustness due to genetic drift or secondary effects of the genetic changes, resulting in strains with reduced or even no growth at previously optimized pH, temperature, and chemical composition of the fermentation media.

14.4 FERMENTATION

High-cell-density cultivation (HCDC) is a relatively new term referring to fermentation processes that aim to produce a high amount of biomass (>100 g/L) and products (i.e., oil) within a short cycle time (<10 days) (Riesenberg and Guthke 1999). HCDC of heterotrophic microalgae shares many of the fundamentals known from already well-studied microorganisms, such as some bacteria, yeast, and fungi (Bumbak et al. 2011). These include organic and mineral feedstocks, supply of oxygen, fermentation equipment, process sensors and instrumentation, and process control strategies.

When using a stirred-tank fermentor, the three typical modes of cultivation are batch, fed-batch, and continuous. Batch mode operation is the simplest but cannot meet the final high-cell-density target. When using glucose as the carbon source, a yield range of 0.3–0.5 g dry weight cells/g glucose means that 200–300 g/L glucose is required in the culture media to support ~100 g/L biomass. Such a high substrate concentration, if added all at once in the initial medium, would inhibit growth through osmotic stress. Other major nutrients, such as ammonium (NH_4^+), can also result in growth limitations at high concentrations. Therefore, a fed-batch operation, having lower initial substrate and nutrient concentrations and further additions of these supplies during the course of the fermentation, is required for HCDC purposes.

Optimization strategies for fermentation with microalgae fall into two areas: composition of culture media and process operations. Both include a number of specific factors that affect the productivities and the costs of the final products. A detailed description of the importance of each ingredient in the fermentation media and a discussion of the factors in process operation that most commonly impact fermentation performance and economics follows.

14.4.1 Equipment and Instrumentation

Many decades of fermentation process and equipment development for nonalgal microorganisms can be used for heterotrophic microalgae production, with the conventional aerated stirred-tank fermentor being the default choice for HCDC processes. These fermentors have been scaled up by industry from lab scale vessels up to 500,000 L industrial fermentors. Other types of bioreactors, such as the membrane dialysis bioreactor and the airlift bioreactor, remain valuable research choices for particular culturing needs but cannot be scaled up to such a level.

Commercially available process sensors and controls, for temperature, pH, foam level, pO_2, and so on, are also applicable to heterotrophic microalgal cultivation. Exhaust gas analyzers are often used for measuring CO_2 production and O_2 consumption, as direct indicators of culture metabolic activity. Although their use requires some caution due to different cell size during the different stages of microbial growth, optical density sensors can be useful for monitoring biomass concentration. High-pressure liquid chromatography (HPLC) and enzyme-based probes can be used for monitoring the presence and consumption of organic substrates. These can be integrated into an online fermentation monitoring system purposely designed for an online sampling and processing system (Park et al. 1992).

14.4.2 Carbon Feedstock

Carbon is major component of microalgal cells, and several organic carbon sources can be used by heterotrophic microalgae, including glucose, fructose, acetic acid, ethanol, and glycerol. In order to support the high carbon demand needed to reach a high biomass or product density, the carbon source needs to be highly concentrated, promoting a reasonable yield and consistently available in bulk. The carbon source is the most expensive ingredient of the culture media because of the large volumes used compared with all the remaining fermentation ingredients. The use of organic carbon from wastewater streams such as food processing operations has some potential but in most cases the dilute nature of the wastewater, limited scale, seasonality and low yields make such approaches unattractive.

Glucose (also called dextrose) is the most widely used carbon source in algal fermentation, and can be obtained for somewhat less than $0.5/kg in large quantities as, for example, in ethanol fermentations. Similar to all heterotrophic microbes, heterotrophic microalgae have highly expressed enzymes that allow them to metabolize glucose at high rates through well-known biochemical pathways. For instance, glycolysis linked to the tricarboxylic acid cycle and then to fatty acid biosynthesis in the case of oil-producing algae. Hence, most initial laboratory tests and fermentation trials are conducted using glucose. Glucose solutions above 75% (w/v) tend to crystallize and clog feeding lines, requiring addition of water, which must be balanced in the overall fermentation media.

Glucose is the most commonly used sugar for heterotrophic production of microalgae, especially for high-value products for human consumption where food grade glucose is generally approved by regulatory agencies. For lower-value products, such as biofuels, glucose would be cost prohibitive. Sucrose, a disaccharide of glucose and fructose, is widely available and the cheapest in Brazil, where many fermentation plants are colocated with the sugarcane industry, thus avoiding the cost of sugar purification and transport. Sucrose generally needs to be converted into the monosaccharides glucose and fructose prior to their fermentation by microalgae. Even if some microalgae have the ability to hydrolyze sucrose and then metabolize the resultant monosaccharides, prehydrolyzing sucrose will reduce fermentation times and thus increase productivities.

Glycerol (or glycerin) is often available in large volumes and at low costs as a by-product of the biodiesel industry (Figure 14.2b). One major challenge is that if untreated, such glycerol sources contain impurities and contaminants (i.e., salts, methanol/ethanol) that can inhibit any microbial or algal fermentation. In comparison to glucose, glycerol can be a more concentrated source of carbon at above 90% (v/v), reducing the required dilution water in the fermentation. Microalgal strains that are highly efficient at using glycerol as a carbon source would still need to be identified and optimized. Furthermore, if not food grade, it would limit its use to fermentations that do not produce biomass for human consumption.

Xylose, a 5-carbon sugar (C5), is the most abundant sugar in nature after C6 glucose, found in all hemicellulosic materials, wheat straw, corn stovers and cobs, forest residues, and so on (Gong et al. 1981). The glucose-to-xylose ratio in these hemicellulosic feedstocks varies depending on the source but, in general, is about 2:1 (glucose/xylose) (Hu et al. 2011; Armenta and Valentine 2013). Some microalgae can use xylose as a carbon feedstock in fermentations (Gupta et al. 2013) along with the glucose available within the 2:1 (glucose/xylose) ratio. However, presently, the main limiting factor is in the deconstruction of lignocellulosic biomass to liberate these sugars and make them available for fuel and other fermentations (Figure 14.1). Large-scale biorefineries are currently being constructed and starting to operate worldwide to achieve the goal of low-cost sugar production for ethanol fermentation. If successful, these substrates would also become available for heterotrophic algae production. Other carbon substrates potentially useful for microalgal fermentations include lactose (composed of galactose and glucose), as a by-product of cheesemaking, and alcohol sugars such as sorbitol/mannitol derived from seaweed (macroalgae), which lack lignin and thus would require milder front-end processing to generate fermentable sugars than lignocellulosic biomass (Wei et al. 2012, 2013).

14.4.3 Nitrogen Fertilizer

Nitrogen typically accounts for 6%–8% of total cellular biomass (Sansawa and Endo 2004; Jakobsen et al. 2008) and is thus a significant input, in volume and cost, for the fermentation media. Most published studies on heterotrophic microalgae use complex nitrogen sources that provide amino acids, vitamins, and trace elements, such as yeast extract, soy peptone and other peptones, monosodium glutamate, or combinations thereof (Armenta and Valentine 2013). These sources of complex nitrogen are expensive, 10–100-fold higher than inorganic N fertilizers, with large volume costs of >$15/kg for yeast extracts and >$25/kg for soy peptone. However, they are used because they increase growth performance and product yield during fermentation and are difficult to replace because besides nitrogen, they provide essential amino acids, vitamins, and trace minerals. Although such complex nitrogen inputs can be reduced significantly, other nutrients, including trace minerals and vitamins, also will need to be readjusted to compensate for their reduction. The cost performance optimum will require a significant R&D effort for each case.

Complex nitrogen sources could also come from waste protein, such as rendered protein (Liang et al. 2011), and even from municipal wastewater (Chi et al. 2011), which have been used to grow heterotrophic microalgae experimentally. However, neither of these materials are food grade, and their potential use would be limited to the production of nonfood products such as biofuel oil and perhaps, though requiring significant regulatory work, for animal feed.

14.4.4 Other Nutrients

Heterotrophic microalgal requirements for mineral nutrients and growth factors (vitamins) vary among strains but are typically a small portion in cost compared to carbon and nitrogen. These trace minerals and vitamins are needed within the fermentation culture media to reach and maintain an optimal microalgal fermentation performance. Trace metals are mostly used as cofactors of enzymes involved in cellular metabolism.

Water-soluble vitamins, mostly from the B complex, are also needed to reach and maintain high performance by microalgal cultures. For example, vitamin B_{12} is needed because it is involved in cell metabolism affecting DNA synthesis and regulation, as well as fatty acid synthesis and energy production (Croft et al. 2006). This B vitamin needs to be added to the fermentation media because only bacteria are able to produce it, and normally, bacteria are completely excluded from the process. One rationale behind the concept of nonaxenic or consortia cultures, where microalgae and bacteria can coexist in nonsterile fermentations, is that bacteria can provide vitamin B_{12} to the algae in exchange for carbon substrates from the microalgae (Croft et al. 2005). However, such bacterial growth must be controlled to avoid bacterial fermentation taking over the culture and slowing down or stopping microalgal growth.

14.4.5 Nutrient Feeding

Many carbon and other key nutrient-feeding strategies have been established by HCDC development for bacteria, yeast, and microalgae. The chosen strategy is the result of a combination of optimum nutrient composition, physiological requirements, nutrient tolerance level, growth rate, and mechanisms of product and by-product formation. The key nutrient is typically the carbon substrate but can also include nitrogen, phosphorus, or other nutrient components.

For processes where products are strictly growth associated or where biomass is the product, there are two feeding strategies:

1. *Open-loop predefined feeding*: These strategies are widely used in heterotrophic cultivation of fast-growing bacteria. In the case of *Escherichia coli* cultivation, excessive carbon leads to the overproduction of acetate, which severely inhibits further growth of the culture (Shiloach and Fass 2005). For *Pseudomonas putida*, unlimited carbon supply often leads to the culture growing too fast and hitting the oxygen supply limit of the system too abruptly, causing excessive foam (Sun et al. 2007). Predefined feeding profiles have been developed to supply carbon at a rate that properly regulates the maximum growth rate of the culture, resulting in limited amounts of carbon substrate being present in the medium at any time. To achieve this, typical feeding equations based on flow rate (Lee 1996) or mass demand (Sun et al. 2007) have been developed.

   Due to the generally slower growth rate of microalgae compared to bacteria and yeast, limiting the carbon supply may not be necessary. However, inhibitory by-products, such as polysaccharides from *C. cohnii* (De Swaaf et al. 2001), might also be produced. In those cases, an open-loop predefined carbon-limited feeding strategy becomes useful.

2. *Closed-loop (feedback) feeding*: This type of feedback includes all feeding methods that apply to predefined actions based on some form of process signal. Commonly used strategies include online/offline-based nutrient monitoring, pH-stat, and dissolved oxygen–stat (DO-stat). Nutrient monitoring and feedback control can be employed when the optimum range of concentration of a key nutrient (usually a carbon source such as glucose) has been established, and quick online/offline analytical methods are available. Prior to ingredients being exhausted in the culture media, typically there is no significant change in fermentation parameters such as pH, temperature, and dissolved oxygen. Although *in situ* sensors for certain soluble compounds do exist (Phelps et al. 1995), frequent dose or feed rate adjustments based on cycles of offline sampling analysis, remain the most practical and widely used strategy for closed-loop feeding. Such a feeding strategy requires a deep understanding of the culturing process but, when executed well, can attain nearly unrestricted, hence maximum, growth potential of the culture without delay and interruption. pH-stat and DO-stat carbon/nitrogen feeding strategies are based on unique physiological responses that a given culturing process exhibits. When an ammonium

salt (ammonium sulfate) is used as a nitrogen source, NH_4^+ assimilation produces protons, lowering the pH of the medium. If ammonium hydroxide is used as the base to maintain the pH, the process becomes a pH-stat with respect to nitrogen feeding (Bailey et al. 2012). When culturing *C. cohnii* (De Swaaf et al. 2003) using acetic acid as a carbon source, the pH would drift upward when acetic acid was depleted, requiring more to be added to maintain the pH at a set point of 6.5. The process is therefore a pH-stat with respect to the carbon source.

14.4.6 SPECIFIC GROWTH RATE AND OXYGEN REQUIREMENTS

One of the first growth kinetic parameters with which to determine the productivity potential of a microorganism, is the maximum specific growth rate (μ_{max}). For heterotrophic microalgae, there is a general perception that they grow slower than bacteria and fungi, as even the fast-growing *Chlorella* sp. has a μ_{max} of only 0.2 h^{-1} (Doucha and Livansky 2012) and most microalgae have a μ_{max} below 0.1 h^{-1} (Schmidt et al. 2005), or doubling times of 7 hours or longer. In contrast, typical heterotrophic bacteria, including *Bacillus*, *Escherichia*, *Alcaligenes*, and *Pseudomonas*, exhibit a much higher μ_{max}, in the range of 0.3–0.5 h^{-1}, corresponding to a doubling time around 2 h or less.

Although lower μ_{max} means a longer doubling time, lower growth rates do not rule out heterotrophic microalgae's potential to achieve high-cell-density cultures. For high-cell-density aerobic cultivation, oxygen supply, typically provided by air only, is often the real limiting factor. Since $dO_2/dt = Q_{O2}X$, where dO_2/dt is the oxygen transfer rate (g/L·h), Q_{O2} is the specific oxygen uptake rate of the culture (g O_2/g cells·h), and X is the biomass yield (g/L) for a given maximum oxygen transfer rate provided by a fermentor, a lower Q_{O2} enables a higher X before the culture reaches oxygen limitation. Compared to *E. coli* with a Q_{O2} of 150–600 mg O_2/g·h (Marr 1991), heterotrophic microalgae have a much lower Q_{O2}. For example, the Q_{O2} for *Chlorella* is 60–140 mg O_2/g·h (Sansawa and Endo 2004) and for the thraustochytrid *Aurantiochytrium* is below 60 mg O_2/g·h (Jakobsen et al. 2008). Once the oxygen uptake rate of a fast-growing culture reaches the maximum oxygen transfer rate of a bioreactor, the rate of biomass increase will change to a constant that is bottlenecked by the oxygen transfer rate of the system. At that point, all aerobic culture proliferation rates become similar.

14.4.7 OPERATING CONDITIONS

Fermentation parameters such as pH, temperature, dissolved oxygen, agitation, and back pressure need to be controlled within specified values under which the selected microalgal strain thrives and reaches targeted productivities. For pH, the goal is to define optimal cell growth and results in pH being typically controlled between 6 and 7. A pH of <6 is used if the microalgal strain can tolerate low pH and still perform well, as at acidic pH bacterial contamination is generally reduced. Agitation or mixing is important to maintain an efficient mass transfer to enable sugars, O_2, and other nutrients to freely diffuse toward and into the microalgal cells. Mixing is not usually a problem in laboratory fermentors but increases in importance as the fermentation scale increases. However, agitation cannot simply be set at a maximum as the algal strain may be susceptible to shearing damage.

Another important factor is dissolved oxygen (DO). Specific DO values in the fermentation media have been associated with levels of unsaturation of the oil produced by heterotrophic microalgae, though this is also affected by other factors, such as temperature and the fatty acid pathways present within the algal strain. DO in the fermentation broth typically responds to the exhaustion of carbon source by rapidly rising. Therefore, a predefined dose of carbon can be fed in response to the DO changes, maintaining a DO-stat process. However, unlike pH, DO signal is also affected by other factors, from temperature and media composition to agitation and pressure, with the risk of feeding excess carbon into the fermentation. CO_2 production, measured by an off-gas CO_2 analyzer, is another indicator of heterotrophic culture activity, responding to carbon exhaustion by a rapid drop in CO_2 production. Therefore, it can also be used as a trigger for carbon feeding.

In general, a higher temperature increases cell growth, while a cooler temperature promotes accumulation of specific fatty acids within the oil, usually unsaturated ones. Often, a fixed temperature of >24°C is used when the goal is to maximize cell and total oil production. For producing oils with increased levels of unsaturated fatty acids for nutritional applications (i.e., DHA), lower temperatures are used. One approach is to use a higher temperature during active microalgal growth and switch to a cooler temperature when nitrogen limitation occurs, and the process has switched mostly to the stage of oil accumulation. High-DHA levels at lower temperatures may be attributed to the microalga trying to maintain proper membrane lipid fluidity by accumulating PUFAs.

Operational conditions can have a great impact on the outcome of large-scale fermentation. Indeed, any fermentation scale-up is challenging, and microalgae are no exception. Even the most improved strain at the laboratory and pilot plant levels presents challenges because scaling-up fermentation is not a linear process. For microalgae that require oxygen to grow in fermentors, operational physical phenomena, such as head pressure, particularly higher pressure at the bottom of a large fermentor and also high shearing that may be needed to assure proper mass transfer for optimal carbon consumption, need to be carefully considered and balanced to avoid reducing productivities. These challenges are expected at large scale because these situations are not fully realized or tested in smaller fermentors used at laboratory and pilot plant setups.

14.5 YIELD COEFFICIENTS AND PRODUCTIVITIES

The yield coefficient for cell biomass ($Y_{biomass/glucose}$) is typically 0.3–0.5 (Ratledge 2006). Also, the maximum theoretical oil yield (yield coefficients) for organic carbon conversion to oil is 0.31 and 0.3 for glucose ($Y_{oil/glucose}$) and glycerol ($Y_{oil/glycerol}$), respectively (Ratledge 1988). Yield coefficients are of paramount importance when assessing microalgal fermentation processes. When biomass to sugar conversion yields appear higher than the 0.5 value indicated, it is likely due to incorrect calculation or unsound data. This also applies to oil yield coefficients, the values above being theoretical maxima and actual fermentation processes will exhibit lower yields.

Volumetric productivities with heterotrophic microalgae are generally <20 g/L·h of biomass and <10 g/L·h of oil. These are the parameters of greatest interest in commercial applications, and it is possible that these productivities could be surpassed with prolific and highly optimized microalgal strains. Both yield coefficient and volumetric productivities must be determined and compared between strains, operating parameters, and production systems to select the most favorable one.

14.6 CONTAMINATION CONTROL

Sterilization of the fermentation media and vessel is expensive due to high capital, operating costs, and energy requirements. Thus, a promising research area is the study of microalgal fermentation under nonsterile environments. Some robust, usually nongenetically modified, strains of algae can exhibit good growth characteristics and product output, under adverse operating conditions such as low or acid pH and high osmotic pressures due to high sugar concentrations that inhibit or control growth of contaminating microorganisms, such as bacteria.

To achieve successful nonsterile fermentations, particularly when using a low-pH culture media, a critical factor is to use a robust or highly concentrated sterile seed culture. This provides a significant initial advantage in terms of cell mass per litre over other microorganisms that may be present in the initial nonsterile fermentation media. Figure 14.2c shows an oil-producing, nonsterile microalgal fermentation arrangement.

14.7 CONCLUSIONS

Heterotrophic culturing of microalgae is gaining increasing academic and industrial interest as a viable biotechnology for synthesizing products of commercial interest, including nutritional lipids, other oils, specialty proteins, and antioxidants. Compared with photoautotrophic cultivation, an advantage of heterotrophic microalgae is that it produces a significantly higher titer of product, an important requirement because fermentation processes are generally associated with high costs of initial capital (fermentor vessels), operational costs (labor and energy input), and carbon feedstocks (organic carbon). Hence, reaching high-cell-density cultures becomes the goal, or at least the prerequisite, for realizing the economic potential of heterotrophic microalgal cultivation.

Currently, the commercial heterotrophic microalgal products are in the categories of nutritional lipids, animal feeds, and cosmetics. Other promising high-value products, including antioxidants, vitamins, and sterols, are likely to achieve commercial-scale production in the near future. Key goals for cost reductions of microalgal fermentation processes are in the ingredients of culture media, especially of the carbon and nitrogen feedstocks, as well as optimization of the operating conditions (i.e., temperature, agitation, aeration). Finally, realizing the potential of nonsterile algal fermentation could reduce operating costs to those currently achieved in fuel ethanol fermentation.

REFERENCES

Armenta, R. E. and M. Valentine. 2013. Single-cell oils as a source of omega-3 fatty acids: An overview of recent advances. *J. Am. Oil Chem. Soc.* 90:167–182.

Bailey, R. B., D. DiMasi, J. M. Hansen et al. 2012. Enhanced production of lipids containing polyenoic fatty acid by very high density cultures of eukaryotic microbes in fermentors. US Patent 8133706 B2.

Bumbak, F., S. Cook, V. Zachleder, S. Hauser, and K. Kovar. 2011. Best practices in heterotrophic high-cell-density microalgal processes: Achievements, potential and possible limitations. *Appl. Microbiol. Biotechnol.* 91:31–46.

Burja, A. M., H. Radianingtyas, A. Windust, and C. J. Barrow. 2006. Isolation and characterization of polyunsaturated fatty acid producing *Thraustochytrium* species: Screening of strains and optimization of omega-3 production. *Appl. Microbiol. Biotechnol.* 72:1161–1169.

Chi, Z., Y. Zheng, A. Jiang, and S. Chen. 2011. Lipid production by culturing oleaginous yeast and algae with food waste and municipal wastewater in an integrated process. *Appl. Biochem. Biotechnol.* 165:442–453.

Commission Implementing Decision (EU) 2015/515/EC of 31 March 2015 authorising the placing on the market of oil from the micro-algae *Schizochytrium* sp. (ATCC PTA-9695) as a novel food ingredient under Regulation (EC) No 258/97 of the European Parliament and of the Council (notified under document C(2015) 2082). Official Journal of the European Union. L90/7-10.

Croft, M. T., A. D. Lawrence, E. Raux-Deery, M. J. Warren, and A. G. Smith. 2005. Algae acquire vitamin B_{12} through a symbiotic relationship with bacteria. *Nature* 438:90–93.

Croft, M. T., M. J. Warren, and A. G. Smith. 2006. Algae need their vitamins. *Euk. Cell* 5:1175–1183.

De Swaaf, M. E., G. J. Grobben, G. Eggink, T. C. de Rijk, P. van der Meer, and L. Sijtsma. 2001. Characterisation of extracellular polysaccharides produced by *Crypthecodinium cohnii. Appl. Microbiol. Biotechnol.* 57:395–400.

De Swaaf, M. E., L. Sijtsma, and J. T. Pronk. 2003. High-cell-density fed-batch cultivation of the docosahexaenoic acid producing marine alga *Crypthecodinium cohnii. Biotechnol. Bioeng.* 81:666–672.

Doncheck, J. A., R. J. Huss, J. A. Running, and T. J. Skatrud. 1996. L-Ascorbic acid containing biomass of *Chlorella pyrenoidosa*. US Patent 5,521,090.

Doucha, J. and K. Livansky. 2012. Production of high-density *Chlorella* culture grown in fermenters. *J. Appl. Phycol.* 24:35–43.

Gong, C. S., M. R. Ladisch, and G. T. Tsao. 1981. Production of ethanol from wood hemicellulose hydrolyzates by a xylose-fermenting yeast mutant, *Candida* sp. XF 217. *Biotechnol. Lett.* 3:657–662.

Gupta, A., S. Wilkens, J. L. Adcock, M. Puri, and C. J. Barrow. 2013. Pollen baiting facilitates the isolation of marine thraustochytrids with potential in omega-3 and biodiesel production. *Biotechnol. J. Ind. Microbiol.* 40:1231–1240.

Hu, C., S. Wu, Q. Wang, G. Jin, H. Shen, and Z. K. Zhao. 2011. Simultaneous utilization of glucose and xylose for lipid production by *Trichosporon cutaneum. Biotechnol. Biofuels* 4:25.

Huang, J., T. Aki, T. Yokochi et al. 2003. Grouping newly isolated docosahexaenoic acid-producing thraustochytrids based on their polyunsaturated fatty acid profiles and comparative analysis of 18S rRNA genes. *Mar. Biotechnol.* 5:450–457.

Jakobsen, A. N., I. M. Aasen, K. D. Josefsen, and A. R. Strom. 2008. Accumulation of docosahexaenoic acid-rich lipid in thraustochytrid *Aurantiochytrium* sp. strain T66: Effects of N and P starvation and O_2 limitation. *Appl. Microbiol. Biotechnol.* 80:297–306.

Lee, S. Y. 1996. High cell-density culture of *Escherichia coli. Trends Biotechnol.* 14:98–105.

Liang, Y., R. Garcia, G. Piazza, and Z. Wen. 2011. Nonfeed application of rendered animal proteins for microbial production of eicosapentaenoic acid by the fungus *Pythium irregulare. J. Agricul. Food Chem.* 59:11990–11996.

Marr, A. G. 1991. Growth rate of *Escherichia coli. Microbiol. Rev.* 55:316–333.

Mercer, P. and R. E. Armenta. 2011. Developments in oil extraction from microalgae. *Eur. J. Lipid Sci. Technol.* 113:539–547.

Nakazawa, A., Y. Kokubun, H. Matsuura et al. 2013. TLC screening of thraustochytrid strains for squalene production. *J. Appl. Phycol.* 26:29–41.

Park, Y. S., K. Kai, S. J. Iijima, and T. Kobayashi. 1992. Enhanced β-galactosidase production by high cell-density culture of recombinant *Bacillus subtilis* with glucose concentration control. *Biotechnol. Bioeng.* 40:686–696.

Phelps, M. R., J. B. Hobbs, D. G. Kilburn, and R. F. B. Turner. 1995. An autoclavable glucose biosensor for microbial fermentation monitoring and control. *Biotechnol. Bioeng.* 46: 514–524.

Prabowo, D. A., O. Hiraishi, and S. Suda. 2013. Diversity of *Crypthecodinium* spp. (Dinophyceae) from Okinawa Prefecture, Japan. *J. Mar. Sci. Technol.* 21:181–191.

Raghukumar, S. 2008. Thraustochytrid marine protists: Production of PUFAs and other emerging technologies. *Mar. Biotechnol.* 10:631–640.

Ratledge, C. 1988. Biochemistry, stoichiometry, substrates and economics. In *Single Cell Oil*, ed. R. S. Moreton, pp. 33–70. Harlow, UK: Longman Scientific & Technical.

Ratledge, C. 2006. Biochemistry and physiology of growth, and metabolism. In *Basic Biotechnology*, 3rd edn., eds. C. Ratledge and B. Kristianseb, pp. 25–71. Cambridge, UK: Cambridge Univ. Press.

Riemann, F. and K. Schaumann. 1993. Thraustochytrid protists in Antarctic fast ice? *Antarctic Sci.* 5:279–280.

Riesenberg, D. and R. Guthke. 1999. High-cell-density cultivation of microorganisms. *Appl. Microbiol. Biotechnol.* 51:422–430.

Running, J. A., D. K. Severson, and K. J. Schneider. 2002. Extracellular production of l-ascorbic acid by *Chlorella protothecoides, Prototheca* species, and mutants of *P. moriformis* during aerobic culturing at low pH. *J. Ind. Microbiol. Biotechnol.* 29:93–98.

Sansawa, H. and H. Endo. 2004. Production of intracellular phytochemicals in *Chlorella* under heterotrophic conditions. *J. Biosci. Bioeng.* 98:437–444.

Schmidt, R. A., M. G. Wiebe, and N. T. Eriksen. 2005. Heterotrophic high cell-density fed-batch cultures of the phycocyanin-producing red alga *Galdieria sulphuraria. Biotechnol. Bioeng.* 90:77–84.

Shah, M. M. R., M. J. Alam, and M. Y. Mia. 2003. *Chlorella sp.*: Isolation, pure culture and small scale culture in brackish-water. *Bangladesh J. Sci. Ind.* 38:165–174.

Shiloach, J. and R. Fass. 2005. Growing *E. coli* to high cell density—A historical perspective on method development. *Biotechnol. Adv.* 23:345–357.

Sun, N., Y. Wang, Y. Lib, J. Huang, and F. Chen. 2008. Sugar-based growth, astaxanthin accumulation and carotenogenic transcription of heterotrophic *Chlorella zofingiensis* (Chlorophyta). *Process. Biochem.* 43:1288–1292.

Sun, Z., J. A. Ramsay, M. Guay, and B. A. Ramsay. 2007. Carbon-limited fed-batch production of medium-chain-length polyhydroxyalkanoates from nonanoic acid by *Pseudomonas putida* KT2440. *Appl. Microbiol. Biotechnol.* 74:69–77.

Wei, N., J. Quarterman, and Y. S. Jin. 2012. Marine microalgae: An untapped resource for producing fuels and chemicals. *Cell* 31:70–77.

Wei, N., J. Quarterman, S. R. Kim, J. H. D. Cate, and Y.-S. Jin. 2013. Enhanced biofuel production through coupled acetic acid and xylose consumption by engineered yeast. *Nat. Commun.* 4:2580.

Yokoyama, R. and D. Honda. 2007. Taxonomic rearrangement of the genus *Schizochytrium* sensu lato based on morphology, chemotaxonomic characteristics, and 18S rRNA gene phylogeny (Thraustochytriaceae, Labyrinthulomycetes): Emendation for *Schizochytrium* and erection of *Aurantiochytrium* and *Oblongichytrium* gen. nov. *Mycoscience* 48:199–211.

Index